AF581679

Genetic and Phenotypic Variation in Tree Crops Biodiversity

Genetic and Phenotypic Variation in Tree Crops Biodiversity

Editor

Gaetano Distefano

MDPI • Basel • Beijing • Wuhan • Barcelona • Belgrade • Manchester • Tokyo • Cluj • Tianjin

Editor
Gaetano Distefano
Università degli Studi di Catania
Italy

Editorial Office
MDPI
St. Alban-Anlage 66
4052 Basel, Switzerland

This is a reprint of articles from the Special Issue published online in the open access journal *Forests* (ISSN 1999-4907) (available at: https://www.mdpi.com/journal/forests/special_issues/Tree_Crops_Biodiversity).

For citation purposes, cite each article independently as indicated on the article page online and as indicated below:

LastName, A.A.; LastName, B.B.; LastName, C.C. Article Title. *Journal Name* **Year**, *Volume Number*, Page Range.

ISBN 978-3-0365-0638-8 (Hbk)
ISBN 978-3-0365-0639-5 (PDF)

Cover image courtesy of Gaetano Distefano.

Contents

About the Editor

Gaetano Distefano is a researcher currently working at the Department of Agriculture, Food and Environment of University of Catania (Italy). During his research career, he has collaborated with several research centers, including the CSIC of University of Zaragoza (Spain) to study aspects of Citrus reproductive biology, the University of New England (Australia) to study HRM fruit tree molecular markers' characterization, and the Oxford Brookes University (UK) and Hunan Agricultural University (China) to study specific techniques of genetic transformation and gene expression analysis. He has been lecturing at the University of Catania since 2012 on fruit tree crops' genetic improvement. Most of his studies have focused on gene identification and expression quantification in Citrus reproductive biology and fruit ripening. He also has a solid background in histology for flower biology studies. He has also worked in studies of biodiversity characterization and phyletic relationship among Citrus species and related genera and on the molecular characterization (AFLP, SSR and SNPs) of several Mediterranean fruit tree species such as carob, pomegranate, olive, almond, pear, cactus pear and grape, developing new markers and new methods of analysis.

MDPI

Editorial

Research and Application of Molecular and Phenotypic Data for Tree Biodiversity Evaluation

Gaetano Distefano

Department of Agriculture, Food and Environment, University of Catania, Via Valdisavoia 5, 95123 Catania, Italy; distefag@unict.it

Citation: Distefano, G. Research and Application of Molecular and Phenotypic Data for Tree Biodiversity Evaluation. *Forests* **2021**, *12*, 564. https://doi.org/10.3390/f12050564

Received: 14 April 2021
Accepted: 25 April 2021
Published: 30 April 2021

The main challenges for tree crop improvement are linked to the sustainable development of agro-ecological habitats, improving the adaptability to limiting environmental factors and resistance to biotic stresses or promoting novel genotypes with improved agronomic traits.

The exploitation of plant genetic resources and their conservation and sustainable management are the key to guaranteeing environmental and food security for future generations, since this will determine the ability of crops to adapt in several conditions. Significant advances in the application of molecular tools to characterize and conserve genetic diversity in tree crops took place in the last decades.

Molecular marker technologies are being increasingly used to explore genetic structure and function, and to provide high-resolution profiling of nucleotide variation within tree crop germplasm collections. Advances in DNA-derived data and innovative phenotyping are bridging the genotype-to-phenotype gap in tree crop selection.

In this Special Issue, fourteen original articles and one review represent a brief overview of the latest genetic and phenotypic trait characterization for tree crop.

Genetic characterization of different tree crops was performed by studying S-allele segregation, epi-markers, and molecular markers.

The study by Bennici et al. [1] was conducted on Italian pear germplasm collection (*Pyrus communis*) composed of varieties selected across the last two centuries, for their traits of agronomic interest and was complemented with wild related species (*P. pyrifolia*, *P. amygdaliformys*), for the S-allele genotyping. The results shaded light on the differences between the traditional varieties, with the wild-related species reflecting a more complex history of hybridization.

Tuisima-Coral et al. [2] used the amplified fragment length polymorphic (AFLP) fingerprints to estimate the genetic diversity of *Guazuma crinita*, a fast-growing timber tree species. The analysis of molecular variation showed higher genetic diversity within rather than among provenances. Cluster analysis and principal coordinate analysis did not show correspondence between genetic and geographic distance and significant genetic differentiation among population types.

Ma et al. [3] detected epimarkers using a methylation-sensitive amplification polymorphism (MSAP) method and performed epimarker–trait association analysis on the basis of nine growth and wood property traits within populations of 432 genotypes of *Populus tomentosa*. Tree height was positively correlated with relative full-methylation level, suggesting that changes in DNA methylation might contribute to regulating tree growth and wood property traits.

The study by Rollo et al. [4] assessed the genetic diversity and structure of the ice-cream-bean (Inga edulis Mart.; Fabaceae) in wild and cultivated populations from the Peruvian Amazon. The average pod length in cultivated trees was significantly higher than that in wild trees. The expected genetic diversity and the average number of alleles was higher in the wild compared to the cultivated populations; thus, a loss of genetic diversity was confirmed in the cultivated populations.

Lee et al. [5] characterized 410 tea accessions collected from South Korea, using 21 simple sequence repeat (SSR) markers. The study revealed the presence of lower diversity and a simpler population structure in Korean tea germplasms. Suggesting, more attention on collecting and conserving the new tea individuals, to broaden the genetic variation of new cultivars in future breeding of the tea plant.

Particular interest was paid to the novel sequencing method of plastid genome combined with molecular markers or phenotypic traits for several forest tree species as a tool to better define the genetic phylogenies and domestication process.

In particular, Su et al. [6] studied the different phenotypic traits of four endemic Ilex species (*I. latifolia*, *I. suaveolens*, *I. viridis*, and *I. micrococca*) on Mount Huangshan, China. A comprehensive comparison of plastomes within eight Ilex revealed the incongruence with the traditional taxonomy, whereas it informed a strong association between clades and geographic distribution.

Zeng et al. [7] used chloroplast DNA sequences and 29 nuclear microsatellite markers to investigate *C.* × *kuchugouzhui* (natural hybrid between *C. sclerophylla* and *C. tibetana*) confirming that it is a rare hybrid between *C. sclerophylla* and *C. tibetana*.

The study by Guo et al. [8] was based on the first collection of cultivated *Paeonia rockii* (flare tree peony, FTP) germplasm, across the main distribution area. Using phenotypic traits, expressed sequence tag (EST)-simple sequence repeat (SSR) markers and chloroplast DNA sequences (cpDNA), the authors showed that the selected accessions could fully reflect the genetic background information of FTP germplasm resources, so their protection and utilization would be of great significance for genetic improvement of woody peonies.

Deng et al. [9] carried out two different studies on *Michelia shiluensis* (Magnoliaceae), a rare and endangered magnolia species found in South China. The first study was dedicated to examining the genetic diversity of *M. shiluensis*, in which high genetic diversity and low differentiation were detected using eight nuclear single sequence repeat (nSSR) markers, and two chloroplast DNA (cpDNA) markers.

The second study relied on the complete chloroplast genome sequencing of *M. shiluensis* [10]. The genetic structure was represented by 160,075 bp in length with two inverted repeat regions (26,587 bp each), a large single-copy region (88,105 bp), and a small copy region (18,796 bp). The genomic information presented in this study was valuable for further classification, phylogenetic studies, and to support ongoing conservation efforts.

Yu et al. [11] reported the complete chloroplast genomes of five species of Acer sect. Platanoidea. The length of Acer sect. Platanoidea cp genomes ranged from 156,262 bp to 157,349 bp and detected the structural variation in the inverted repeat (IRs) boundaries. Platanoidea, with high resolutions for nearly all identified nodes, suggests a promising opportunity to resolve infrasectional relationships of the most species-rich section Platanoidea of Acer.

On the other hand, phenotypic traits were evaluated to select improved individuals or traits.

The study by Kim et al. [12] was conducted to select plus trees of two evergreen oaks, *Quercus salicina* and *Q. glauca*, in Korea. To select the candidate trees, they developed a subjective grading system with six characteristics in three categories and introduced a weighted generalized value (GVIw) to compare the superiority of the candidate trees. Through this process, 44 candidate trees of *Q. salicina* and 41 candidate trees of *Q. glauca* were selected as plus trees.

The study by Baniulis et al. [13] monitored the constitutive protein expression differences detected during active growth associated with cell metabolism and stress response, and conveyed a population-specific adaptation to the distinct climatic conditions in geographically distant Scots pine (*Pinus sylvestris* L.) populations adapted to specific photoperiods and temperature gradients, and which markedly vary in the timing of growth patterns and adaptive traits.

Dinulică et al. [14] measured the acoustic properties of wood for identifying simple, valid criteria for diagnosis, which remains an exciting challenge when selecting materials for manufacturing musical instruments. The results showed that the spruce trees with acoustic and structural features that match the requirements for the manufacture of violins have a bark phenotype distinguishable by color—as well as by scale shape. The south-facing side of the trunk and the external side of the scale were best for identifying resonance trees by their bark. The authors concluded that the differences among bark phenotypes were noticeable to the naked eye.

Zhao et al. [15] monitored the leaf color mutation as ideal materials for studying pigment metabolism, chloroplast development and differentiation, photosynthesis, and other pathways that could also provide important information for improving varietal selection. In this review, the authors summarized the research on leaf color mutants, such as the functions and mechanisms of leaf color mutant-related genes, which affect chlorophyll synthesis, chlorophyll degradation, chloroplast development, and anthocyanin metabolism. The review provides a reference for the study and application of leaf color mutants in the future.

This Special Issue end points were to contribute to the growth of this area of research, trigger research interest on biodiversity conservation, and valorization, by adding genetic and phenotypic information scientifically substantiated with new data.

We would like to thank all authors and reviewers of the papers published in this Special Issue for their great contributions and efforts. We are also grateful to the editorial board members and to the staff of the Journal for their kind support in the preparation steps of this Special Issue.

Funding: This research received no external funding.

Informed Consent Statement: Not applicable.

Conflicts of Interest: The author declares no conflict of interest.

References

1. Bennici, S.; Di Guardo, M.; Distefano, G.; Las Casas, G.; Ferlito, F.; De Franceschi, P.; Dondini, L.; Gentile, A.; La Malfa, S. Deciphering *S*-RNase Allele Patterns in Cultivated and Wild Accessions of Italian Pear Germplasm. *Forests* **2020**, *11*, 1228. [CrossRef]
2. Tuisima-Coral, L.L.; Hlásná Čepková, P.; Weber, J.C.; Lojka, B. Preliminary Evidence for Domestication Effects on the Genetic Diversity of *Guazuma crinita* in the Peruvian Amazon. *Forests* **2020**, *11*, 795. [CrossRef]
3. Ma, K.; Song, Y.; Ci, D.; Zhou, D.; Tian, M.; Zhang, D. Genome Cytosine Methylation May Affect Growth and Wood Property Traits in Populations of *Populus tomentosa*. *Forests* **2020**, *11*, 828. [CrossRef]
4. Rollo, A.; Ribeiro, M.M.; Costa, R.L.; Santos, C.; Clavo, P.Z.M.; Mandák, B.; Kalousová, M.; Vebrová, H.; Chuqulin, E.; Torres, S.G.; et al. Genetic Structure and Pod Morphology of *Inga edulis* Cultivated vs. Wild Populations from the Peruvian Amazon. *Forests* **2020**, *11*, 655. [CrossRef]
5. Lee, K.J.; Lee, J.-R.; Sebastin, R.; Shin, M.-J.; Kim, S.-H.; Cho, G.-T.; Hyun, D.Y. Assessment of Genetic Diversity of Tea Germplasm for Its Management and Sustainable Use in Korea Genebank. *Forests* **2019**, *10*, 780. [CrossRef]
6. Su, T.; Zhang, M.; Shan, Z.; Li, X.; Zhou, B.; Wu, H.; Han, M. Comparative Survey of Morphological Variations and Plastid Genome Sequencing Reveals Phylogenetic Divergence between Four Endemic *Ilex* Species. *Forests* **2020**, *11*, 964. [CrossRef]
7. Zeng, X.; Chen, R.; Bian, Y.; Qin, X.; Zhang, Z.; Sun, Y. Identification of a Natural Hybrid between *Castanopsis sclerophylla* and *Castanopsis tibetana* (Fagaceae) Based on Chloroplast and Nuclear DNA Sequences. *Forests* **2020**, *11*, 873. [CrossRef]
8. Guo, X.; Cheng, F.; Zhong, Y. Genetic Diversity of *Paeonia rockii* (Flare Tree Peony) Germplasm Accessions Revealed by Phenotypic Traits, EST-SSR Markers and Chloroplast DNA Sequences. *Forests* **2020**, *11*, 672. [CrossRef]
9. Deng, Y.; Liu, T.; Xie, Y.; Wei, Y.; Xie, Z.; Shi, Y.; Deng, X. High Genetic Diversity and Low Differentiation in *Michelia shiluensis*, an Endangered Magnolia Species in South China. *Forests* **2020**, *11*, 469. [CrossRef]
10. Deng, Y.; Luo, Y.; He, Y.; Qin, X.; Li, C.; Deng, X. Complete Chloroplast Genome of *Michelia shiluensis* and a Comparative Analysis with Four Magnoliaceae Species. *Forests* **2020**, *11*, 267. [CrossRef]
11. Yu, T.; Gao, J.; Huang, B.-H.; Dayananda, B.; Ma, W.-B.; Zhang, Y.-Y.; Liao, P.-C.; Li, J.-Q. Comparative Plastome Analyses and Phylogenetic Applications of the *Acer* Section *Platanoidea*. *Forests* **2020**, *11*, 462. [CrossRef]
12. Kim, I.S.; Lee, K.M.; Shim, D.; Kim, J.J.; Kang, H.-I. Plus Tree Selection of *Quercus salicina* Blume and *Q. glauca* Thunb. and Its Implications in Evergreen Oaks Breeding in Korea. *Forests* **2020**, *11*, 735. [CrossRef]
13. Baniulis, D.; Sirgėdienė, M.; Haimi, P.; Tamošiūnė, I.; Danusevičius, D. Constitutive and Cold Acclimation-Regulated Protein Expression Profiles of Scots Pine Seedlings Reveal Potential for Adaptive Capacity of Geographically Distant Populations. *Forests* **2020**, *11*, 89. [CrossRef]

14. Dinulică, F.; Albu, C.-T.; Vasilescu, M.M.; Stanciu, M.D. Bark Features for Identifying Resonance Spruce Standing Timber. *Forests* **2019**, *10*, 799. [CrossRef]
15. Zhao, M.-H.; Li, X.; Zhang, X.-X.; Zhang, H.; Zhao, X.-Y. Mutation Mechanism of Leaf Color in Plants: A Review. *Forests* **2020**, *11*, 851. [CrossRef]

Article

Deciphering *S*-RNase Allele Patterns in Cultivated and Wild Accessions of Italian Pear Germplasm

Stefania Bennici [1,†], Mario Di Guardo [1,†], Gaetano Distefano [1,*], Giuseppina Las Casas [2], Filippo Ferlito [2], Paolo De Franceschi [3], Luca Dondini [3], Alessandra Gentile [1,4] and Stefano La Malfa [1]

1 Dipartimento di Agricoltura, Alimentazione ed Ambiente, Università di Catania, Via Valdisavoia 5, 95123 Catania, Italy; stefania.bennici@hotmail.it (S.B.); mario.diguardo@unict.it (M.D.G.); gentilea@unict.it (A.G.); slamalfa@unict.it (S.L.M.)

2 Consiglio per la ricerca in agricoltura e l'analisi dell'economia agraria, Centro di ricerca Olivicoltura, Frutticoltura e Agrumicoltura, Corso Savoia, 190-95024 Acireale (CT), Italy; giuseppina.lascasas@outlook.it (G.L.C.); filippo.ferlito@crea.gov.it (F.F.)

3 Dipartimento di Scienze e Tecnologie Agroalimentari (DISTAL), Università degli Studi di Bologna, 40126 Bologna, Italy; paolodefranceschi@gmail.com (P.D.F.); luca.dondini@unibo.it (L.D.)

4 College of Horticulture and Landscape, Hunan Agricultural University, Changsha 410128, China

* Correspondence: distefag@unict.it

† Both authors contributed equally to this work.

Received: 4 November 2020; Accepted: 19 November 2020; Published: 22 November 2020

Abstract: The genus *Pyrus* is characterized by an *S*-RNase-based gametophytic self-incompatibility (GSI) system, a mechanism that promotes outbreeding and prevents self-fertilization. While the *S*-genotype of the most widely known pear cultivars was already described, little is known on the *S*-allele variability within local accessions. The study was conducted on 86 accessions encompassing most of the local Sicilian varieties selected for their traits of agronomic interest and complemented with some accessions of related wild species (*P. pyrifolia* Nakai, *P. amygdaliformis* Vill.) and some national and international cultivars used as references. The employment of consensus and specific primers enabled the detection of 24 *S*-alleles combined in 48 *S*-genotypes. Results shed light on the distribution of the *S*-alleles among accessions, with wild species and international cultivars characterized by a high diversity and local accessions showing a more heterogeneous distribution of the *S*-alleles, likely reflecting a more complex history of hybridization. The *S*-allele distribution was largely in agreement with the genetic structure of the studied collection. In particular, the "wild" genetic background was often characterized by the same *S*-alleles detected in *P. pyrifolia* and *P. amygdaliformis*. The analysis of the *S*-allele distribution provided novel insight into the contribution of the wild and international cultivars to the genetic background of the local Sicilian or national accessions. Furthermore, these results provide information that can be readily employed by breeders for the set-up of novel mating schemes.

Keywords: *S*-genotyping; *S*-locus; *P. communis*; *P. pyrifolia*; *P. amygdaliformis*; genetic structure

1. Introduction

European pear (*Pyrus communis* L.) is an economically important fruit tree species belonging to the Rosaceae family. Like the majority of the Rosaceae, the genus *Pyrus* exhibits the *S*-RNase-based gametophytic self-incompatibility (GSI) system, evolved by flowering plants to prevent self-fertilization and promote outbreeding [1]. The GSI system prevents self-fertilization through a specific pollen–pistil recognition that selectively inhibits the growth of those pollen tubes recognized by the pistil as "self" (i.e., pollen from the same plant or from individuals that are genetically related) [2]. The GSI system is controlled by the single, multigenic, and multiallelic *S*-locus expressing a female determinant in the style, the stylar ribonuclease (*S*-RNase) [3], and a male determinant in the pollen, a pool of F-box proteins known as SFBB (*S*-locus F-box brothers) [4]. In the GSI system, a match between the male *S*-determinant carried by the haploid genome of the pollen grain and one of the two *S*-alleles carried by the diploid genome of the stylar tissue (self-recognition) results in the arrest of the pollen tube growth triggered by the "self" *S*-RNase, which acts as a cytotoxin and possibly activates programmed cell death. Conversely, in non-self-recognition, *S*-RNase is inactivated by a specific F-box protein, through a proteolytic degradation mechanism [5]. Self-incompatibility is generally considered an undesired trait, especially for those cultivated species in which the success of the fertilization process is essential for fruit set. An understanding of the *S*-genotype is crucial for the choice of pollinators during the set-up of novel orchards and for novel breeding programs. Traditionally, the degree of compatibility between cultivars was determined directly in the field via the set-up of controlled crosses; however, this approach is time-consuming and often not reliable in the discrimination between fully and semi-compatible combinations. The advent of molecular biology techniques enabled both the sequencing of the gene coding for the *S*-RNase and the development of molecular markers, allowing a fast and relatively inexpensive screening of the *S*-genotypes in germplasms of interest [6]. The *S*-RNase gene is composed of five consensus conserved regions (C1, C2, C3, RC4, and C5) and the highly conserved noncanonical hexapeptide (IIWPNW); moreover, between C2 and C3 is located the hypervariable region (RHV) harboring a highly polymorphic intron. This RHV region has been largely exploited to assess the *S*-locus diversity in European pear by cloning and sequencing PCR products amplified with universal primers on the basis of conserved regions [6–16]. Sanzol [14] developed a PCR-based method for the detection of 20 *S*-RNase alleles in European pear by using consensus primers simultaneously amplifying a large number of alleles characterized by different intron sizes, plus a set of allele-specific primers. Then, further primer pairs were developed by Nikzad Gharehaghaji and colleagues [16], allowing the detection of additional alleles of European pear, some of which were highly similar and possibly derived from other *Pyrus* species.

Sicily is characterized by a wide pear biodiversity; to this extent, Mount Etna represents an ideal reservoir of different local accessions due to the occurrence of different microclimates, soils, and orographic conditions combined with the ancient history of cultivation and natural seed-based propagation. Moreover, the geographic location of Sicily in the Mediterranean Sea and its historical involvement in commercial exchanges may have favored the introgression of several traits of agronomical interest from different pear species. This wide pear biodiversity includes autochthonous and wild pears such as *P. amygdaliformis* Vill. and *P. pyraster* (L.) Burgsd. (*P. communis* ssp. *pyraster* L.) that were largely employed as rootstocks to increase the hardiness and longevity of the trees in past centuries [17]. *P. amygdaliformis* is native to the Mediterranean region and is highly tolerant to drought stress [18]. *P. pyraster* comes from the western Black Sea region, with a distribution area spanning from the British Isles to Latvia, and it is believed to be one of the most probable ancestors that gave rise to European pear [19,20].

A previous study of genetic structure, conducted largely on the same accessions of the present study, revealed the presence of two subpopulations that can be reconducted to a "wild" and "cultivated" genetic status. Within the Sicilian local germplasm, only a small number of accessions were characterized by a high admixture of the two subpopulations, while the majority were characterized by a clear prevalence of one of the two subpopulations [17]. In such a genetic background, identifying the *S*-allele distribution within the subpopulations can provide valuable insights into the relationship and gene flow between wild and cultivated Sicilian pears.

In the present work, the PCR-based *S*-genotyping method described by Sanzol [14] was used to (i) ascertain the *S*-RNase composition in local Sicilian pear varieties and native wild accessions collected from the Mount Etna area in comparison with national and international varieties used as a reference, (ii) determine the distribution pattern of the *S*-alleles among the different groups of accessions, and (iii) compare the S-allele distribution in genotypes characterized by "wild" and "cultivated" genetic backgrounds.

2. Materials and Methods

2.1. Plant Material and DNA Extraction

The germplasm in analysis consisted of 86 accessions composed of 43 local varieties (LV), 16 individuals belonging to wild related species (RS) (nine *P. pyraster* and seven *P. amydgaliformis*, all collected from Mount Etna area, Italy), 18 nationally cultivated varieties (NCV), and nine internationally cultivated varieties (ICV) (Table 1). Accessions were located in two ex situ germplasm collections located in Catania district (Sicily, South Italy): the experimental farm of Catania University (UNICT, 10 m above sea level (a.s.l.)) and the Germplasm Bank of "Parco dell'Etna" (Mt. Etna, 850 m a.s.l.).

Genomic DNA was extracted from fresh leaves using ISOLATE II Plant DNA Kit (Bioline, Meridian Life Science, Memphis, TN, USA) following the protocol provided by the manufacturer. The concentration and the purity of the extracted DNA were assessed using a Nanodrop 2000 spectrophotometer (Thermo Scientific, Waltham, MA, USA) and agarose gel electrophoresis.

2.2. S-Genotyping Assay

The S-genotyping assay was performed according to the protocol described by Sanzol [14] using a pair of consensus primers, PycomC1F and PycomC5R [12], and 17 specific primers able to discriminate among 23 different S-RNase alleles (Table S1, Supplementary Materials). The S-genotyping assay was implemented with additional allele-specific primer pairs for alleles PcS_{126} and PcS_{127} (Table S1, Supplementary Materials). Primer pairs were ad hoc developed in this study on the basis of genomic sequences of PcS_{126} (accession number KF588567) and PcS_{127} (accession number KF588568) using Primer-BLAST [21]. The nine ICVs and the NCVs "Bella di Giugno", "Coscia", and "Gentile", which have a known S-genotype, were used as controls (Table S2, Supplementary Materials).

Consensus PCR was performed in a 20 µL volume containing 40 ng of genomic DNA, 1× PCR buffer II, 2 mM magnesium chloride, 0.2 mM dNTPs, 0.6 µM each primer (PycomC1F and PycomC5R), and 1 U of MyTaq DNA polymerase (Bioline, Meridian Life Science, Memphis, TN, USA). Amplification was conducted using a program with an initial denaturation at 94 °C for 10 min, followed by 35 cycles at 94 °C for 30 s, 57 °C for 45 s, and 72 °C for 2 min, with a final cycle of 72 °C for 7 min.

Allele-specific PCR was performed in a 20 µL volume containing 40 ng of genomic DNA, 1× PCR buffer II, 2 mM magnesium chloride, 0.2 mM dNTPs, from 0.3 to 0.6 µM each primer (Table S1, Supplementary Materials), and 1 U of MyTaq DNA polymerase (Bioline, Meridian Life Science, Memphis, TN, USA). Amplification was conducted using a program with an initial denaturation at 94 °C for 10 min, followed by 35 cycles at 94 °C for 30 s, 58 °C for 45 s, and 72 °C for 1 min, with a final cycle of 72 °C for 7 min.

Amplicons were separated by gel electrophoresis in 1% agarose stained with SYBR Safe DNA gel stain (Invitrogen, Carlsbad, CA, USA). Image acquisition and fragment size estimation were performed using Image LabTM software with the GelDOCTM XR+ system (BIO-RAD Molecular Imager®, Hercules, CA, USA).

2.3. DNA Sequencing and Allele Identification

Consensus PCR products were excised from agarose gel and purified using UPGRADE TO ISOLATE II Nucleic Acid Isolation Kits (Bioline, Meridian Life Science, Memphis, TN, USA) following the protocol provided by the manufacturer. Purified products were sequenced in the forward and reverse directions starting from primers PycomC1F1 and PycomC5R1 sequenced using an ABI310 genetic analyzer (Applied Biosystems, Foster City, CA, USA).

Table 1. *S*-Genotypes and features of accessions employed in this study.

Species	Accession Name	Origin	Status [a]	Collection [b]	Q1	Q2	Subpopulation [c]	*S*-Genotype
Pyrus communis	Adamo	Italy (Sicily)	LV	UNICT	0.053	0.947	Wild	PcS_{105}/PcS_{126}
P. communis	Alessio	Italy (Sicily)	LV	Mt Etna	0.97	0.03	Cultivated	$PcS_{107}/-$
P. communis	Angelico	Italy (Sicily)	NCV	Mt Etna	0.989	0.011	Cultivated	PcS_{104}/PcS_{103}
P. communis	Azzone di Cassone	Italy (Sicily)	LV	UNICT	0.972	0.028	Cultivated	PcS_{103}/PcS_{127}
P. communis	Bella di Giugno	Italy (Sicily)	NCV	Mt Etna	0.054	0.946	Wild	PcS_{104}/PcS_{120} *
P. communis	Bergamotto	Italy (Sicily)	NCV	UNICT	0.984	0.016	Cultivated	PcS_{103}/PcS_{111}
P. communis	Bianchetto	Italy (Sicily)	NCV	UNICT	0.012	0.988	Wild	PcS_{109}/PcS_{117}
P. communis	Bianchettone	Italy (Sicily)	LV	Mt Etna	0.032	0.968	Wild	PcS_{104}/PcS_{105}
P. communis	Bruttu Beddu	Italy (Sicily)	LV	Mt Etna	0.929	0.071	Cultivated	$PcS_{103}/-$
P. communis	Buona Luisa	Italy (Sicily)	NCV	Mt Etna	0.993	0.007	Cultivated	PcS_{101}/PcS_{103}
P. communis	Butirra	Italy (Sicily)	NCV	UNICT	0.991	0.009	Cultivated	PcS_{101}/PcS_{103}
P. communis	Campana	Italy (Sicily)	NCV	UNICT			NA	PcS_{103}/PcS_{122}
P. communis	Catanese	Italy (Sicily)	LV	Mt Etna	0.019	0.981	Wild	PcS_{104}/PcS_{108}
P. communis	Cavaliere	Italy (Sicily)	LV	UNICT			NA	PcS_{104}/PcS_{115}
P. communis	Chiuzzu	Italy (Sicily)	LV	UNICT	0.215	0.785	Admixed	PcS_{103}/PcS_{127}
P. communis	Coscia	Italy (Sicily)	NCV	UNICT	0.99	0.01	Cultivated	PcS_{104}/PcS_{103} *
P. communis	Duchessa d'Angio'	Italy (Sicily)	LV	Mt Etna	0.98	0.02	Cultivated	PcS_{103}/PcS_{105}
P. communis	Faccia Donna	Italy (Sicily)	LV	UNICT	0.009	0.991	Wild	$PcS_{124}/-$
P. communis	Faccibedda	Italy (Sicily)	LV	UNICT	0.018	0.982	Wild	PcS_{126}/PcS_{127}
P. communis	Franconello	Italy (Sicily)	LV	Mt Etna			NA	PcS_{126}/PcS_{127}
P. communis	Garibaldi	Italy (Sicily)	LV	Mt Etna			NA	PcS_{104}/PcS_{115}
P. communis	Garofalo	Italy (Sicily)	NCV	Mt Etna	0.009	0.991	Wild	$PcS_{124}/-$
P. communis	Gentile	Italy (Sicily)	NCV	UNICT	0.644	0.356	Admixed	PcS_{101}/PcS_{106} *
P. communis	Ialufaru	Italy (Sicily)	LV	Mt Etna	0.014	0.986	Wild	$PcS_{124}/-$
P. communis	Ianculiddu	Italy (Sicily)	LV	Mt Etna	0.012	0.988	Wild	PcS_{125}/PcS_{127}
P. communis	Iazzuleddu	Italy (Sicily)	LV	UNICT	0.984	0.016	Cultivated	PcS_{103}/PcS_{127}
P. communis	Mezza Campana	Italy (Sicily)	LV	Mt Etna			NA	PcS_{104}/PcS_{108}
P. communis	Moscatello 1	Italy (Sicily)	NCV	Mt Etna	0.018	0.982	Wild	PcS_{104}/PcS_{123}
P. communis	Moscatello 2	Italy (Sicily)	NCV	UNICT	0.027	0.973	Wild	PcS_{103}/PcS_{116}
P. communis	Moscatello Maiolino	Italy (Sicily)	LV	Mt Etna			NA	PcS_{126}/PcS_{127}
P. communis	Moscatello Nero	Italy (Sicily)	LV	Mt Etna	0.014	0.986	Cultivated	PcS_{101}/PcS_{106}
P. communis	Paradiso/Confittaru	Italy (Sicily)	LV	Mt Etna	0.081	0.919	Wild	PcS_{103}/PcS_{126}
P. communis	Pasqualino	Italy (Sicily)	LV	UNICT	0.968	0.032	Cultivated	PcS_{103}/PcS_{111}
P. communis	Pauluzzo	Italy (Sicily)	LV	Mt Etna	0.018	0.982	Wild	PcS_{126}/PcS_{127}
P. communis	Pergolesi	Italy (Sicily)	LV	Mt Etna	0.989	0.011	Cultivated	PcS_{101}/PcS_{103}

Table 1. *Cont.*

Species	Accession Name	Origin	Status [a]	Collection [b]	Q1	Q2	Subpopulation [c]	S-Genotype
P. communis	Piccola Dolce	Italy (Sicily)	LV	Mt Etna			NA	PcS_{104}/PcS_{117}
P. communis	Piridda	Italy (Sicily)	LV	Mt Etna	0.007	0.993	Wild	PcS_{109}/PcS_{117}
P. communis	Piru Mulinciana	Italy (Sicily)	LV	Mt Etna	0.99	0.01	Cultivated	PcS_{101}/PcS_{108}
P. communis	Piru Pizzu	Italy (Sicily)	LV	Mt Etna	0.139	0.861	Wild	PcS_{105}/PcS_{127}
P. communis	Pisciazzanu	Italy (Sicily)	LV	Mt Etna	0.056	0.944	Wild	PcS_{101}/-
P. communis	Pistacchino	Italy (Sicily)	LV	Mt Etna	0.699	0.301	Admixed	PcS_{117}/PcS_{121}
P. communis	Putiru d'Estate	Italy (Sicily)	LV	Mt Etna	0.985	0.015	Cultivated	PcS_{101}/PcS_{103}
P. communis	Putiru d'Inverno	Italy (Sicily)	LV	UNICT	0.992	0.008	Cultivated	PcS_{103}/-
P. communis	Razzuolo Rosata	Italy (Sicily)	LV	Mt Etna	0.017	0.983	Wild	PcS_{103}/PcS_{106}
P. communis	Regina	Italy (Sicily)	LV	UNICT			NA	PcS_{101}/PcS_{122}
P. communis	Rosa	Italy (Sicily)	LV	Mt Etna	0.01	0.99	Wild	PcS_{104}/PcS_{103}
P. communis	San Cono	Italy (Sicily)	LV	Mt Etna	0.651	0.349	Admixed	PcS_{102}/PcS_{103}
P. communis	San Giovanni	Italy (Sicily)	NCV	UNICT	0.009	0.991	Wild	PcS_{102}/PcS_{109}
P. communis	San Giovannino	Italy (Sicily)	LV	Mt Etna	0.01	0.99	Wild	PcS_{105}/-
P. communis	San Pietro	Italy (Sicily)	NCV	UNICT	0.015	0.985	Wild	PcS_{120}/PcS_{126}
P. communis	Santa Caterina	Italy (Sicily)	LV	UNICT	0.01	0.99	Wild	PcS_{104}/PcS_{125}
P. communis	Sciaduna	Italy (Sicily)	LV	Mt Etna	0.05	0.95	Wild	PcS_{111}/PcS_{117}
P. communis	Spadona	Italy (Sicily)	NCV	Mt Etna	0.029	0.971	Wild	PcS_{104}/PcS_{105}
P. communis	Spineddu	Italy (Sicily)	NCV	UNICT	0.039	0.961	Wild	PcS_{111}/PcS_{117}
P. communis	Tabaccaro	Italy (Sicily)	LV	Mt Etna	0.029	0.971	Wild	PcS_{101}/PcS_{127}
P. communis	Ucciarduni	Italy (Sicily)	NCV	UNICT			NA	PcS_{103}/PcS_{108}
P. communis	Urzi'	Italy (Sicily)	LV	UNICT			NA	PcS_{103}/PcS_{125}
P. communis	Villalba	Italy (Sicily)	LV	Mt Etna	0.276	0.724	Admixed	PcS_{101}/PcS_{109}
P. communis	Virgolese	Italy (Sicily)	NCV	UNICT	0.99	0.01	Cultivated	PcS_{101}/PcS_{103}
P. communis	Zio Pietro	Italy (Sicily)	LV	UNICT	0.986	0.014	Cultivated	PcS_{101}/PcS_{108}
P. communis	Zuccareddu	Italy (Sicily)	LV	Mt Etna	0.034	0.966	Wild	PcS_{101}/PcS_{107}
P. amygdaliformis	1	Italy (Sicily)	RS	North (N) 37° 47′ 692″ east (E) 14° 50′ 925″	0.013	0.987	Wild	PcS_{120}/PcS_{122}
P. amygdaliformis	2	Italy (Sicily)	RS	N 37° 51′ 189″ E 14° 50′ 795″	0.008	0.992	Wild	PcS_{126}/-
P. amygdaliformis	3	Italy (Sicily)	RS	N 37° 51′ 147″ E 14° 50′ 739″	0.007	0.993	Wild	PcS_{101}/PcS_{116}
P. amygdaliformis	4	Italy (Sicily)	RS	N 37° 51′ 109″ E 14°50′ 809″	0.007	0.993	Wild	PcS_{116}/PcS_{125}
P. amygdaliformis	5	Italy (Sicily)	RS	N 37° 51′ 116″ E 14° 50′ 806″			NA	PcS_{120}/PcS_{122}
P. amygdaliformis	7	Italy (Sicily)	RS	N 37° 51′ 139″ E 14° 50′ 765″	0.009	0.991	Wild	PcS_{108}/PcS_{125}
P. amygdaliformis	10	Italy (Sicily)	RS	N 37° 51′ 137″ E 14° 50′ 748″			NA	PcS_{106}/-

Table 1. *Cont.*

Species	Accession Name	Origin	Status [a]	Collection [b]	Q1	Q2	Subpopulation [c]	*S*-Genotype
P. pyraster	2	Italy (Sicily)	RS	N 37° 48′ 143″ E 14° 51′ 203″			NA	PcS_{111}/-
P. pyraster	3	Italy (Sicily)	RS	N 37° 51′ 796″ E 14° 52′ 409″	0.036	0.964	Wild	PcS_{122}/-
P. pyraster	4	Italy (Sicily)	RS	N 37° 52′ 808″ E 14° 52′ 571″			NA	PcS_{127}/-
P. pyraster	5	Italy (Sicily)	RS	N 37° 52′ 868″ E 14° 52′ 565″			NA	PcS_{105}/PcS_{123}
P. pyraster	7	Italy (Sicily)	RS	N 37° 47′ 691″ E 14° 50′ 901″	0.097	0.903	Wild	PcS_{101}/PcS_{111}
P. pyraster	8	Italy (Sicily)	RS	N 37° 52′ 006″ E 14° 52′ 969″	0.007	0.993	Wild	PcS_{108}/PcS_{127}
P. pyraster	9	Italy (Sicily)	RS	N 37° 51′ 865″ E 14° 52′ 672″	0.009	0.991	Wild	PcS_{105}/PcS_{109}
P. pyraster	10	Italy (Sicily)	RS	N 37° 48′ 276″ E 14° 51′ 269″	0.477	0.523	Admixed	PcS_{104}/PcS_{110}
P. pyraster	11	Italy (Sicily)	RS	N 37° 53′ 038″ E 14° 52′ 578″	0.057	0.943	Wild	PcS_{108}/-
P. communis	Abbé Fétel	France	ICV	UNIBO	0.967	0.033	Cultivated	PcS_{104}/PcS_{105} *
P. communis	Beurre Hardy	USA	ICV	UNIBO	0.391	0.609	Admixed	PcS_{108}/PcS_{114} *
P. communis	Cascade	USA	ICV	UNIBO			NA	PcS_{101}/PcS_{104} *
P. communis	Dr. Jules Guyot	France	ICV	UNIBO			NA	PcS_{101}/PcS_{105} *
P. communis	Kaiser	France	ICV	UNIBO	0.914	0.086	Cultivated	PcS_{107}/PcS_{125} *
P. communis	Max Red Barlet	USA	ICV	UNIBO	0.993	0.007	Cultivated	PcS_{101}/PcS_{102} *
P. communis	Old Home	USA	ICV	UNIBO			NA	PcS_{101}/PcS_{113} *
P. communis	Harrow Sweet	Canada	ICV	UNIBO	0.987	0.013	Cultivated	PcS_{102}/PcS_{105} *
P. communis	Williams	England	ICV	UNIBO	0.993	0.007	Cultivated	PcS_{101}/PcS_{102} *

[a] Status: (LV) local varieties; (RS) wild related species; (NCV) nationally cultivated varieties; (ICV) internationally cultivated varieties. [b] Collection: (UNICT) University of Catania, Catania, Italy, experimental field; (Mt Etna) the Germplasm Bank of "Parco dell'Etna", Nicolosi, Italy; (UNIBO) University of Bologna, Bologna, Italy, experimental field. [c] Subpopulation: for the 68 accessions genotyped in the work published by Bennici et al. 2018; https://doi.org/10.1371/journal.pone.0198512, the structure results for K = 2 were reported. Samples showing a prevalence (≥0.8) of subpopulations Q1 or Q2 were classified as "cultivated" or "wild", respectively, accessions showing a not clear predominance of one of the two subpopulations were classified "admixed", while samples that were not genotyped were scored as not applicable (NA). * Reference genotypes.

2.4. Clustering

S-RNase alleles identified for each genotype were converted into a binary matrix and used to compute a principal component analysis (PCA) on the basis of a dissimilarity matrix, performed using the statistical package R [22].

3. Results

The germplasm was genotyped using the consensus primers PycomC1F1 and PycomC5R [12]. The analysis of the PCR products allowed the identification of six alleles: PcS_{101} (1300 bp), PcS_{102} (1700 bp), PcS_{104} (750 bp), PcS_{110} (2200 bp), PcS_{113} (2000 bp), and PcS_{120} (800 bp), while 16 *S*-alleles were identified through the use of specific primers (Table S1, Supplementary Materials). The consensus PCR products and the amplification for each *S*-RNase allele tested for every accession are shown in the Table S2 (Supplementary Materials).

The accession "Iazzuleddu" showed a consensus amplicon of 1650 bp, positive to PcS_{103} primers and a new PCR product size of approximately 850 bp, which could not be identified by any of the tested allele-specific primers. The same amplicon was detected in the LVs "Azzone di Cassone", "Chiuzzu", "Faccibedda", "Franconello", "Ianculiddu", "Moscatello maiolino", "Pauluzzo", "Piru Pizzu", and "Tabaccaro" and in the two RS genotypes of *P. pyraster* (no. 4 and no. 8). The sequencing of the 850 bp amplicon of "Iazzuleddu" showed a 100% similarity to the *S*-RNase-PcS_{127} allele of *Pyrus communis* (Sequence ID: KF588568.1). A new pair of primers was designed to selectively amplify the *S*-RNase-PcS_{127} allele (Table S1, Supplementary Materials) and allowed its detection in all genotypes carrying the band of 850 bp producing an amplicon of 214 bp.

The LV "Pauluzzo", in addition to the band of 850 bp (PcS_{127} allele), was characterized by a smaller band of 680 bp that was not identified by any of the allele-specific primers. The sequencing of the 680 bp amplicon revealed a 99% similarity to the *S*-RNase-PcS_{126} allele of *Pyrus communis* (Sequence ID: KF588567.1). This amplicon was also detected in the LVs "Adamo", "Faccibedda", "Franconello", "Moscatello maiolino", and "Paradiso Confittaru", in the NCV "S. Pietro", and in the RS *P. amygdaliformis* (no. 2). A specific primer pair was designed to selectively amplify the *S*-RNase-PcS_{126} allele (Table S1, Supplementary Materials) producing an amplicon of 100 bp in all genotypes carrying the initial band of 680 bp.

Summing up, the use of the consensus, the specific, and the two ad hoc designed primers allowed the detection of 24 *S* alleles; for 72 accessions, both *S* alleles were detected (resulting in 48 different *S*-genotypes), while, for the remaining 14 accessions, only a single allele was detected (Table 1). The relative frequencies of the *S*-RNase alleles identified for each group (ICV, NCV, LV, and RS) are shown in Table 2.

The *S*-allele showing the highest absolute frequency in the germplasm was PcS_{103}, detected in 23 accessions (Table 2). Looking at the distribution of the *S*-RNase allele among the four pear groups, PcS_{103} was detected only among Italian varieties (14 LVs and nine NCVs) (Table 2). The *S*-RNase alleles PcS_{101}, PcS_{104}, PcS_{105}, and PcS_{108} (identified in 21, 16, 11, and nine accessions, respectively) were detected, although with different frequencies, in all four groups; in contrast, five *S*-RNase alleles (PcS_{110}, PcS_{113}, PcS_{114}, PcS_{115}, and PcS_{121}) were group-specific (PcS_{110} for RS, PcS_{113} and PcS_{114} for ICV, and PcS_{115} and PcS_{121} for LV) (Table 2). None of the *S*-RNase alleles detected in more than two samples were found in only one of the four classes presented (Table 2).

A number of 68 out of the 86 accessions here characterized were previously SSR-genotyped, and genetic structure analysis detected two subpopulations defined as "wild" and "cultivated" [17] (Table 1). The "wild" subpopulation largely characterized the RS group (contributing for an average of 93.4% on the genetic makeup of such accessions), while the "cultivated" subpopulation was predominantly detected in the ICVs (average of 87.4%). A more complex pattern was detected for the accessions deemed as LVs or NCVs; in both cases most of the accessions showed a clear prevalence (more than the 80%) of one of the two subpopulations with only five accessions showing a more balanced presence of the "wild" and "cultivated" subpopulations ("admixed" [17]). Figure 1 showed the relative frequency of the different S-alleles according to the structure analysis (subpopulations "wild", "cultivated", and "admixed").

Table 2. S-Allele frequencies among analyzed accessions.

S-Allele	Count	RS	LV	NCV	ICV
PcS_{101}	21	0.1	0.48	0.19	0.24
PcS_{102}	5	0	0.2	0.2	0.6
PcS_{103}	23	0	0.61	0.39	0
PcS_{104}	16	0.06	0.5	0.31	0.13
PcS_{105}	11	0.18	0.45	0.09	0.27
PcS_{106}	4	0.25	0.5	0.25	0
PcS_{107}	3	0	0.67	0	0.33
PcS_{108}	9	0.33	0.44	0.11	0.11
PcS_{109}	5	0.2	0.4	0.4	0
PcS_{110}	1	1	0	0	0
PcS_{111}	6	0.33	0.33	0.33	0
PcS_{113}	1	0	0	0	1
PcS_{114}	1	0	0	0	1
PcS_{115}	2	0	1	0	0
PcS_{116}	3	0.67	0	0.33	0
PcS_{117}	6	0	0.67	0.33	0
PcS_{120}	4	0.5	0	0.5	0
PcS_{121}	1	0	1	0	0
PcS_{122}	5	0.6	0.2	0.2	0
PcS_{123}	2	0.5	0	0.5	0
PcS_{124}	3	0	0.67	0.33	0
PcS_{125}	6	0.33	0.5	0	0.17
PcS_{126}	8	0.13	0.75	0.13	0
PcS_{127}	12	0.17	0.83	0	0

For each of the S-alleles detected, the absolute frequency is reported together with the relative frequency according to the four classes: RS (wild related species), LV (local varieties), NCV (nationally cultivated varieties), and ICV (internationally cultivated varieties).

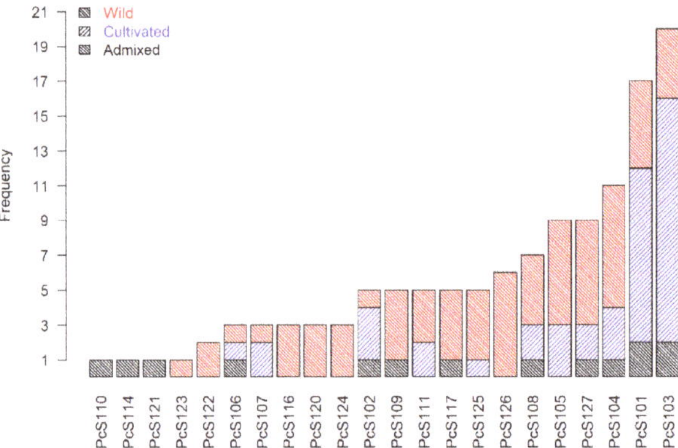

Figure 1. S-Allele frequency among analyzed accessions according to their population structure information shown in Bennici et al. (2018) [17]. Samples without an assigned subpopulation are excluded.

Results indicated that the two most abundant S-alleles, PcS_{103} and PcS_{101}, were largely detected in accessions characterized by a clear predominance of the "cultivated" subpopulation (60% and 56%, respectively; blue color in Figure 1), whereas the other S-alleles were mostly associated with individuals characterized by the "wild" subpopulation (red color in Figure 1). The alleles PcS_{116}, PcS_{120}, PcS_{122},

PcS_{123}, PcS_{124}, and PcS_{126} (18 accessions in total) were detected only in samples showing a "wild" genetic background (Figure 1).

The results presented in Table S2 (Supplementary Materials) were converted into a binary matrix and used to compute a PCA (Figure 2).

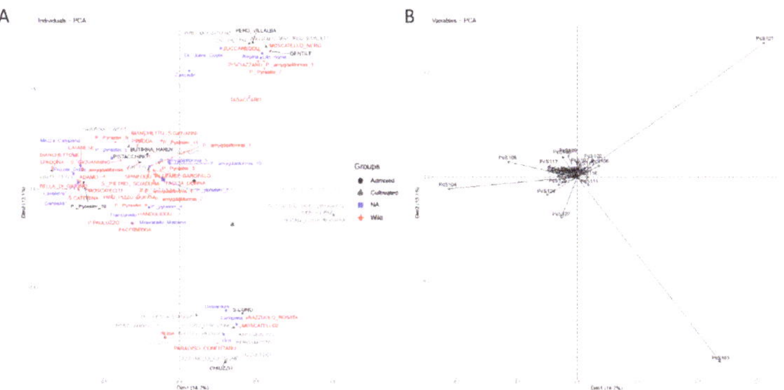

Figure 2. Principal component analysis (PCA) depicting the distribution of accessions over the first two PCs (Dim1 and Dim2) performed on the genetic data scored for the presence/absence of each S-allele. (**A**) PCA biplot: different colors indicate the subpopulations detected with structure analysis in Bennici et al. (2018) [17]: "wild" (red); "cultivated" (blue); "admixed" (black); not available (gray). (**B**) Loading projections of the variables employed.

The first two PCs explained 27.3% of the whole variability (14.2% and 13.1% for PC1 and PC2, respectively). The analysis of the first two principal components (Dim1 and Dim2) allowed the definition of four clusters composed of five (Dim1 > 0, Dim2 ~ 0, S-genotype = $PcS_{101,}$ PcS_{103}), 16 (Dim1 > 0, Dim2 > 0, one PcS_{101}), 18 (Dim1 ~ 0, Dim2 < 0, one PcS_{103}), and 47 (Dim1 < 0, Dim2 ~ 0, other *S*-alleles) accessions. The five accessions characterized by a PcS_{101}/PcS_{103} *S*-genotype included the NCVs "Butirra", "Buona Luisa", and "Virgolese", and the LVs "Putiru d'Estate" and "Pergolesi" (Figure 2). Even though the PCA was computed with *S*-allele data, the first principal component (Dim1) was highly predictive for the genetic structure results, with individuals plotted in the upper right and lower-right quadrants (Dim1 > 0) showing a predominance of the "cultivated" subpopulation (blue color), while samples characterized by Dim1 negative values were largely "wild" (red color).

4. Discussion

Despite its importance for breeding and as an agronomic trait, information on GSI genetic background is mainly known for the most commonly used varieties. The germplasm collection herein analyzed encompassed local cultivars selected through the last two centuries for their traits of agronomical interest such as chilling requirements, resistance to biotic/abiotic stress, and fruit quality. The present work aimed to decipher the *S*-genotype of such local varieties and to assess similarity and differences with the close wild accessions *P. pyraster* and *P. amygdaliformis* and with some national and international cultivars of *P. communis*. Analyses were carried out employing the PCR-based *S*-genotyping method described by Sanzol [14], resulting in the identification of 24 *S*-alleles [14,16]. *S*-RNase alleles identified in the reference ICVs and the NCVs "Bella di Giugno", "Coscia", and "Gentile" agreed with previous reports [6,8,10,11,14,23,24], confirming the reliability of the protocol for the detection of the known European pear *S*-RNases.

S-Allele genotyping allowed the definition of the complete *S*-genotype for 84% of the accessions, while, for the remaining 14 samples, only one allele was identified. Given the forced heterozygosity at the *S*-locus, the detection of a single amplicon (Table S2, Supplementary Materials) implied the presence of additional alleles that were undetected suggesting the occurrence of sequence diversity and/or of insertions/deletions (INDELs) at the probe site [25]. These 14 genotypes included mostly LVs and RSs, reinforcing the assumption that the gene pool of pears from the Etna region is, to a certain extent, different from that of the widely employed cultivars.

S-Alleles were not uniformly distributed across the germplasm; in particular, the most abundant *S*-alleles were PcS_{103}, detected in 23 accessions (61% LV; 39% NCV), and PcS_{101} detected in 21 accessions (10% RS; 48% LV; 19% NCV; 24% ICV) (Table 2). Fourteen *S*-alleles were detected in five or fewer accessions each (Table 2). Such variability reflected the different history and utilization of the accessions, with NCV or ICV cultivars widely employed both for cultivation and in breeding plans, such as "Coscia", "Gentile", "Cascade", "Dr. Jules Guyot", "Max Red Bartlett", "Old home", and "Williams", contributing to the spread of either PcS_{103} or PcS_{101}. Conversely, only two RS (*P. amygdaliformis* n.3 and *P. pyraster* n.7) showed PcS_{101} and none showed PcS_{103} (Table 1).

None of the most abundant *S*-alleles were exclusively detected in one of the four classes in which the germplasm was subdivided according to its diffusion, suggesting a high heterogeneity within each of the groups, with LV and NCV accessions being highly similar to either RS or ICV.

This result agreed with the findings of the structure analysis performed by Bennici et al. [17] in which the presence of two subpopulations ("wild" and "cultivated") was revealed and that, although characterizing the RSs and ICVs, respectively, coexisted in the LVs and NCVs. Within these two groups, accessions were either "wild" or "cultivated" with very few accessions showing an admixed genetic configuration. The high genetic diversity between RSs and ICVs was here confirmed by the analysis of the *S*-alleles; in fact, 10 *S*-alleles (PcS_{106}, PcS_{109}, PcS_{110}, PcS_{111}, PcS_{116}, PcS_{120}, PcS_{122}, PcS_{123}, PcS_{126}, and PcS_{127}), detected in at least one RS accession, were absent in the ICVs, and four *S*-alleles (PcS_{102}, PcS_{107}, PcS_{113}, PcS_{114}) were detected in ICVs and absent in RS accessions.

Interestingly, when the *S*-allele genotypic data were matched with the population structure results, the most frequent *S*-alleles, PcS_{103} and PcS_{101}, were found largely present in the "cultivated" group (Figure 1), while the remaining *S*-alleles were exclusively or predominantly detected in the "wild" accessions. The PCA analysis confirmed the close relationship between *S*-alleles and the genetic stratification of the germplasm collection with accessions that largely clustered according to their "wild" or "cultivated" nature (Figure 2).

Among the *S*-alleles largely present in the "wild" subpopulation, PcS_{126} and PcS_{127}, were detected in 20 accessions through the design of specific primers (Table 2), the same *S*-alleles were already detected among Iranian local varieties [16]. Nikzad Gharehaghaji and colleagues [16] highlighted the high sequence homology of PcS_{126} and PcS_{127} with the *S*-alleles of the Asian pear species *P. korshinskyi* (S9) and *P.* × *bretschneideri* (S19), respectively, suggesting that these alleles might have been introgressed by hybridization. In fact, *P. korshinskyi* Litv. is native to Central Asia, and it was proposed to have originated from the hybridization between *P. communis* and *P. regelii* Rehder., a species native of Afghanistan [26–28]. The Chinese white pear (*P.* × *bretschneideri* Rehder.) is native of East Asia [28]. Molecular studies using RAPD and SSR markers suggested that *P.* × *bretschneideri* might share a common ancestor with *P. pyrifolia* [29,30]. Recently, a genotyping-by-sequencing (GBS) study proposed that the genomes of the Ussurian pear (*P. ussuriensis* Maxim.) and the Chinese sand pear (*P. pyrifolia* Nakai) could have both contributed to the origin of *P.* × *bretschneideri* [31]. Interestingly, four LVs ("Faccibedda", "Pauluzzo", "Moscatello Maiolino", and "Franconello") showed both alleles Pc_{S127} and Pc_{S126}, suggesting a possible contribution of Asian pear species on the genetic backgrounds of these accessions.

The *S*-RNase allele PcS_{117} (detected in the NCV "Bianchetto" and the LVs "Piccola Dolce", "Piridda", "Pistacchino", "Sciaduna", and "Spineddu", Table 1) was amplified using the primer pairs developed for the PpS_9 allele of the Japanese pear *P. pyrifolia*. PpS_9 is one of the most common *S*-alleles characterizing Japanese pear; nevertheless, the high similarity between PpS_9 and PcS_{117} from *P. communis* was already described in [16]. The occurrence of hybridization events between *P. communis* and other wild species is also confirmed by the high degree of homology between the *S*-RNase alleles of *P. pyraster* and those of *P. communis* [20]. Furthermore, many plants identified as *P. pyraster* likely represent various stages of hybridization between *P. pyraster* and *P. communis* [32]. Such close genetic proximity of the wild accessions to most of the LVs could also be explained by their wide use as rootstock to propagate selected varieties, as well as increase plant vigor and adaptability in different pedoclimatic conditions [33].

Collectively, the *S*-genotyping results confirmed the existence of genetic distinctness between the "wild" and "cultivated" subpopulations which emerged from previous SSR analyses. While natural and human selections indeed shaped the population genetic structure differently, forced allogamy and insect-mediated pollination favored gene flow between wild and cultivated populations. It is reasonable to hypothesize that at least some of the ICVs did not come in contact with the Sicilian pear populations, preventing gene exchange with local genotypes; however, those cultivars and genotypes which were introduced in Sicily in historical times offered "new" *S*-alleles that had the chance to spread into the local gene pool. It should be considered that, unlike other loci, the *S*-locus is subject to frequency-dependent balancing selection; i.e., pollen harboring rare alleles has increased chances to be accepted by pistils with respect to more frequent ones [34], making the frequency of a rare allele increase across generations until an equilibrium is reached. In such a scenario, an *S*-allele introduced in Sicily through foreign cultivars would not only rapidly spread in the local population (thanks to the ability of its pollen to be accepted by 100% of local pistils), but would then have a great chance to become a stable part of the local gene pool and to be maintained for long times in the population, as frequency-dependent balancing selection makes it very unlikely to loose *S*-alleles due to random frequency fluctuations or genetic drift [35]. Natural selection, therefore, might have favored the introgression of new *S*-alleles from cultivated to wild populations; however, on the other hand, the opposite path (from wild to cultivated material) would be theoretically more unlikely to occur, as human selection tends to eliminate wild-related detrimental traits, which in most cases affect hybrid progenies. On the basis of these assumptions, wild populations are expected to maintain a greater allelic diversity at the *S*-locus than cultivated ones. The SSR-based data on population structure previously described, combined with the *S*-genotypes determined in this study, support the following hypothesis: when the two groups supposed to correspond to "wild" and "cultivated" subpopulations according to SSR data are analyzed separately, the former shows a higher number of *S*-alleles than the latter (19 vs. 11; Figure 1). Moreover, allele frequencies in the "wild" subgroup are less skewed, with none of the alleles reaching 10%, while, in the "cultivated" group, only two alleles accounted for more than 50% of the allelic composition of the entire subpopulation (PcS_{101} and PcS_{103}; Figure 1). The *S*-allele composition of the "wild" group is, therefore, less distant from an equilibrium state, in which natural balancing selection tends to maintain comparable frequencies for all the *S*-alleles present in the population.

5. Conclusions

The *S*-allele genotyping analysis was conducted on an ex situ collection encompassing most of the local Sicilian varieties selected for their traits of agronomic interest complemented with national/international cultivars and with related wild species. Results shed light on the distribution of the *S*-alleles among accessions and revealed that RSs display a high diversity from ICVs, in terms of the S-allele composition. On the other hand, LSs and NCVs showed a more heterogeneous distribution of the *S*-alleles as the results of a more complex history of hybridization. The analysis of the *S*-allele distribution provided novel insight into the contribution of RSs and ICVs to the genetic background of the LVs and NCVs. Furthermore, these results provide information that can be readily employed by breeders for the set-up of novel mating schemes, both for rootstocks and varieties, and they are the ideal completion of the phenotypic and genotypic evaluation of the Mount Etna pear germplasm described by Ferlito et al. (under review) and Bennici et al. [17].

Supplementary Materials: The following are available online at http://www.mdpi.com/1999-4907/11/11/1228/s1: Table S1. Allele-specific primer pairs used for *S*-genotyping; Table S2. *S*-Genotypes assigned to the accessions in analysis combining consensus and allele-specific PCR primers.

Author Contributions: Conceptualization, G.D. and L.D.; methodology: G.D. and P.D.F.; validation G.L.C. and P.D.F.; formal analysis, S.B. and M.D.G.; resources F.F.; data curation, M.D.G.; writing—original draft preparation S.B. and M.D.G.; writing—review and editing G.D., A.G., and S.L.M. All authors read and agreed to the published version of the manuscript.

Funding: This research received no external funding.

Conflicts of Interest: The authors declare no conflict of interest.

References

1. McClure, B.; Haring, V.; Ebert, P.; Anderson, M.; Simpson, R.; Sakiyama, F.; Clarke, A. Style self-incompatibility gene products of *Nicotiana alata* are ribonucleases. *Nature* **1989**, *342*, 955–957. [CrossRef] [PubMed]
2. De Nettancourt, D. *Incompatibility and Incongruity in Wild and Cultivated Plants*; Springer: Berlin, Germany, 2001; Volume 3.
3. Broothaerts, W.; Janssens, G.A.; Proost, P.; Broekaert, W.F. cDNA cloning and molecular analysis of two self-incompatibility alleles from apple. *Plant Mol. Biol.* **1995**, *27*, 499–511. [CrossRef] [PubMed]
4. Sassa, H.; Kakui, H.; Miyamoto, M.; Suzuki, Y.; Hanada, T.; Ushijima, K.; Kusaba, M.; Hirano, H.; Koba, T. S Locus F-Box Brothers: Multiple and Pollen-Specific F-Box Genes With S Haplotype-Specific Polymorphisms in Apple and Japanese Pear. *Genetics* **2007**, *175*, 1869–1881. [CrossRef] [PubMed]
5. De Franceschi, P.; Dondini, L.; Sanzol, J. Molecular bases and evolutionary dynamics of self-incompatibility in the *Pyrinae (Rosaceae)*. *J. Exp. Bot.* **2012**, *63*, 4015–4032. [CrossRef] [PubMed]
6. Zuccherelli, S.; Tassinari, P.; Broothaerts, W.; Tartarini, S.; Dondini, L.; Sansavini, S. S-Allele characterization in self-incompatible pear (*Pyrus communis* L.). *Sex. Plant Reprod.* **2002**, *15*, 153–158. [CrossRef]
7. Sanzol, J.; Herrero, M.B. Identification of self-incompatibility alleles in pear cultivars (*Pyrus communis* L.). *Euphytica* **2002**, *128*, 325–331. [CrossRef]
8. Zisovich, A.; Stern, R.; Sapir, G.; Shafir, S.; Goldway, M. The RHV region of S-RNase in the European pear (Pyrus communis) is not required for the determination of specific pollen rejection. *Sex. Plant Reprod.* **2004**, *17*, 151–156. [CrossRef]
9. Sanzol, J.; Sutherland, B.G.; Robbins, T.P. Identification and characterization of genomic DNA sequences of the S-ribonuclease gene associated with self-incompatibility alleles S1 to S5 in European pear. *Plant Breed.* **2006**, *125*, 513–518. [CrossRef]
10. Takasaki, T.; Moriya, Y.; Okada, K.; Yamamoto, K.; Iwanami, H.; Bessho, H.; Nakanishi, T. cDNA cloning of nine S alleles and establishment of a PCR-RFLP system for genotyping European pear cultivars. *Theor. Appl. Genet.* **2006**, *112*, 1543–1552. [CrossRef]
11. Moriya, Y.; Yamamoto, K.; Okada, K.; Iwanami, H.; Bessho, H.; Nakanishi, T.; Takasaki, T. Development of a CAPS marker system for genotyping European pear cultivars harboring 17 S alleles. *Plant Cell Rep.* **2007**, *26*, 345–354. [CrossRef]
12. Sanzol, J.; Robbins, T.P. Combined Analysis of S-Alleles in European Pear by Pollinations and PCR-based S-Genotyping; Correlation between S-Phenotypes and S-RNase Genotypes. *J. Am. Soc. Hortic. Sci.* **2008**, *133*, 213–224. [CrossRef]
13. Goldway, M.; Takasaki-Yasuda, T.; Sanzol, J.; Mota, M.; Zisovich, A.; Stern, R.A.; Sansavini, S. Renumbering the S-RNase alleles of European pears (*Pyrus communis* L.) and cloning the S109 RNase allele. *Sci. Hortic.* **2009**, *119*, 417–422. [CrossRef]
14. Sanzol, J. Genomic characterization of self-incompatibility ribonucleases (S-RNases) in European pear cultivars and development of PCR detection for 20 alleles. *Tree Genet. Genomes* **2009**, *5*, 393–405. [CrossRef]
15. Sanzol, J. Pistil-function breakdown in a new S-allele of European pear, S21°, confers self-compatibility. *Plant Cell Rep.* **2008**, *28*, 457–467. [CrossRef] [PubMed]
16. Gharehaghaji, A.N.; Arzani, K.; Abdollahi, H.; Shojaeiyan, A.; Dondini, L.; De Franceschi, P. Genomic characterization of self-incompatibility ribonucleases in the Central Asian pear germplasm and introgression of new alleles from other species of the genus Pyrus. *Tree Genet. Genomes* **2014**, *10*, 411–428. [CrossRef]
17. Bennici, S.; Casas, G.L.; Distefano, G.; Di Guardo, M.; Continella, A.; Ferlito, F.; Gentile, A.; La Malfa, S. Elucidating the contribution of wild related species on autochthonous pear germplasm: A case study from Mount Etna. *PLoS ONE* **2018**, *13*, e0198512. [CrossRef]

18. Matsumoto, K.; Tamura, F.; Chun, J.-P.; Tanabe, K. Native Mediterranean Pyrus Rootstock, P. amygdaliformis and P. elaeagrifolia, Present Higher Tolerance to Salinity Stress Compared with Asian Natives. *J. Jpn. Soc. Hortic. Sci.* **2006**, *75*, 450–457. [CrossRef]
19. Łukasz, W.; Antkowiak, W.; Lenartowicz, E.; Bocianowski, J. Genetic diversity of European pear cultivars (*Pyrus communis* L.) and wild pear (*Pyrus pyraster* (L.) Burgsd.) inferred from microsatellite markers analysis. *Genet. Resour. Crop. Evol.* **2010**, *57*, 801–806. [CrossRef]
20. Łukasz, W.; Antkowiak, W.; Sips, M.; Slomski, R. Self-incompatibility alleles in Polish wild pear (*Pyrus pyraster* (L.) Burgsd.): A preliminary analysis. *J. Appl. Genet.* **2010**, *51*, 33–35. [CrossRef]
21. Ye, J.; Coulouris, G.; Zaretskaya, I.; Cutcutache, I.; Rozen, S.G.; Madden, T.L. Primer-BLAST: A tool to design target-specific primers for polymerase chain reaction. *BMC Bioinform.* **2012**, *13*, 134. [CrossRef]
22. R Core Team. *R: A Language and Environment for Statistical Computing*; R Foundation for Statistical Computing: Vienna, Austria, 2017.
23. Tassinari, P. Analisi molecolari ed agronomiche dell'incompatibilita' gemetofitica in pero (*Pyrus communis*) e ciliegio (*Prunus avium* e *P. cerasus*): Ricerche sul locus S. Ph.D. Thesis, University of Bologna, Bologna, Italy, 2005.
24. Mota, M.; Tavares, L.; Oliveira, C.M. Identification of S-alleles in pear (*Pyrus communis* L.) cv. 'Rocha' and other European cultivars. *Sci. Hortic.* **2007**, *113*, 13–19. [CrossRef]
25. Takasaki-Yasuda, T.; Nomura, N.; Moriya-Tanaka, Y.; Okada, K.; Iwanami, H.; Bessho, H. Cloning anS-RNaseallele, including the longest intron, from cultivars of European pear (*Pyrus communis* L.). *J. Hortic. Sci. Biotechnol.* **2013**, *88*, 427–432. [CrossRef]
26. Rehder, A. *A Manual of Cultivated Trees and Shrubs Hardy in North American Exclusive of the Subtropical and Warmer Temperate Regions*, 2nd ed.; Macmillan: New York, NY, USA, 1940; p. 403.
27. Katayama, H.; Tachibana, M.; Iketani, H.; Zhang, S.-L.; Uematsu, C. Phylogenetic utility of structural alterations found in the chloroplast genome of pear: Hypervariable regions in a highly conserved genome. *Tree Genet. Genomes* **2011**, *8*, 313–326. [CrossRef]
28. Volk, G.M.; Henk, A.D.; Richards, C.M.; Bassil, N.; Postman, J. Chloroplast sequence data differentiate Maleae, and specifically Pyrus, species in the USDA-ARS National Plant Germplasm System. *Genet. Resour. Crop. Evol.* **2018**, *66*, 5–15. [CrossRef]
29. Liu, Q.; Song, Y.; Liu, L.; Zhang, M.; Sun, J.; Zhang, S.; Wu, J. Genetic diversity and population structure of pear (*Pyrus* spp.) collections revealed by a set of core genome-wide SSR markers. *Tree Genet. Genomes* **2015**, *11*, 128. [CrossRef]
30. Teng, Y.; Tanabe, K.; Tamura, F.; Itai, A. Genetic Relationships of Pyrus Species and Cultivars Native to East Asia Revealed by Randomly Amplified Polymorphic DNA Markers. *J. Am. Soc. Hortic. Sci.* **2002**, *127*, 262–270. [CrossRef]
31. Kumar, S.; Kirk, C.; Deng, C.; Wiedow, C.; Knaebel, M.; Brewer, L. Genotyping-by-sequencing of pear (*Pyrus* spp.) accessions unravels novel patterns of genetic diversity and selection footprints. *Hortic. Res.* **2017**, *4*, 1–10. [CrossRef]
32. Dolatowski, J.; Podyma, W.; Szymańska, M.; Zych, M. Molecular studies on the variability of Polish semi-wild pears (*Pyrus*) using AFLP. *J. Fruit Ornam. Plant Res.* **2004**, *12*, 331–337.
33. Volk, G.M.; Cornille, A. Genetic Diversity and Domestication History in *Pyrus*. In *The Pear Genome*; Korban, S.S., Ed.; Springer: Cham, Germany, 2019; pp. 51–62.
34. Wright, S. The Distribution of Self-Incompatibility Alleles in Populations. *Evolution* **1964**, *18*, 609. [CrossRef]
35. Ioerger, T.R.; Clark, A.G.; Kao, T.H. Polymorphism at the self-incompatibility locus in Solanaceae predates speciation. *Proc. Natl. Acad. Sci. USA* **1990**, *87*, 9732–9735. [CrossRef]

Publisher's Note: MDPI stays neutral with regard to jurisdictional claims in published maps and institutional affiliations.

Article

Comparative Survey of Morphological Variations and Plastid Genome Sequencing Reveals Phylogenetic Divergence between Four Endemic *Ilex* Species

Tao Su [1,2], Mengru Zhang [1,2], Zhenyu Shan [1], Xiaodong Li [1], Biyao Zhou [1,2], Han Wu [1] and Mei Han [1,*]

[1] Co-Innovation Center for Sustainable Forestry in Southern China, College of Biology and the Environment, Nanjing Forestry University, Nanjing 210037, China; tao.su@cos.uni-heidelberg.de (T.S.); zhangmengru@njfu.edu.cn (M.Z.); njfuszy1007@163.com (Z.S.); xiaodong.li@chinahaisheng.com (X.L.); zhoubiyao@njfu.edu.cn (B.Z.); wh010510@163.com (H.W.)

[2] Key Laboratory of State Forestry Administration on Subtropical Forest Biodiversity Conservation, Nanjing Forestry University, Nanjing 210037, China

* Correspondence: sthanmei@njfu.edu.cn; Tel.: +86-158-9598-9551

Received: 3 August 2020; Accepted: 1 September 2020; Published: 3 September 2020

Abstract: Holly (*Ilex* L.), from the monogeneric Aquifoliaceae, is a woody dioecious genus cultivated as pharmaceutical and culinary plants, ornamentals, and industrial materials. With distinctive leaf morphology and growth habitats, but uniform reproductive organs (flowers and fruits), the evolutionary relationships of *Ilex* remain an enigma. To date, few contrast analyses have been conducted on morphology and molecular patterns in *Ilex*. Here, the different phenotypic traits of four endemic *Ilex* species (*I. latifolia*, *I. suaveolens*, *I. viridis*, and *I. micrococca*) on Mount Huangshan, China, were surveyed through an anatomic assay and DNA image cytometry, showing the unspecified link between the examined morphology and the estimated nuclear genome size. Concurrently, the newly-assembled plastid genomes in four *Ilex* have lengths ranging from 157,601 bp to 157,857 bp, containing a large single-copy (LSC, 87,020–87,255 bp), a small single-copy (SSC, 18,394–18,434 bp), and a pair of inverted repeats (IRs, 26,065–26,102 bp) regions. The plastid genome annotation suggested the presence of numerable protein-encoding genes (89–95), transfer RNA (tRNA) genes (37–40), and ribosomal RNA (rRNA) genes (8). A comprehensive comparison of plastomes within eight *Ilex* implicated the conserved features in coding regions, but variability in the junctions of IRs/SSC and the divergent hotspot regions potentially used as the DNA marker. The *Ilex* topology of phylogenies revealed the incongruence with the traditional taxonomy, whereas it informed a strong association between clades and geographic distribution. Our work herein provided novel insight into the variations in the morphology and phylogeography in Aquifoliaceae. These data contribute to the understanding of genetic diversity and conservation in the medicinal *Ilex* of Mount Huangshan.

Keywords: *Ilex* species; Aquifoliaceae; morphological traits; DNA C-value; plastid genome

1. Introduction

The *Ilex* L. (holly) is the only living woody dioecious angiosperm genus, accounting for approximately 700 species within the monogeneric family of Aquifoliaceae [1]. The *Ilex* species are evergreen and deciduous trees, prostrate shrubs, and climbers with a broad distribution from tropics to temperate regions [2,3]. Over 200 species have been documented in the center of East Asia and South America, whereas several species can grow in Europe, tropical Africa, and northern Australia [4]. The unexpected spreading in the oceanic islands as patchy populations prompted the assumption of unspecialized pollination and efficient seed dispersal by birds [5]. Recently, a report in phylogeny implicated the origin of the Aquifoliaceae as subtropical lowlands with a mesic climate type. This study indicated that the eternal existence of *Ilex* species in humid and warm subtropical monsoon forests of southern China could trace back to the middle Eocene [6,7].

For tens of decades, various *Ilex* species have been used as herbal infusions in traditional Chinese medicine. Some large-leaved species of *Ilex* (e.g., *I. kudingcha* and *I. latifolia*) are processed by brewing to produce a bitter-tasting "Kuding" tea, used for its nutritional merit and nourishment for the daily health stimulation, consumption, and pharmaceutical drugs [8,9]. A traditional Chinese herb, *I. pubescens*, is used particularly to treat cardiovascular diseases [10]. Originating from southern areas of South America, *I. paraguariensis* is consumed commercially as a popular beverage, known as "yerba mate" in America and the Middle East, highly appreciated for its peculiar flavor and health maintenance effects [11,12]. An alternative mate tea, *I. dumosa*, has a similar taste and mild effects; it is available to satisfy the different consumers as a low-caffeine and xanthine-containing substitute [13]. As the natural resources for pharmaceuticals and dietary food, progress in phytochemistry and pharmacology has led to discoveries of terpenoids, saponins, flavonoids, glycosides, amino acids, and other bioactive compounds in many additional *Ilex* species, reflecting the worldwide economic, medicinal, and clinical value of *Ilex* [13–15]. Some larger native *Ilex* species have been developed for timber production and garden use. Other holly species (e.g., *I. aquifolium*, *I. opaca*, and *I. crenata*) retaining the green foliage and bright red drupes are grown as traditional Christmas decorations and ornamentals during the winter holidays [16,17].

Enabling the estimation of absolute genome size, flow cytometry (FCM) has provided relevant clues for taxonomy in various plant species [18]. Independent contrast studies suggested a significant relationship between the phenotypic traits (e.g., cell size and stomata density) and the genome size in angiosperms [19]. However, a proposed correlation between cellular architecture and organelle DNA content remains poorly understood [20]. As all plastids possess the same DNA and a few functional features, the plastid genome comparison between different species of plant branches, integrating basic backgrounds with gene content and bar codes, was significantly relevant to the understanding of the evolution of plastid DNA and the adapted ancient environments [21]. The plastid phylogeny offers an alternative strategy for the investigation of phylogeography due to its maternal inheritance and conserved circular structure as well as small sufficient population size [22]. Increasing evidence suggests that plastid genomes provide more suitable material for species identification and conservation, enabling a significant improvement in the resolution of branches under the frame of the nuclear phylogenies [23,24]. To date, emerging advances in sequencing technologies have increased the availability of plastid genomes to approximately 3000 in the GenBank database (https://www.ncbi.nlm.nih.gov/genbank/). Therefore, contrasting analyses of the plastid genome and nuclear genome using the molecular genetic methodologies may provide the theoretical bases for the phylogenetic reconstruction and exploration of genetic diversity and plant systematics.

The unique subtropical monsoon pattern in Mount Huangshan (Anhui, China) led to high forest diversity and broad distribution of evergreen broad-leaved forest [25]. Based on a dynamic field survey in a large-scale forest plot, a total of 12 documented endemic *Ilex* species in Aquifoliaceae were found to be predominant [26]. However, comparative reports are lacking on the morphology traits in combination with molecular and genomic patterns within the *Ilex* genus. In this work, the four *Ilex* species were chosen to conduct a contrast analysis in terms of their phenotypic variation, DNA-C values, and the newly assembled plastid genomes. The main objectives were: (1) to investigate the possible relationships between the morphological traits and nuclear genome size between the four selected *Ilex* species, (2) to discover the variation and highly divergent regions of the plastome that could be used in the classification and identification of various *Ilex* species, and (3) to analyze the plastid genome structures and reconstruct the plastome-derived phylogenetic relationships of *Ilex* species in Aquifoliaceae.

2. Materials and Methods

2.1. Plant Materials

The respective organs (e.g., leaves and seeds) of *I. latifolia*, *I. suaveolens*, *I. viridis*, and *I. micrococca* were harvested from 10.24 ha (320 m × 320 m) forest plot (30°8′26″ N, 118°6′38″ E) in Mount Huangshan, Anhui, China [26]. The plot ranges altitude from 430 to 565 m with an annual average temperature of 7.8 °C and annual precipitation of 2394.5 mm. The voucher specimens of four *Ilex* species (accession numbers: YL20190417014, YL20190417015, YL20190417016, and YL20190417017) were preserved in the herbarium of Nanjing Forestry University, Nanjing, China.

2.2. Phenotype Quantification and Determination of DNA Content

For each species, more than 60 healthy mature leaves and 300 mature seeds of 5 independent trees of similar age were randomly sampled for the morphological analyses. In total, thirty anatomic sections of epidermis were prepared from the cut leaves (5 mm × 5 mm) macerated by H_2O_2-HAC solutions [27]. To quantitate the size of the upper leaf epidermal cell (LEC), the stomata aperture (STA), and the stomata density (STD), twenty visual fields were captured using the same scale during optical microscopy. The leaf area (LA) was measured from 50 randomly selected leaves by Image J v1.53c (https://imagej.nih.gov/ij/) after scanning with Expression 11000XL (EPSON, Beijing, China). The specific leaf area (SLA) was calculated based on the ratio of leaf area to leaf dry mass. We used 100 air-dried seeds to determine the seed weight (SW). The variance of three perpendicular seed dimensions (VSD) was calculated using the average of 50 seeds according to a previous report [28]. Spherical seeds have a variance of 0, and elongated or flattened seeds have a variance of up to 0.33. Flower size (FS) was measured from the average diameters of 12 female flowers. The significance of phenotypic variation between four *Ilex* samples was statically analyzed using SPSS 24.0 (https://www.ibm.com) based on the ANOVA ($p < 0.05$) and Duncan's multiple range tests.

For the determination of the DNA C-value, the young leaves were chopped to isolate the crude nuclear DNA with the addition of woody plant buffer followed by RNase digest. DNA staining of propidium iodide (PI) and FCM analysis were performed based on a previous report [29]. Together with the internal standard (*Solanum lycopersicum*, 2C = 2.00 pg), the resulting suspensions were analyzed with BD Influx^{TM} cell sorter (BD, Piscataway, NJ, USA). The histograms of FCM were generated by the software BD FACSTM 1.0.0.650. The coefficients of variation (CV) of DNA peaks below 5% were considered as reliable. The chromosome numbers in four diploid *Ilex* species were retrieved from the IPCN database (http://legacy.tropicos.org/Project/IPCN). The genome size (DNA C-value or the haploid DNA content) was calibrated by multiplying the standard by the ratio of the mean fluorescent intensity of each sample to that of the standard [30]. The DNA-C value is represented as means ± standard error (±SE) of at least four independent biological replicates.

2.3. Plastome Sequencing, Assembly, Annotation, Codon Usage, and Repeat Analyses

The fresh leaves of *I. suaveolens*, *I. viridis*, and *I. micrococca* were harvested and flash-frozen in liquid nitrogen. DNA was extracted using a method modified from a previous report [31]. The plastid genome sequencing data of *I. latifolia* were obtained from the NCBI database (https://www.ncbi.nlm.nih.gov/) based on the new accession MN688228 [32]. The next-generation sequencing of the whole-plastid genomes was performed by Biodata Biotechnologies Inc. (Hefei, China) for *I. suaveolens*, *I. viridis*, and *I. micrococca* on the BGISEQ-500 platform (BGI, Shenzhen, China). Approximately 50 MB of high-quality clean paired-end reads was generated, and the filtered sequences were assembled by SPAdes assembler 3.14.1 (http://cab.spbu.ru/software/spades/) with default parameters [33]. The genome was annotated using the CPGAVAS (http://47.96.249.172:16014/analyzer/home) program [34] and checked further by DOGMA (https://dogma.ccbb.utexas.edu/) following with nBLAST searches in NCBI based on the reference genome to identify the specific genes [1]. The circular graphical maps of the plastome were drawn using the program OGDRAWv1.3.1 (https://chlorobox.mpimp-golm.mpg.de/OGDraw.html) [35]. The complete plastome sequences of *I. suaveolens*, *I. viridis*, and *I. micrococca* were submitted to the GenBank with accession numbers MN830249, MN830250, and MN830251, respectively. The relative synonymous codon usage for all protein-encoding genes was analyzed using MEGA X (https://www.megasoftware.net/) [36]. The use frequency of amino acids was calculated by the percentage of the codons divided by the total codons. The simple sequence repeats (SSRs) were analyzed by MISA (http://pgrc.ipk-gatersleben.de/misa/) [37]. The parameters of categorized SSRs (mono-, di-, tri-, tetra-, pena-, and hexanucleotide) sequence length were set up with a minimum number of repeats of 8, 5, 4, 3, 3, and 3, respectively. The long repeat sequences, including the forward, palindrome, reverse, and complement repeats, were analyzed using the online program REPuter (https://bibiserv.cebitec.uni-bielefeld.de/reputer) with adjusted parameters [38].

2.4. Plastome Divergence and Phylogenetic Analyses

Using the annotation of *I. suaveolens* as the reference, we compared of the entire plastid genomes of eight *Ilex* species in Aquifoliaceae using the program GView (https://www.gview.ca/wiki/GView/WebHome) [39] and the mVISTA program (http://genome.lbl.gov/vista/mvista/submit.shtml) in the Shuffle-LAGAN mode [40,41]. For the inverted repeat (IR) expansion and contraction of border genes, eight *Ilex* plastomes were aligned to analyze the variations in the junctions of LSC, IRs, and SSC using IRscope (https://irscope.shinyapps.io/irapp/) [42]. In total, nineteen plastid genome sequences (15 *Ilex* species) were retrieved from the GenBank. *Populus trichocarpa*, *Populus deltoides*, *Quercus acutissima*, and *Helwingia himalaica* were used as outgroups. The multiple sequences were aligned using MAFFT v7.471 (https://mafft.cbrc.jp/alignment/server/index.html) [43]. The phylogenetic topology was constructed using the software MEGA X software by the methods of maximum likelihood (ML) and maximum parsimony (MP) using the nucleotide substitution model of Tamura-Neighbour. The bootstrap values are shown on the branches of the phylogenetic tree based on 1000 replicates.

3. Results

3.1. Variation of the Morphological Trait and Nuclear DNA Content

To investigate the variation in morphology between the four *Ilex* species, we initially performed a comparative analysis of phenotypic traits of the major vegetative and reproductive organs, including LA, LEC, SLA, STA, STD, SW, VSD, and FS (Table 1). Based on the anatomic analyses, the significant variation in LA was observed in all *Ilex* species, with *I. latifolia* having the highest value (79.19 cm^2) (Table 1). Significant differences were also observed in SLA, ranging the values from 40.88 cm^2/g (*I. latifolia*) to 165.4 cm^2/g (*I. micrococca*). Regarding the LEC, the highest value (938.65 μm^2) was found in *I. suaveolens*; however, the value varied insignificantly as *I. viridis* and *I. micrococca* have a similar-sized LEC (Figure 1b). Image analyses of stomata-related traits revealed that STD is drastically different in four *Ilex* species, and *I. micrococca* has the highest density (257.64/mm^2; Table 1 and Figure 1c). The significantly different values in SLA appeared to be related to the STD values in the four *Ilex* species, as both traits showed similar patterns (Table 1). Analyses of VSD indicated that *I. suaveolens* seeds are closer to spherical shape (0.1137). However, for STA, VSD, and FS, significantly statistical distinctions were not found within the four *Ilex* species.

To determine the DNA-C value of four *Ilex* species by FCM, *S. lycopersicum* was used as the internal standard to calculate the genome size [44]. The DNA-associated fluorescence on FCM histograms showed that the CV values for G0/G1 peaks were between 2.92% and 4.67% (Figure 1d). The low CV (< 5%) indicated a constant quality considered reliable for FCM assessments [45]. The inspection of the FCM fluorescent peak revealed that *I. micrococca* has the most abundant nuclear DNA content (3.053 pg), followed by *I. viridis* (2.519 pg). *I. suaveolens* (2.242 pg) and *I. latifolia* (1.910 pg) exhibited similar levels of DNA 2C-value. Further calculation of nuclear genome size (NG) revealed that the mean levels vary with a range from 955 Mb (*I. latifolia*) to 1493 Mb (*I. micrococca*) in the four *Ilex* species (Table 1).

Table 1. Comparison of the morphological traits between four subtropical *Ilex* species.

Traits	*I. latifolia* Thunb.	*I. suaveolens* (H. Lév.) Loes.	*I. viridis* Champ.ex Benth.	*I. micrococca* Maxim.
LA (cm^2)	79.19 ± 2.24 [a]	19.81 ± 1.84 [c]	8.89 ± 0.89 [d]	32.61 ± 1.26 [b]
SLA (cm^{-2}/g)	40.88 ± 2.42 [d]	93.02 ± 3.47 [c]	100.81 ± 3.91 [b]	165.4 ± 3.62 [a]
LEC (μm^2)	848.21 ± 94 [b]	938.65 ± 103 [a]	732.42 ± 85 [cd]	734.21 ± 81 [c]
STA (μm)	32.76 ± 0.35 [a]	30.55 ± 0.29 [c]	31.73 ± 0.29 [b]	32.65 ± 0.27 [ab]
STD (n/mm^2)	118.31 ± 17 [d]	216.90 ± 12 [c]	232.39 ± 11 [b]	257.64 ± 29 [a]
SW (g/100)	6.29 ± 0.46 [b]	3.04 ± 0.36 [c]	8.35 ± 0.69 [a]	2.12 ± 0.10 [d]
VSD	0.083 ± 0.03 [b]	0.114 ± 0.02 [a]	0.118 ± 0.01 [a]	0.136 ± 0.01 [a]
FS (mm)	5.20 ± 0.13 [cd]	7.10 ± 0.21 [a]	5.50 ± 0.11 [bc]	6.50 ± 0.17 [b]
DNA 2C (pg)	1.910 ± 0.021	2.242 ± 0.022	2.519 ± 0.038	3.053 ± 0.047
NG (≈Mb)	955	1121	1232	1493

LA, leaf area; LEC, upper leaf epidermal cells; SLA, specific leaf area; STA, stomata aperture; STD, stomata density; SW, weight of 100 seeds; VSD, variance of seed dimensions; FS, flower size in diameter; NG, nuclear genome size. The data represent mean values ± SE of the independent biological replicates (see materials and methods). According to Duncan's multiple range tests, the superscripted letters indicate statistical significance per ANOVA ($p < 0.05$) between four *Ilex* species.

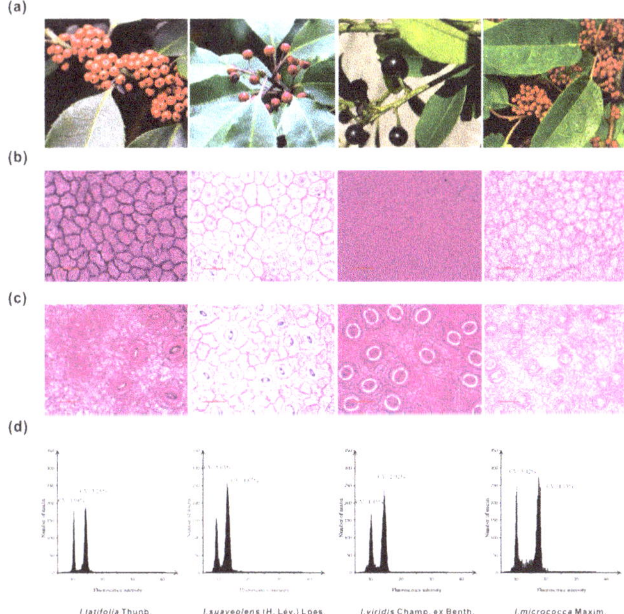

Figure 1. Analyses of the morphological traits and the fluorescent histograms of DNA content in four *Ilex* species. (**a**) The phenotypes of leaves and fruits; (**b**) LEC images; (**c**) STA and STD images; (**d**) flow cytometry (FCM) histograms obtained from leaves of four *Ilex* species. Nuclear DNA was stained with PI, using tomato (*S. lycopersicum*) as the internal standard.

3.2. The Plastid Genome Features and Sequence Divergence

The de novo sequencing and reference-guided assembling of the plastid genome of four *Ilex* species revealed a double-stranded circular DNA with a range of length from 157,601 (*I. latifolia*) to 157,857 bp (*I. suaveolens*) (Figure 2). The typical combined quadripartite structures of the plastid genome contain two inverted repeats (IR) regions, IRA and IRB, between 26,065 and 26,102 bp, each separated by a large single-copy (LSC, 87,020–87,255 bp) and a small single-copy (SSC, 18,394–18,434 bp) sections. Except for *I. latifolia*, the plastomes encode 134 genes comprised of 89 putative protein-encoding (PE) genes, 37 transfer RNA (tRNA) genes, and 8 ribosomal RNA (rRNA) genes in the other three *Ilex* species (Table 2). Among the annotated genes in *Ilex* plastome, eight PE genes (*rps7*, *ndhB*, *ycf15*, *ycf2*, *rpl23*, *rpl2*, *rpl12*, and *ycf1*), five tRNA genes (*trnV-GAC*, *trnL-GAU*, *trnA-UGC*, *trnR-ACG*, and *trnN-GUU*), and all four rRNA genes (*rrn4.5*, *rrn5*, *rrn23*, and *rrn16*) are anchored in the IR regions. Twelve PE genes and one tRNA gene (*trnL-UAG*) characterize the locations in the SSC sections (Figure 2). Most of the gene sequences assembled as a single-copy, whereas 18 genes occur with duplication in IR regions, including seven PE genes, four rRNA genes, and six tRNA genes (Table 2). Interestingly, *ycf1* is the only variable gene identified with an incomplete duplication in the junction of SSC and IRB regions (Figure 2 and Table 2). In *Arabidopsis*, translocon complex of *ycf1/Tic214* and *Tic20* has been proposed as the central component for plastid proteins accumulation [46]; however, a recent report stated that *ycf1/Tic214* might not be involved in the general import machinery [47]. All four *Ilex* plastomes uniformly contain *ycf2*, a duplicated gene, showing a maximum length of 6894 bp in IR regions. Another single-copy gene *rpoC2* has a relatively shorter length (4155 bp) in the LSC section. Besides, a total of 19 genes, including 11 PE genes and 7 tRNA genes, were identified with two exons, while three of PE genes (*clpP*, *ycf3*, and *rps12*) contain three exons. Contrast analyses revealed a slight variation in GC content in various regions of four *Ilex* plastomes (Table S1). Additionally, analyses of the total numbers of universal genetic code for the coding genes showed a similar range from 26,729 (*I. viridis*) to 27,121 (*I. latifolia*) (Table S2). Based on the statistical analysis of the codon usage, leucine (Leu) represents the most abundant amino acid with a frequency of 10.5%, whereas cysteine (Cys) shows the lowest abundance of 1.1%. Overall, the codon usage in all identified genes exhibits similar patterns across the four *Ilex* species. The almost identical gene numbers, annotation, and plastid genome length prompted us to further explore the sequence variation and divergence through the analyses of SSRs and long repeat sequences.

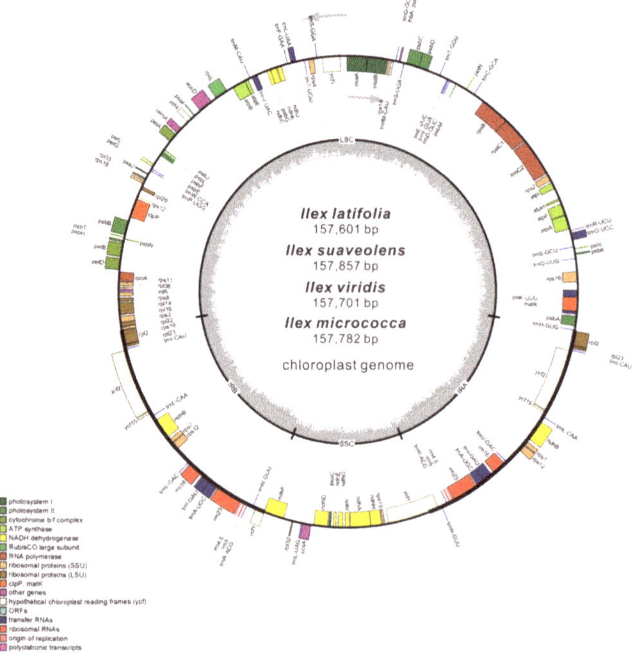

Figure 2. The plastome map of four *Ilex* species. The genes are shown by the gray arrowheads. Genes inside the circle indicate clockwise transcription, and those outsides are transcribed counterclockwise. The protein-encoding genes are marked in different colors. The GC content graphs are included as dark gray bars toward the center of the diagram.

Table 2. List of annotated genes in the plastid genome of *Ilex*.

Category	Gene Family	Gene Name	Numbers
Photosynthesis	Photosystem I	*psaA, psaB, psaC, psaI, psaJ*	5
	Photosystem II	*psbA, psbB, psbC, psbD, psbE, psbF, psbG, psbH, psbI, psbJ, psbK, psbL, psbM, psbN, psbT, psbZ/lhbA*	16
	Cytochrome b/f	*petA, petB*[2], *petD*[2], *petL, petG, petN*	6
	ATP synthase	*atpA, atpB, atpE, atpF*[2], *atpH, atpI*	6
	Cytochrome c	*ccsA*	1
	photosystem I assembly	*ycf3*[3], *ycf4*	2
	NADH dehydrogenase	*ndhA*[2], *ndhB*[2d], *ndhC, ndhD, ndhE, ndhF, ndhG, ndhH, ndhI, ndhJ, ndhK*	12
	Rubisco subunit L	*rbcL*	1
Transcription and translation	RNA polymerase β-subunit	*rpoC1* [2], *rpoC* [2], *rpoA, rpoB*	4
	30S ribosomal protein S	*rps2, rps3, rps4, rps7* [d], *rps8, rps11, rps12* [d+], *rps14, rps15, rps16* [2], *rps18, rps19*	14
	50S ribosomal protein L	*rpl2* [2d], *rpl14, rpl16* [2], *rpl20, rpl22, rpl23* [d], *rpl32, rpl33, rpl36*	11
RNA	Ribosomal RNA	*rrn4.5* [d], *rrn5* [d], *rrn16* [d], *rrn23* [d]	8
	Transfer RNA	*trnP-U/GGG, trnH-GUG, trnK-UUU2, trnQ-UUG, trnS-GCU, trnG-GCC2, trnG-UCC, trnR-UCU, trnC-GCA, trnD-GUC, trnY-GUA, trnE-UUC, trnT-GGU, trnM-CAU, trnS-UGA, trnfM-CAU, trnS-GGA, trnT-UGU, trnL-UAA2, trnF-GAA, trnV-UAC2, trnW-CCA, trnP-UGG, trnL-CAA* [d], *trnV-GAC* [d], *trnI-GAU2* [d], *trnR-ACG* [d], *trnL-UAG, trnN-GUU* [d], *trnA-UGC2* [d], *trnI-CAU* [d]	37
Other	RNA splicing	*matK*	1
	Plastid envelope	*cemA*	1
	Acetyl-CoA carboxylase β	*accD*	1
	Serine Protease	*clpP3*	1
	Translational initiation factor 1	*infA*	1
Unknown function	Conserved hypothetical gene	*ycf1* [d], *ycf2* [d], *ycf15* [d]	6
		orf56 [d], *orf42* [d], *ycf68* [d], *orf188*	7 *

[2] Genes containing two exons; [3] genes containing three exons; [d] two gene copies in the IRs; [d+] two gene copies with one shared exon 1 in the LSC, whereas exon 2 and 3 are in the IRs; * pseudogene.

3.3. *Analyses of SSRs and Long Repeat Sequences*

Using the MISA program, a total of 221 SSRs (mono- /di- /trinucleotide repeats) were characterized in the plastid genomes of *I. latifolia* (46/2/1), *I. suaveolens* (46/3/1), *I. viridis* (53/3/1), and *I. micrococca* (52/2/1). Among the SSRs, the mononucleotide repeats are the most abundant type with a location in the non-coding regions, which is related to AT richness. The proportions of A/T sequences in the mononucleotide repeats appear to be identical in the four *Ilex* species, varying from 92.00% (*I. suaveolens*) to 92.98% (*I. viridis*), whereas C/G only exists in the plastid genomes of *I. latifolia* and *I. micrococca*. The AT/AT sequence of dinucleotide repeats showed similar levels (1.75%–4.08%); however, the AAT/ATT sequence of trinucleotide repeats showed significant variability, ranging in proportion from 1.82% to 6.00% (Figure S1). Other types of SSRs (e.g., tetra- and pentanucleotide repeats) were not identified in any *Ilex* plastid genomes. The long repeats (forward, reverse, complement, and palindrome repeats) were concurrently analyzed by REPuter. A total of 196 unique long repeats were detected in *I. latifolia* (24/0/0/22), *I. suaveolens* (38/9/3/0), *I. viridis* (41/8/1/0), and *I. micrococca*

(38/9/3/0) (Figure S2). The forward repeats appear to be the most common type in the four plastid genomes. We found no reverse and complement repeats identified in *I. latifolia*, but it uniquely contains the palindrome repeats (Figure S2a). Further analyses of various types of long repeats showed that the 20–30 bp sequence length is the typical pattern (Figures S2b–e). In four plastid genomes, the sequence length in reverse and complement repeats usually has a limitation of 30 bp, whereas this length can be extended to 60 bp in the forward and palindrome repeats.

3.4. Comparative Analyses of Complete Plastomes in Ilex Species

We used *I. suaveolens* as a reference to conduct a BLAST comparison of eight *Ilex* species, showing that the entire plastid genomes are well conserved across eight selected species. The LSC and SSC regions were found to be more substantially divergent than the IR regions (Table 3, Figure 3). The non-coding regions appear to have more significant variation compared with the coding regions, in which some genes are relatively conserved. The VISTA analysis resulted in the findings of 12 hotspot regions for genome divergence and variable genes (e.g., *rpoC1*, *rbcL*, *ndhF*, *clpP*, and *psbA*). The highly divergent hotspot regions are particularly located in the intergenic regions than in coding regions, including *trnH-psbA*, *matK-rps16*, *psbK-psbI-trnS-trnG*, *petN-psbM*, *trnE-psbD-trnT*, *trnS-psbZ-psaB*, *trnL-ycf3*, *rbcL-accD-ycf4*, *clpP-rpl33*, *rpl16-rpoA*, *rpl32-ccsA*, and *ycf15-rps12-rrn16* (Figure 4). These divergent sites may facilitate the development of potential DNA markers for the species identification and reconstruction of phylogeny in the genus *Ilex*.

Table 3. List of plastid genome features in *Ilex* species.

Species	Plastome (bp)	LSC (bp)	SSC (bp)	IRs (bp)	PE Genes	tRNA Genes	rRNA Genes	GC (%)
I. szechwanensis	157,900	87,204	18,513	52,183	96	40	8	37.6
I. pubescens	157,741	87,109	18,436	52,196	96	40	8	37.7
I. paraguariensis	157,614	87,144	18,307	52,154	86	37	8	37.6
I. wilsonii	157,918	87,266	18,432	52,220	96	40	8	37.6
I. latifolia	157,610	87,020	18,427	52,154	95	40	8	37.7
I. suaveolens	157,857	87,255	18,398	52,204	89	37	8	37.6
I. viridis	157,701	87,177	18,394	52,130	89	37	8	37.7
I. micrococca	157,782	87,200	18,434	52,148	89	37	8	37.6
I. integra	157,548	86,935	18,426	52,186	86	37	8	37.6

Further sequence comparison of quadripartite borders revealed that the IR regions are extremely conserved in eight *Ilex* species (Figure 5). The various sites of the *rps19*, *rpl2*, *ycf1*, and *ndhF* are generally located in the junction of IRs/SSC (JSA/B) and IRs/LSC (JLA/B). The *rps19* locates within the LSC region, showing 8–13 bp gaps to the JLB in *I. latifolia*, *I. pubescens*, *I. wilsonii*, *I. szechwanensis*, and *I. micrococca*. *rps19* was identified across the JLB with the addition of 4 bp in *I. paraguariensis*, *I. viridis*, and *I. suaveolens*. At the JLA, gene *rpl2* shows the same location of 55 bp away from the LSC in all *Ilex* species. A tRNA gene (*trnH*) showed 11 bp shifts in the LSC region (Figure 5). The JLB and JLA are moderately conserved, whereas the JSA and JSB are strikingly different in all *Ilex* species. One short copy gene, *ycf1* in IRB, shows a significant sequence variation, ranging in length from 1038 bp (*I. wilsonii*) to 1085 bp (*I. suaveolens*). The other long copy gene of *ycf1* in IRB shows an identical length of 5690 bp in most *Ilex* species, excluding *I. paraguariensis* (5693 bp), *I. wilsonii* (5684 bp), and *I. micrococca* (5684 bp). Within the SSC, the gene *ndhF*, which is correlated with photosynthesis, differs in length from 2231 bp, 2258 bp (15 bp shift), and 2270 bp (40 bp shift) from the JSB. The overlapping sequences of the short copy gene *ycf1* and *ndhF* are also commonly found in *I. latifolia*, *I. paraguariensis*, *I. micrococca*, *I. viridis*, and *I. suaveolens*. Other variable sequences were also identified between the gene *psbA* and the JLA (Figure 5). Overall, the compared sequence information of the IRs/SC boundaries suggested that the contractions and expansions in IRs regions have relatively stable patterns in the eight *Ilex* plastid genomes.

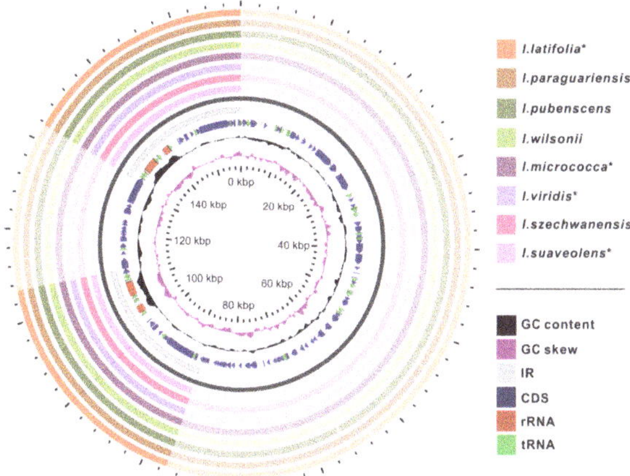

Figure 3. Plastome comparison between eight *Ilex* species in GView. The eight outer circles represent the BLAST results for *I. suaveolens* vs. *I. latifolia*, *I. paraguariensis*, *I. pubescens*, *I. wilsonii*, *I. micrococca*, *I. viridis*, *I. szechwanensis*, and itself, respectively. The clockwise inner cycle shows the CDS, ribosomal RNA (rRNA) genes, and transfer RNA (tRNA) genes in the plastome of *I. suaveolens*. The GC skew in a purple color indicates either G > C or G < C, and the GC content is shown in black.

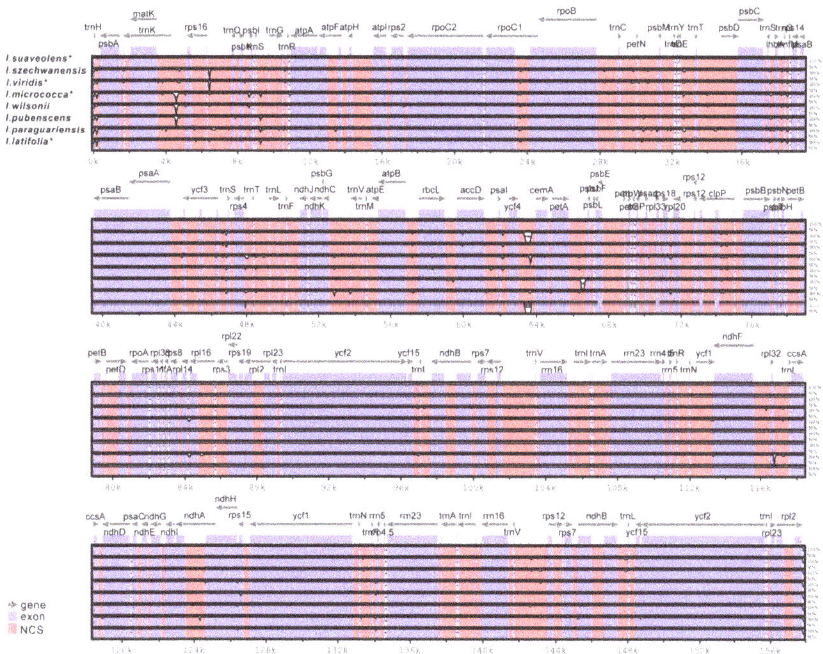

Figure 4. VISTA visualization of the alignment between the eight *Ilex* plastomes. The plots show the sequence identity with the *I.suaveolens* plastome, used as a reference. The vertical and horizontal axes represent the sequence consistency degree (50%–100%) and the sequence length, respectively. The locations of divergent hotspot regions are labeled along the top of the alignment, where the gray arrows indicate the orientation of the annotated genes. The red bars indicate the non-coding sequences (NCS), and white peaks represent differences in chloroplast genomes.

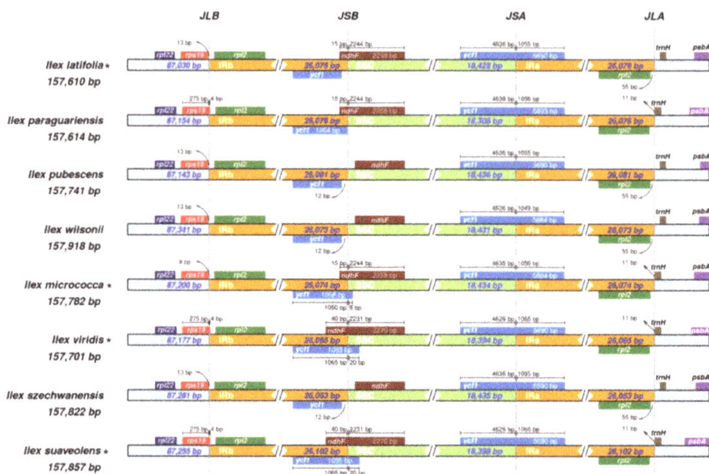

Figure 5. Comparative analyses of the junctions for the SSC/IRs and LSC/IRs regions among eight *Ilex* plastomes. The colored boxes above the strip scaled with sequence length indicate the denoted genes. The gaps in base length (bp) are indicated between boxed genes and boundaries. JLA, the junction of IRa/LSC; JLB, the junction of IRb/LSC; JSA, the junction of IRa/SSC; JSB, the junction of IRb/SSC.

3.5. Phylogenetic Analyses of Ilex in Aquifoliaceae

To investigate the phylogenies, the entire plastid genome sequences of 15 *Ilex* accessions were aligned using MAFFT. The phylogenetic topology was generated by MEGA X using the ML method supported with 1000 bootstrap values. *P. trichocarpa*, *P. deltoides*, *Q. acutissima*, and *H. himalaica* were used as out groups in the phylogenetic analyses. In Figure 6, the phylogenetic tree shows that four clades were deduced from all examined *Ilex* species. *I. latifolia* and *I. integra* cluster together with an identical high value in clade I, which also includes three additional *Ilex* species (*I. delavayi*, *I.sp. XY-2016*, and *I. cornuta*). All of them are evergreen trees with leathery leaves and broad distribution in subtropical Asia. The endemic *I. viridis* and *I. suaveolens* show close relationships in the clade III. *I. micrococca* clusters around *I. wilsonii* and *I. asprella* in clade IV, and the latter two are deciduous shrubs or trees. Yerba mate *I. paraguariensis* and *I. dumosa*, originating from South America, are located in clade II. The plastid phylogeny revealed that the species might not have originated from a single ancestor in Aquifoliaceae. The full encoding sequences were used to construct the phylogenetic tree, showing a distinctive clade (Figure S3).

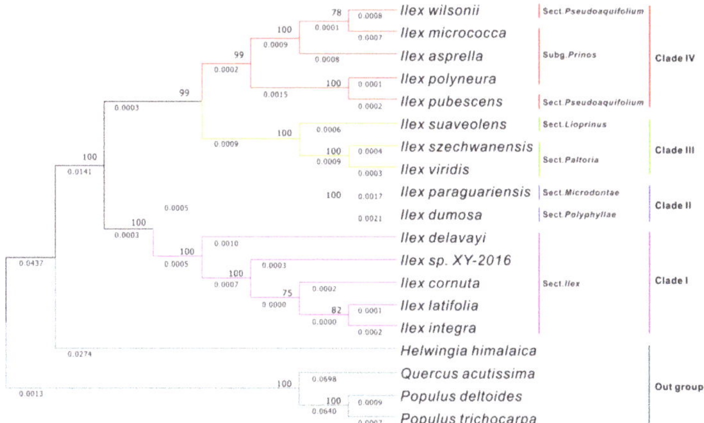

Figure 6. The phylogenetic analyses of 15 *Ilex* species using the entire plastid genomes. The evolutionary tree was constructed using MEGA X with the maximum likelihood (ML) method and the Tamura-Neighbour model. The percentage of trees in which the associated taxa clustered together is shown next to the branches. This analyses involved 19 plastid entire plastomes of *I. latifolia*, *I. suaveolens*, *I. viridis*, *I. micrococca*, *I. paraguariensis* (KP016928), *I. dumosa* (KP016927), *I. integra* (MK335537), *I. cornuta* (MK335536), *I. sp. XY-2016* (KX426469), *I. delavayi* (KX426470), *I. szechwanensis* (KX426466), *I. wilsonii* (KX426471), *I. polyneura* (KX426468), *I. asprella* (NC_045274.1), and *I. pubescens* (KX426467). The plastomes of *P. trichocarpa* (NC_009143), *P. deltoides* (MK267316), *Q. acutissima* (MF593895), and *H. himalaica* (KX434807) were used as outgroups.

4. Discussion

The holly *Ilex* mostly grows in mesic environments with global distribution. A recent report on phylogeny and biogeography suggested the origin of *Ilex* being in subtropical Asia [6]. Subsequently, it colonized other areas (e.g., South and North America, Australia, Europe, Africa, and some ocean islands) with divergence time from 4 to 30 million years ago [6]. Approximately 204 *Ilex* species (149 endemic species) have been described in China [48]. Located in a transition zone of north–south flora of eastern China, Mount Huangshan is considered as a priority spot for biodiversity conservation, in which all of 20 recorded *Ilex* species exhibited many medicinal properties and economic importance for garden and industry use [8,49,50]. Among these *Ilex* species, *I. suaveolens* is the most abundantly native species. The deciduous *I. micrococca* is locally grown as popular ornamental trees and for providing superior materials for paper and tannin production, in addition to herbal medicine use. The "Kuding" tea represented by *I. latifolia*, and pharmaceutical *I. viridis*, show potential clinical functions for scavenging heat, anti-inflammation, and detoxification, are moderately distributed in the dynamic forest plot [26].

The plasticity and diversity in leaf traits, including venation, anatomy, stomatal distribution, and stomatal conductance, were drastically affected by both cellular factors and environmental cues [51–53]. Hence, this extensively phenotypic survey of the vegetative and reproductive organs in plants contributes to the understanding of plant physiological and ecological adaptation. In our work, initial comparative analyses of leaf phenotypes revealed the significantly different values in LA, SLA, STD, and SW between four *Ilex* species, suggesting that these morphological traits could be potentially used as candidates to distinguish different *Ilex* species (Table 1). The persistent fruits and distinctive leave morphology in *Ilex* result from complicated genetic variation and ecological adaptability to various environments; therefore, the underlying molecular mechanisms require an extensive understanding of genome architecture and genome size diversity [54]. The transcriptome assembly was previously only conducted in *I. paraguariensis* [13]. Unfortunately, the draft of the full nuclear genome sequencing and annotation remains unavailable in *Ilex*. Using the tomato as the internal standard, analyses of DNA image cytometry in four *Ilex* species revealed that the calculated NG showed approximately 955 Mb in *I. latifolia*. This NG value is higher than *I.cornuta* (642 Mb) [55]. The examined NG in *I. suaveolens* and *I. viridis* showed similar levels to *I. mucronata* (1073 Mb) [56]. The most abundant NG was characterized in *I. micrococca*, slightly lower than *I. paraguariensis* (1078 Mb) [29]. The estimated DNA C-value (NG) in four *Ilex* species had not been previously reported. We found that *I. micrococca* had the largest size in NG but showed the lowest SW levels in comparison with other *Ilex*. A similar phenomenon was reported in a recent study, which proposed a negative association between the seed mass and nuclear genome size in the diploid *Aesculus* species [57]. In contrast with all detected morphological traits, the calculated NG appears to have a linear relationship with SLA and STD in four *Ilex* species, which is inconsistent with the associated NG/STD patterns from a large-scale comparative analysis in angiosperms [19]. We hypothesized that the statistical associations between the NG, LEC, and STD might not be evaluated adequately with small sample capacity, which could reason the discordance above. Perhaps, the relationships of NG size and various leaf traits appeared to be not conservative among angiosperms due to the diversified environmental adaptation [58–60].

The advances in novel biotechnologies, particularly in whole-genome sequencing, created the opportunity to explore the phylogenies, species identification, and molecular markers among taxa [23,61]. In our work, using high-throughput sequencing technologies combined with de novo and reference-guided assembly, the complete plastomes of four *Ilex* species were constructed, showing the conserved quadripartite circular structure, comprising of LSC, SSC, and two copies of IR regions. A total of 144 putative genes (96 PE, 40 tRNA, and 8 rRNA genes) were previously annotated in *Ilex* plastid genomes [1]. The number of identified PE genes and tRNA appeared to be variable in various *Ilex* species [62,63]. Further sequence analyses indicated that *ycf68*, *orf42*, and *ycf68* were located between two exons of *trnA-UGC*, and *orf188* sequence overlap with one exon of *ndhA*. *trnP-GGG* shows a sequence overlapping with *trnP-UGG*. Thus, the total numbers of genes identified were normalized into 134 (89 PE, 37 tRNA, and 8 rRNA genes) in the plastomes of *I. suaveolens*, *I. viridis*, and *I. micrococca*. The most distinctive length in plastome is 370 bp between various *Ilex* species, whereas the maximum sequence difference of LSC is 331 bp, suggesting that the divergence in the LSC region may cause the varied plastome lengths, which might depend on the IR contraction and expansion [64]. Plastid genome sequences exhibit extensive variations in the length, number, and distribution of SSRs. These variations have potential importance for the outcome of genomic diversity [65]. Plastid SSRs have been used extensively in the taxonomy, phylogenies, and the maternal structures in the community, diversity, and differentiation [66]. In total, 221 SSRs and 196 unique long repeats were characterized in four *Ilex* plastomes. The mononucleotide repeats (A/T) in all SSRs represent the dominant type distributed in the non-coding regions, which is related to the AT abundance of the nucleotide composition. This typical pattern was also identified in previous reports [67–69]. The forward repeat and 20–30bp in sequence length were the most general features within the different types of long repeats.

In our study, both GView and mVISTA analyses of eight plastid genomes illustrated numerous divergent hotspots that are primarily located in the SSC regions, suggesting more variable sites in intergenic non-coding sequences than in coding genes. These variable sequences could potentially be used to develop new molecular markers for the identification and taxonomy in *Ilex*. The findings of these hotspot regions are compatible with a previous study that reported the presence of at least 11 divergent regions [1]. Although several divergent genes (*rbcL*, *atpB*, and *matK*) helped inform the reconstruction of the phylogenetic tree among distantly-related species, they might provide suitable resolution for studies within *Ilex* due to the relative lower divergence than the hotspot regions [2,6,22,70]. Recently, *trnH-psbA*, *rbcL-accD*, and *trnS-trnG* have been developed as genetic markers in other species [71,72]. The plastome-divergent hotspot regions were considered powerful for species-level identification. However, whether these divergence hotspot regions could be extensively used for classification and taxonomy in *Ilex* remains to be assessed experimentally.

The comparative detection of sequence variability in the junctions of IRs/SSC and IRs/LSC showed relative fluid patterns in eight *Ilex* species. Mostly, gene *rps19* in LSC regions is nearest to the JLB; however, it was found to span the JLB into IRB with a 4 bp extension in *I. paraguariensis*, *I. viridis*, and *I. suaveolens*. The typical variability in JSB was also observed for *ycf1* in the IRB region and *ndhF* in the SSC region. A similar phenomenon of the extent in IR regions has been recognized in other *Ilex* species [1]. The sequence variations in four junctions of SC/IRs appeared to frequently occur during genome evolution, which resulted in alteration of the plastome size [73]. Hence, the contraction and expansion at the boundaries of IR regions could explain the variability in sequence length between different plastid genomes [74].

Although the Aquifoliaceae has good fossil records, several systematic studies revealed a high incongruity of phylogenies between the nuclear trees and plastid trees, suggesting that the evolutionary and phylogenetic patterns of Aquifoliaceae remain to be elucidated [1,6,22]. Some molecular genetic factors, including inter-lineage hybridization, lineage sorting, introgression, and gene duplication and loss, were reported to significantly influence the phylogenetic incongruence, which broadly occurred in Angiosperms [75]. In contrast to the nuclear trees, the plastid trees strongly reflect the biogeographic distribution of extant species [22]. In our work, the phylogenetic topology revealed the presence of four clades within *Ilex*, in agreement with the fossil record [6], and also fit well with recent reports on plastid phylogenetic analysis [1,32,63]. All five *Ilex* species in the clade I belong to section *Ilex*, showing leathery morphology in the leaf, which is consistent with the traditional classification [48]. *I. paraguariensis* and *I. dumosa* are both located in clade II, suggesting a similar geographic distribution in Southern America. However, *I. suaveolens* in section *Lioprinus* clusters together with *I. szechwanensis* and *I. viridis* in clade III, while the latter two species are categorized into section *Paltoria*. The findings that evergreen species surprisingly have a close relationship with deciduous species in clade IV, further reflects the incongruence between the plastid phylogeny and traditional taxonomy [1,3,67]. Nevertheless, three endemic *Ilex* species in clade III show a relative similarity of the leaf morphology and growth pattern (600–1600 m altitude). Both subtropical deciduous *I. micrococca* and *I. asprella* in clade IV have very similar leaf margins, growing at an altitude of 400–1000 m. Additionally, the topology constructed using the entire plastomes significantly improved the quality of phylogenies compared to the use of full coding sequences (Figure S3). Therefore, the plastid topology provides more reliable clues to infer the species' geographic patterns and origins. Overall, when more complete plastome sequences of *Ilex* species become accessible, plastid trees combined with the nuclear markers will contribute to the resolution of the deeper branches of the biogeography and phylogeny in Aquifoliaceae.

5. Conclusions

The unspecific pollination and weakness of reproductive isolation resulted in the frequently intraspecific genetic exchange, confusing the phylogenies and biogeography in the *Ilex* genus. Dioecious *Ilex* species have similar flowers and fruits, but the variable morphology in leaves is commonly affected by climate and season, creating an obstacle to the identification and classification of the *Ilex* specimens [22,75]. Hitherto, the anatomical features and plastid genomes of four endemic *Ilex* species in Mount Huangshan were undocumented. In this work, we performed comparative analyses of a variety of phenotypic traits (e.g., leaves, flowers, and seeds) and the molecular profiles, including nuclear genome size, plastid genome structures, variable patterns, and phylogenetic relationships, in four *Ilex* species. The reconstruction of the plastid phylogenies verified the significant usage of the complete plastomes in identifying the phylogeography and phylogenetic evolution of *Ilex*. However, the evolved phenotypic variation and variable molecular patterns might be tightly related to the adapted ecology and environments. The limited accessions in the plastid genome restrained the extensive exploration of the phylogenies in *Ilex*, which necessitates additional sequencing samples. In summary, the morphological and molecular data in the present study provided informative resources for the in-depth mining of the phylogenies, biogeography, and genetic diversity in the family Aquifoliaceae.

Supplementary Materials: The following are available online at http://www.mdpi.com/1999-4907/11/9/964/s1, Figure S1: Analysis of SSRs in four *Ilex* plastid genomes, Figure S2: Analysis of the long repeated sequences in four *Ilex* plastid genomes, Figure S3: The phylogenetic analysis of 15 *Ilex* species using the coding sequences, Table S1: Base composition and proportions of the four *Ilex* plastid genomes, Table S2: Statistical analysis of codon usage, RSCU values, codon-anticodon recognition patterns of four *Ilex* plastid genomes.

Author Contributions: T.S. and M.H. designed the experiment, analyzed all of the data, and prepared the initial draft. M.H. developed the concept and approved of the final manuscript. M.Z. and Z.S. collected all genome sequencing data and performed the statistical analysis. Z.S. and X.L. conducted the section preparation, image visualization, and FCM assessment. B.Z. and H.W. assisted M.Z. with the sequence comparison, experiments conduction, and DNA calculation. All authors have read and agreed to the published version of the manuscript.

Funding: This research was funded by the National Natural Science Foundation of China (NSFC) (31870589; 31700525); The Natural Science Foundation of Jiangsu Province (NSFJ) (BK20170921); the Scientific Research Foundation for High-Level Talents of Nanjing Forestry University (SRFNFU) (GXL2017011; GXL2017012). The Priority Academic Program Development of Jiangsu Higher Education Institutions.

Acknowledgments: The authors would like to thank the NSFC, NSFJ, and SRFNFU for funding this work and the Co-Innovation Center for Sustainable Forestry in Southern China and PAPD for the instrument use. Thanks go to Yao Li, who assisted with the species identification and Mingzhi Li for the technical support with sequence processing. We also appreciate the critical comments from Yanming Fang.

Conflicts of Interest: The authors declare no conflict of interest.

References

1. Yao, X.; Tan, Y.-H.; Liu, Y.-Y.; Song, Y.; Yang, J.-B.; Corlett, R.T. Chloroplast genome structure in Ilex (Aquifoliaceae). *Sci. Rep.* **2016**, *6*, 28559. [CrossRef] [PubMed]
2. Cuénoud, P.; Spichiger, R.; Andrews, S.; Manen, J.-F.; Martinez, M.A.D.P.; Loizeau, P.-A. Molecular Phylogeny and Biogeography of the Genus Ilex L. (Aquifoliaceae). *Ann. Bot.* **2000**, *85*, 111–122. [CrossRef]
3. Manen, J.-F.; Barriera, G.; Loizeau, P.-A.; Naciri, Y. The history of extant Ilex species (Aquifoliaceae): Evidence of hybridization within a Miocene radiation. *Mol. Phylogenetics Evol.* **2010**, *57*, 961–977. [CrossRef] [PubMed]
4. Gottlieb, A.M.; Giberti, G.C.; Poggio, L. Molecular analyses of the genus Ilex (Aquifoliaceae) in southern South America, evidence from AFLP and ITS sequence data. *Am. J. Bot.* **2005**, *92*, 352–369. [CrossRef] [PubMed]
5. Tsang, A.C.; Corlett, R.T. Reproductive biology of the Ilex species (Aquifoliaceae) in Hong Kong, China. *Can. J. Bot.* **2005**, *83*, 1645–1654. [CrossRef]
6. Yao, X.; Song, Y.; Yang, J.; Tan, Y.; Corlett, R.T. Phylogeny and biogeography of the hollies (*Ilex*, L., Aquifoliaceae). *J. Syst. Evol.* **2020**. [CrossRef]
7. Xie, Y.; Wu, F.; Fang, X. Middle Eocene East Asian monsoon prevalence over southern China: Evidence from palynological records. *Glob. Planet. Change* **2019**, *175*, 13–26. [CrossRef]
8. Hao, D.; Gu, X.; Xiao, P.; Liang, Z.; Xu, L.; Peng, Y. Research progress in the phytochemistry and biology of Ilex pharmaceutical resources. *Acta Pharm. Sin. B* **2013**, *3*, 8–19. [CrossRef]
9. Zhao, X.; Pang, L.; Li, J.; Song, J.-L.; Qiu, L.-H. Apoptosis Inducing Effects of Kuding Tea Polyphenols in Human Buccal Squamous Cell Carcinoma Cell Line BcaCD885. *Nutrients* **2014**, *6*, 3084–3100. [CrossRef]
10. Cao, D.; Xu, C.; Xue, Y.; Ruan, Q.; Yang, B.; Liu, Z.; Cui, H.; Zhang, L.; Zhao, Z.; Jin, J. The therapeutic effect of Ilex pubescens extract on blood stasis model rats according to serum metabolomics. *J. Ethnopharmacol.* **2018**, *227*, 18–28. [CrossRef]
11. Berté, K.A.S.; Beux, M.R.; Spada, P.K.W.D.S.; Salvador, M.; Hoffmann-Ribani, R. Chemical Composition and Antioxidant Activity of Yerba-Mate (*Ilex paraguariensis* A.St.-Hil., Aquifoliaceae) Extract as Obtained by Spray Drying. *J. Agric. Food Chem.* **2011**, *59*, 5523–5527. [CrossRef] [PubMed]
12. Bracesco, N.; Sanchez, A.G.; Contreras, V.; Menini, T.; Gugliucci, A. Recent advances on Ilex paraguariensis research: Minireview. *J. Ethnopharmacol.* **2011**, *136*, 84–378. [CrossRef] [PubMed]
13. Debat, H.J.; Grabiele, M.; Aguilera, P.M.; Bubillo, R.E.; Otegui, M.B.; Ducasse, D.A.; Zapata, P.D.; Marti, D.A. Exploring the Genes of Yerba Mate (*Ilex paraguariensis* A. St.-Hil.) by NGS and De Novo Transcriptome Assembly. *PLoS ONE* **2014**, *9*, e109835. [CrossRef] [PubMed]
14. Hao, D.C.; Gu, X.-J.; Xiao, P.G. Phytochemistry and biology of Ilex pharmaceutical resources. In *Medicinal Plants*; Elsevier: Amsterdam, The Netherlands, 2015; pp. 531–585.

15. Kim, H.K.; Saifullah; Khan, S.; Wilson, E.G.; Kricun, S.D.P.; Meissner, A.; Goraler, S.; Deelder, A.M.; Choi, Y.H.; Verpoorte, R. Metabolic classification of South American Ilex species by NMR-based metabolomics. *Phytochemistry* **2010**, *71*, 773–784. [CrossRef]
16. Evens, Z.; Stellpflug, S. Holiday Plants with Toxic Misconceptions. *West. J. Emerg. Med.* **2012**, *13*, 538–542. [CrossRef]
17. Rendell, S.; Ennos, R.A. Chloroplast DNA diversity of the dioecious European tree *Ilex aquifolium* L. (English holly). *Mol. Ecol.* **2003**, *12*, 2681–2688. [CrossRef]
18. Spooner, D.M.; Van Den Berg, R.G.; Rivera-Peña, A.; Velguth, P.; Del Rio, A.; Salas-López, A. Taxonomy of Mexican and Central American Members of Solanum series Conicibaccata (sect. Petota). *Syst. Bot.* **2001**, *26*, 743–756.
19. Beaulieu, J.M.; Leitch, I.J.; Patel, S.; Pendharkar, A.; Knight, C.A. Genome size is a strong predictor of cell size and stomatal density in angiosperms. *New Phytol.* **2008**, *179*, 975–986. [CrossRef]
20. Smith, D.R. Does Cell Size Impact Chloroplast Genome Size? *Front. Plant Sci.* **2017**, *8*. [CrossRef]
21. Sabater, B. Evolution and Function of the Chloroplast. Current Investigations and Perspectives. *Int. J. Mol. Sci.* **2018**, *19*, 3095. [CrossRef]
22. Manen, J.F.; Boulter, M.C.; Naciri-Graven, Y. The complex history of the genus Ilex, L. (Aquifoliaceae): Evidence from the comparison of plastid and nuclear DNA sequences and from fossil data. *Plant. Syst. Evol.* **2002**, *235*, 79–98. [CrossRef]
23. Nock, C.J.; Waters, D.L.E.; Edwards, M.A.; Bowen, S.G.; Rice, N.; Cordeiro, G.M.; Henry, R.J. Chloroplast genome sequences from total DNA for plant identification. *Plant. Biotechnol. J.* **2011**, *9*, 328–333. [CrossRef] [PubMed]
24. Parks, M.; Cronn, R.; Liston, A. Increasing phylogenetic resolution at low taxonomic levels using massively parallel sequencing of chloroplast genomes. *BMC Biol.* **2009**, *7*, 84. [CrossRef] [PubMed]
25. Xu, H.; Cao, M.; Wu, Y.; Cai, L.; Cao, Y.; Ding, H.; Cui, P.; Wu, J.; Wang, Z.; Le, Z.; et al. Optimized monitoring sites for detection of biodiversity trends in China. *Biodivers. Conserv.* **2017**, *26*, 1959–1971. [CrossRef]
26. Ding, H.; Fang, Y.; Yang, X.; Yuan, F.; He, L.; Yao, J.; Wu, J.; Chi, B.; Li, Y.; Chen, S.; et al. Community characteristics of a subtropical evergreen broad-leaved forest in Huangshan, Anhui Province, East China. *Biodivers. Sci.* **2016**, *24*, 875–887. [CrossRef]
27. Da Silva, N.R.; Oliveira, M.W.D.S.; Filho, H.A.D.A.; Pinheiro, L.F.S.; Rossatto, D.R.; Kolb, R.M.; Bruno, O.M. Leaf epidermis images for robust identification of plants. *Sci. Rep.* **2016**, *6*, 25994. [CrossRef]
28. Thompson, K.; Band, S.R.; Hodgson, J.G. Seed Size and Shape Predict Persistence in Soil. *Funct. Ecol.* **1993**, *7*, 236–241. [CrossRef]
29. Gottlieb, A.M.; Poggio, L. Quantitative and qualitative genomic characterization of cultivated Ilex, L. species. *Plant. Genet. Resour.* **2015**, *13*, 142–152. [CrossRef]
30. Doležel, J.; Greilhuber, J.; Suda, J. Estimation of nuclear DNA content in plants using flow cytometry. *Nat. Protoc.* **2007**, *2*, 2233–2244. [CrossRef]
31. Su, T.; Han, M.; Min, J.; Cao, D.; Pan, H.; Liu, Y. The complete chloroplast genome sequence of Populus deltoides 'Siyang-2'. *Mitochondrial DNA Part. B* **2020**, *5*, 283–285. [CrossRef]
32. Shi, Y.; Liu, B. Complete chloroplast genome sequence of Ilex latifolia (Aquifoliaceae), a traditional Chinese tea. *Mitochondrial DNA Part. B* **2020**, *5*, 190–191. [CrossRef]
33. Bankevich, A.; Nurk, S.; Antipov, D.; Gurevich, A.A.; Dvorkin, M.; Kulikov, A.S.; Lesin, V.M.; Nikolenko, S.I.; Pham, S.; Prjibelski, A.D.; et al. SPAdes: A New Genome Assembly Algorithm and Its Applications to Single-Cell Sequencing. *J. Comput. Biol.* **2012**, *19*, 455–477. [CrossRef] [PubMed]
34. Liu, C.; Shi, L.; Zhu, Y.; Chen, H.; Zhang, J.; Lin, X.; Guan, X. CpGAVAS, an integrated web server for the annotation, visualization, analysis, and GenBank submission of completely sequenced chloroplast genome sequences. *BMC Genomics* **2012**, *13*, 715. [CrossRef] [PubMed]
35. Greiner, S.; Lehwark, P.; Bock, R. OrganellarGenomeDRAW (OGDRAW) version 1.3.1: Expanded toolkit for the graphical visualization of organellar genomes. *Nucleic Acids Res.* **2019**, *47*, W59–W64. [CrossRef] [PubMed]
36. Kumar, S.; Stecher, G.; Li, M.; Knyaz, C.; Tamura, K. MEGA X: Molecular Evolutionary Genetics Analysis across Computing Platforms. *Mol. Biol. Evol.* **2018**, *35*, 1547–1549. [CrossRef]
37. Beier, S.; Thiel, T.; Münch, T.; Scholz, U.; Mascher, M. MISA-web: A web server for microsatellite prediction. *Bioinformatics* **2017**, *33*, 2583–2585. [CrossRef] [PubMed]

38. Kurtz, S. REPuter: The manifold applications of repeat analysis on a genomic scale. *Nucleic Acids Res.* **2001**, *29*, 4633–4642. [CrossRef]
39. Petkau, A.; Stuart-Edwards, M.; Stothard, P.; Van Domselaar, G. Interactive microbial genome visualization with GView. *Bioinformatics* **2010**, *26*, 3125–3126. [CrossRef]
40. Brudno, M.; Malde, S.; Poliakov, A.; Do, C.B.; Couronne, O.; Dubchak, I.; Batzoglou, S. Glocal alignment: Finding rearrangements during alignment. *Bioinformatics* **2003**, *19*, i54–i62. [CrossRef]
41. Frazer, K.A.; Pachter, L.; Poliakov, A.; Rubin, E.M.; Dubchak, I. VISTA: Computational tools for comparative genomics. *Nucleic Acids Res.* **2004**, *32*, W273–W279. [CrossRef]
42. Amiryousefi, A.; Hyvönen, J.; Poczai, P. IRscope: An online program to visualize the junction sites of chloroplast genomes. *Bioinformatics* **2018**, *34*, 3030–3031. [CrossRef] [PubMed]
43. Katoh, K.; Rozewicki, J.; Yamada, K.D. MAFFT online service: Multiple sequence alignment, interactive sequence choice and visualization. *Brief. Bioinform.* **2019**, *20*, 1160–1166. [CrossRef] [PubMed]
44. Obermayer, R. Nuclear DNA C-values in 30 Species Double the Familial Representation in Pteridophytes. *Ann. Bot.* **2002**, *90*, 209–217. [CrossRef] [PubMed]
45. Dolezel, J. Plant DNA Flow Cytometry and Estimation of Nuclear Genome Size. *Ann. Bot.* **2005**, *95*, 99–110. [CrossRef]
46. Kikuchi, S.; Bédard, J.; Hirano, M.; Hirabayashi, Y.; Oishi, M.; Imai, M.; Takase, M.; Ide, T.; Nakai, M. Uncovering the Protein Translocon at the Chloroplast Inner Envelope Membrane. *Sci.* **2013**, *339*, 571–574. [CrossRef]
47. Bölter, B.; Soll, J. Ycf1/Tic214 Is Not Essential for the Accumulation of Plastid Proteins. *Mol. Plant.* **2017**, *10*, 219–221. [CrossRef]
48. Chen, S.K.; Ma, H.; Feng, Y.; Barriera, G.; Loizeau, P.A. Aquifoliaceae. In *Flora of China*; Wu, Z.Y., Raven, P.H., Hong, D.Y., Eds.; Science Press: Beijing, China; Missouri Botanical Garden Press: St. Louis, MO, USA, 2008; Volume 11, pp. 359–438.
49. Qian, Y.; Tian, R. Research Advance of Ilex Germplasm Resources and Their Application to Landscape. *World For. Res.* **2016**, *29*, 40–45.
50. Yi, F.; Zhao, X.; Peng, Y.; Xiao, P. *Genus Ilex* L.: Phytochemistry, Ethnopharmacology, and Pharmacology. *Chinese Herb. Med.* **2016**, *8*, 209–230. [CrossRef]
51. Hetherington, A.M.; Woodward, F.I. The role of stomata in sensing and driving environmental change. *Nature* **2003**, *424*, 901–908. [CrossRef]
52. Maricle, B.R.; Koteyeva, N.K.; Voznesenskaya, E.V.; Thomasson, J.R.; Edwards, G.E. Diversity in leaf anatomy, and stomatal distribution and conductance, between salt marsh and freshwater species in the C4 genus Spartina (Poaceae). *New Phytol.* **2009**, *184*, 216–233. [CrossRef]
53. Scoffoni, C.; Kunkle, J.; Pasquet-Kok, J.; Vuong, C.; Patel, A.J.; Montgomery, R.A.; Givnish, T.J.; Sack, L. Light-induced plasticity in leaf hydraulics, venation, anatomy, and gas exchange in ecologically diverse Hawaiian lobeliads. *New Phytol.* **2015**, *207*, 43–58. [CrossRef]
54. Dodsworth, S.; Leitch, A.R.; Leitch, I.J. Genome size diversity in angiosperms and its influence on gene space. *Curr. Opin. Genet. Dev.* **2015**, *35*, 73–78. [CrossRef]
55. Zhang, L.; Cao, B.; Bai, C. New reports of nuclear DNA content for 66 traditional Chinese medicinal plant taxa in China. *Caryologia* **2013**, *66*, 375–383. [CrossRef]
56. Bai, C.; Alverson, W.S.; Follansbee, A.; Waller, D.M. New reports of nuclear DNA content for 407 vascular plant taxa from the United States. *Ann. Bot.* **2012**, *110*, 1623–1629. [CrossRef]
57. Krahulcová, A.; Trávníček, P.; Krahulec, F.; Rejmánek, M. Small genomes and large seeds: Chromosome numbers, genome size and seed mass in diploid Aesculus species (Sapindaceae). *Ann. Bot.* **2017**, *119*, 957–964.
58. Bertolino, L.T.; Caine, R.S.; Gray, J.E. Impact of Stomatal Density and Morphology on Water-Use Efficiency in a Changing World. *Front. Plant. Sci.* **2019**, *10*. [CrossRef]
59. Jordan, G.J.; Carpenter, R.J.; Koutoulis, A.; Price, A.; Brodribb, T.J. Environmental adaptation in stomatal size independent of the effects of genome size. *New Phytol.* **2015**, *205*, 608–617. [CrossRef]
60. Morgan, H.D.; Westory, M. The Relationship between Nuclear DNA Content and Leaf Strategy in Seed Plants. *Ann. Bot.* **2005**, *96*, 1321–1330. [CrossRef]
61. Zong, D.; Gan, P.; Zhou, A.; Zhang, Y.; Zou, X.; Duan, A.; Song, Y.; He, C. Plastome Sequences Help to Resolve Deep-Level Relationships of Populus in the Family Salicaceae. *Front. Plant. Sci.* **2019**, *10*, 5. [CrossRef]

62. Cascales, J.; Bracco, M.; Garberoglio, M.; Poggio, L.; Gottlieb, A. Integral Phylogenomic Approach over Ilex, L. Species from Southern South America. *Life* **2017**, *7*, 47. [CrossRef]
63. Park, J.; Kim, Y.; Nam, S.; Kwon, W.; Xi, H. The complete chloroplast genome of horned holly, Ilex cornuta Lindl. & Paxton (Aquifoliaceae). *Mitochondrial DNA Part. B* **2019**, *4*, 1275–1276.
64. Kim, K.; Lee, S.-C.; Lee, J.; Yu, Y.; Yang, K.; Choi, B.-S.; Koh, H.-J.; Waminal, N.E.; Choi, H.-I.; Kim, N.-H.; et al. Complete chloroplast and ribosomal sequences for 30 accessions elucidate evolution of Oryza AA genome species. *Sci. Rep.* **2015**, *5*, 15655. [CrossRef]
65. George, B.; Bhatt, B.S.; Awasthi, M.; George, B.; Singh, A.K. Comparative analysis of microsatellites in chloroplast genomes of lower and higher plants. *Curr. Genet.* **2015**, *61*, 665–677. [CrossRef]
66. Provan, J. Novel chloroplast microsatellites reveal cytoplasmic variation in Arabidopsis thaliana. *Mol. Ecol.* **2000**, *9*, 2183–2185. [CrossRef] [PubMed]
67. Li, D.-M.; Zhu, G.-F.; Xu, Y.-C.; Ye, Y.-J.; Liu, J.-M. Complete Chloroplast Genomes of Three Medicinal Alpinia Species: Genome Organization, Comparative Analyses and Phylogenetic Relationships in Family Zingiberaceae. *Plants* **2020**, *9*, 286. [CrossRef] [PubMed]
68. Nguyen, V.B.; Linh Giang, V.N.; Waminal, N.E.; Park, H.-S.; Kim, N.-H.; Jang, W.; Lee, J.; Yang, T.-J. Comprehensive comparative analysis of chloroplast genomes from seven Panax species and development of an authentication system based on species-unique single nucleotide polymorphism markers. *J. Ginseng Res.* **2020**, *44*, 135–144. [CrossRef] [PubMed]
69. Yang, Y.; Zhou, T.; Duan, D.; Yang, J.; Feng, L.; Zhao, G. Comparative Analysis of the Complete Chloroplast Genomes of Five Quercus Species. *Front. Plant. Sci.* **2016**, *07*, 959. [CrossRef]
70. Rohwer, J.G. Toward a Phylogenetic Classification of the Lauraceae: Evidence from matK Sequences. *Syst. Bot.* **2000**, *25*, 60–71. [CrossRef]
71. Shaw, J.; Shafer, H.L.; Leonard, O.R.; Kovach, M.J.; Schorr, M.; Morris, A.B. Chloroplast DNA sequence utility for the lowest phylogenetic and phylogeographic inferences in angiosperms: The tortoise and the hare IV. *Am. J. Bot.* **2014**, *101*, 1987–2004. [CrossRef]
72. Yang, J.; Vázquez, L.; Chen, X.; Li, H.; Zhang, H.; Liu, Z.; Zhao, G. Development of Chloroplast and Nuclear DNA Markers for Chinese Oaks (Quercus Subgenus Quercus) and Assessment of Their Utility as DNA Barcodes. *Front. Plant. Sci.* **2017**, *8*, 816. [CrossRef]
73. Liu, X.; Chang, E.-M.; Liu, J.-F.; Huang, Y.-N.; Wang, Y.; Yao, N.; Jiang, Z.-P. Complete Chloroplast Genome Sequence and Phylogenetic Analysis of Quercus bawanglingensis Huang, Li et Xing, a Vulnerable Oak Tree in China. *Forests* **2019**, *10*, 587. [CrossRef]
74. Wang, R.-J.; Cheng, C.-L.; Chang, C.-C.; Wu, C.-L.; Su, T.-M.; Chaw, S.-M. Dynamics and evolution of the inverted repeat-large single copy junctions in the chloroplast genomes of monocots. *BMC Evol. Biol.* **2008**, *8*, 36. [CrossRef]
75. Shi, L.; Li, N.; Wang, S.; Zhou, Y.; Huang, W.; Yang, Y.; Ma, Y.; Zhou, R. Molecular Evidence for the Hybrid Origin of Ilex dabieshanensis (Aquifoliaceae). *PLoS ONE* **2016**, *11*, e0147825. [CrossRef] [PubMed]

Article

Identification of a Natural Hybrid between *Castanopsis sclerophylla* and *Castanopsis tibetana* (Fagaceae) Based on Chloroplast and Nuclear DNA Sequences

Xiaorong Zeng, Risheng Chen, Yunxin Bian, Xinsheng Qin, Zhuoxin Zhang and Ye Sun *

Guangdong Key Laboratory for Innovative Development and Utilization of Forest Plant Germplasm, College of Forestry and Landscape Architecture, South China Agriculture University, Guangzhou 510642, China; zengxiaorong1211@126.com (X.Z.); sam.chengrisheng@gmail.com (R.C.); bianyunx@163.com (Y.B.); qinxinsheng@scau.edu.cn (X.Q.); zxzhang@scau.edu.cn (Z.Z.)

* Correspondence: sun-ye@scau.edu.cn; Tel.: +86-136-4265-9676

Received: 30 June 2020; Accepted: 7 August 2020; Published: 11 August 2020

Abstract: *Castanopsis* × *kuchugouzhui*Huang et Y. T. Chang was recorded in Flora Reipublicae Popularis Sinicae (FRPS) as a hybrid species on Yuelushan mountain, but it is treated as a hybrid between *Castanopsis sclerophylla* (Lindl.) Schott. and *Castanopsis tibetana* Hance in Flora of China. After a thorough investigation on Yuelushan mountain, we found a population of *C. sclerophylla* and one individual of *C.* × *kuchugouzhui*, but no living individual of *C. tibetana*. We collected *C.* × *kuchugouzhui*, and we sampled 42 individuals of *C. sclerophylla* from Yuelushan and Xiushui and 43 individuals of *C. tibetana* from Liangyeshan and Xiushui. We used chloroplast DNA sequences and 29 nuclear microsatellite markers to investigate if *C.* × *kuchugouzhui* is a natural hybrid between *C. sclerophylla* and *C. tibetana*. The chloroplast haplotype analysis showed that *C.* × *kuchugouzhui* shared haplotype H2 with *C. sclerophylla* on Yuelushan. The STRUCTURE analysis identified two distinct genetic pools that corresponded well to *C. sclerophylla* and *C. tibetana*, revealing the genetic admixture of *C.* × *kuchugouzhui*. Furthermore, the NewHybrids analysis suggested that *C.* × *kuchugouzhui* is an F2 hybrid between *C. sclerophylla* and *C. tibetana*. Our results confirm that *C.* × *kuchugouzhui* recorded in FRPS is a rare hybrid between *C. sclerophylla* and *C. tibetana*.

Keywords: *Castanopsis* × *kuchugouzhui*; natural hybrid; molecular identification; chloroplast DNA sequence; microsatellite

1. Introduction

Hybridization is a process through which there is interbreeding of individuals from two genetically distinct populations or species [1]. It was estimated that at least 25% of plant species, mostly the youngest species, were involved in hybridization and potential introgression with other species [2]. A large number of studies showed that natural hybridization is ubiquitous in different taxa [3–6], and it plays a significant role in generating genetic diversity, even the origin of new ecotypes or species [7–10]. Research on natural hybridization has become a hot spot in the field of plant systematics and evolution in recent years [11–13].

Hybrid identification is the first step in exploring the intricate evolutionary history of natural hybridization. The morphological characteristics of natural hybrids are usually intermediate or more similar to one of their parents, and they form a gradually morphological transition that often causes the blurred and indistinguishable boundary of the species [14]. Chloroplast DNA is uniparentally inherited (maternal inheritance in most angiosperm) and nuclear DNA is biparentally inherited; thus, a comparative analysis of nuclear DNA and chloroplast DNA would provide complementary and often contrasting information on the genetic structure and phylogenies. Interspecific hybrids are commonly identified by cytonuclear discordance that may indicate different parental contribution to the hybrid genome [4,15–17].

Castanopsis sclerophylla (Lindl.) Schott. and *Castanopsis tibetana* Hance are dominant species in mid-subtropical evergreen broad-leaved forest in China. *Castanopsis sclerophylla* is mainly distributed in the north of Nanling mountain, south of the Yangtze River, and east of Sichuan and Guizhou provinces [18]. *Castanopsis tibetana* is widely distributed in subtropical China and overlaps with the range of *C. sclerophylla*. *Castanopsis tibetana* grows together with *C. sclerophylla* in a forest on Yuelushan mountain of Changsha City in Hunan Province, where a specimen of *Castanopsis* × *kuchugouzhui* Huang et Y. T. Chang was collected according to Flora Reipublicae Popularis Sinicae (FRPS) [18]. The leaves and cupules of *C.* × *kuchugouzhui* show intermediate morphologies between *C. sclerophylla* and *C. tibetana*; thus, it is recognized as a hybrid species in FRPS [18]. However, *C.* × *kuchugouzhui* is assumed to be a putative natural hybrid between *C. sclerophylla* and *C. tibetana* in Flora of China [19]. Up to now, there is no molecular evidence for this putative hybrid. The purpose of this study was to verify whether *C.* × *kuchugouzhui* is a natural hybrid between *C. sclerophylla* and *C. tibetana* by using chloroplast DNA sequences and nuclear microsatellite markers.

2. Materials and Methods

2.1. Plant Materials and DNA Extraction

After a thorough investigation of this forest on Yuelushan mountain in 2017–2019, we found a population of *C. sclerophylla* and one individual of *C.* × *kuchugouzhui*, but no living individual of *C. tibetana*. According to the information obtained from Yuelushan Mountain Scenic Area Administration Bureau and Hunan Normal University, this *C.* × *kuchugouzhui* is more than 100 years old and it is the same individual reported by FRPS. We sampled *C.* × *kuchugouzhui* and 20 individuals of *C. sclerophylla* on Yuelushan. We further sampled 22 individuals of *C. sclerophylla* and 19 individuals of *C. tibetana* from Xiushui, as well as 24 individuals of *C. tibetana* from Liangyeshan (Table 1). For each population, eight to 10 fresh leaves per tree were chosen and quickly dried with silica gel, and the individuals were sampled at least 20 m apart from each other. Voucher specimens were made for *C.* × *kuchugouzhui* and each population, and they were stored in the Dendrological Herbarium of South China Agricultural University (CANT).

Table 1. Location, sample size, and genetic diversity of the investigated populations.

Species	Location/Population Code	Latitude (N)	Longitude (E)	Sample Size	A	A_R	H	F_{IS}
Castanopsis sclerophylla	Xiushui/KZ-XS	28°55′23″	114°43′12″	22	5.6	5.426	0.629	0.103 *
	Yuelushan/KZ-YLS	28°10′49″	112°56′28″	20	5.5	5.430	0.580	0.052
	Average			21	5.6	5.428	0.605	0.078
Castanopsis × *kuchugouzhui*	Yuelushan/KZGK-YLS	28°10′49″	112°56′28″	1				
Castanopsis tibetana	Liangyeshan/GK-LYS	25°12′35″	116°10′48″	24	3.3	3.219	0.481	−0.057
	Xiushui/GK-XS	28°55′23″	114°43′12″	19	2.9	2.862	0.427	−0.068
	Average			22	3.1	3.041	0.454	−0.063

N: north, E: east, A: total number of alleles detected, A_R: allelic richness for 19 diploid individuals, H: gene diversity, F_{IS}: fixation index. * Deviated from Hardy–Weinberg equilibrium significantly ($p < 0.01$).

Total genomic DNA was extracted using the DNeasy Plant Mini Kit (QIAGEN, Hilden, Germany) according to the manufacturer's instructions. The quality and concentration of the genomic DNA were evaluated by electrophoresis on 1% agarose gels and NanoDropTM 2000 Spectrophotometers (Thermo Scientific, Waltham, MA, USA).

2.2. Chloroplast DNA Sequencing and Nuclear Microsatellite Genotyping

The chloroplast DNA of psbA-trnH and trnM-trnV intergenic spacers was amplified and sequenced with primers described in References [20,21]. The total volume of the polymerase chain reaction (PCR) was 30 µL, which contained 1 × ES Tag Master Mix (Cwbiotech, Beijing, China), 0.2 µM each of forward and reverse primers, and 20 ng of DNA. PCR amplification was conducted for 33 cycles in a TaKaRa PCR Thermal Cycler Dice™ Gradient (TP600) (TAKARA, Kyoto, Japan). Each cycle included 45 s at 95 °C, 45 s at annealing temperature, and 45 s at 72 °C. An initial pre-denaturation for 5 min at 95 °C and a final extension for 7 min at 72 °C were added.

Polymorphic nuclear microsatellites were screened from 173 simple sequence repeat (SSR) markers originally developed in *C. sclerophylla*, *Castanopsis fargesii*, *Castanopsis sieboldii*, *Castanopsis tribuloides*, *Castanea sativa*, and *Castanea mollissima* [22–28]. One sample in each population and *C.* × *kuchugouzhui* were used in a preliminary experiment to test SSR amplification. PCR was performed in a total volume of 10 μL that consisted of 1 × ES Tag Master Mix, 0.5 μM each of forward and reverse primers, and 20 ng of DNA. The PCR program was the same as above apart from the annealing temperature. PCR products were separated by electrophoresis on 3% agarose gels.

A total of 75 pairs of primers that generated a clear electrophoretic band were applied to the multiple PCR, in which the forward primers were labeled with fluorochromes TAMRA, HEX, 6-FAM, and ROX. The primers with different fluorescence were combined, and we tried to keep their predicted products 30–50 bp apart. The Type-it Microsatellite PCR Kit (QIAGEN, Hilden, Germany) was used to prepare the multiple PCR with a total volume of 10 μL, which contained 1 × PCR Master Mix, 1 × Q-Solution, 0.2 μM each of forward and reverse primers, and 20 ng of DNA. Two samples in each population and *C.* × *kuchugouzhi* were used in this experiment. The PCR programs included an initial pre-denaturation of 5 min at 95 °C, followed by 28 cycles of 30 s at 95 °C, 90 s at 57 °C, 30 s at 72 °C, and a final extension of 30 min at 60 °C. PCR products were visualized on an ABI-3730XL fluorescence sequencer (Applied Biosystems, Foster City, CA, USA) using LIZ500 as an internal size standard. Alleles (Table S1) were scored using GeneMarker v 2.2.0 [29]. Finally, 32 pairs of primers with high polymorphism and stability were applied to genotype all samples.

2.3. Data Analyses

DNA sequences of psbA-trnH (GenBank accession numbers: MT635060-MT635092) and trnM-trnV (GenBank accession numbers: MT635093-MT635125) were manually checked using BioEdit [30]. Multiple alignments were carried out using MEGA v 7 [31] with *Castanopsis fabri* (GenBank accession numbers: MF592976, MF592882) as the outgroup. Haplotypes were retrieved using DnaSP v 6.12.03 [32], and a reduced median-joining network was constructed using NETWORK v 5.0 [33] to infer haplotype relationships.

Genetic diversity parameters including number of alleles detected (A), allelic richness (A_R), gene diversity (H), observed heterozygosity (H_O), gene diversity within populations (H_S), gene diversity in total population (H_T), fixation index (F_{IS}), and genetic differentiation among populations (F_{ST}) under an infinite allele model were calculated per locus using FSTAT v 2.9.4 [34]. The Hardy–Weinberg equilibrium was tested by permuting alleles and comparing the fixation index calculated from randomized datasets to that obtained from the observed dataset; p-values were subjected to Bonferroni correction for multiple comparisons. Three loci significantly deviated from the Hardy–Weinberg equilibrium both in *C. sclerophylla* and in *C. tibetana*; they were excluded from all subsequent analyses.

Genetic differentiation and exchange of *C. sclerophylla* and *C. tibetana* were assessed with a model-based Bayesian clustering method implemented in the program STRUCTURE v 2.3.4 [35] by choosing the admixture model and correlated allele frequencies between populations. Ten independent runs were conducted for each K value (from 1 to 5) with 100,000 MCMC (Markov chain Monte Carlo) iterations after 50,000 burn-in period. The optimal number of clusters (K) was determined through the statistic ΔK based on the second-order rate of change in the log probability of data between successive K values [36]. The average matrix of ancestry membership proportions was calculated over the 10 runs using CLUMPP v 1.1.2 [37].

Each individual was assigned to a genotype category with posterior probability by using the program NewHybrids v 1.1 beta [38]. We considered six genotype categories: Parent 1, Parent 2,

F1 hybrids, F2 hybrids, backcross generation to Parent 1, and backcross generation to Parent 2 [39]. The analysis was run for 100,000 rounds after a burn-in of 50,000 iterations.

3. Results

3.1. Chloroplatst DNA Variation

After alignment using *C. fabri* as outgroup, the total length of chloroplast sequences was 1158 bp, from which seven variable sites and six haplotypes were identified. The sequence variants and their positions are shown in Table 2. Within the lineage of *C. sclerophylla/C. tibetana*, five variable sites and five haplotypes were identified. Haplotypes H1 and H2 were detected in populations of *C. sclerophylla*, while haplotypes H3, H4, and H5 were found only in *C. tibetana*. *Castanopsis* × *kuchugouzhui* shared haplotype H2 with *C. sclerophylla* at the Yuelushan. Population GK-XS possessed two haplotypes H4 and H5, while the other populations were fixed for one unique haplotype. There was no shared common haplotype between *C. sclerophylla* and *C. tibetana*, nor between *C.* × *kuchugouzhui* and *C. tibetana*. Relationships among haplotypes are shown in Figure 1. Haplotype H2 was closely related to H1, and they diverged from haplotypes H3, H4, and H5, which constituted a haplogroup.

Table 2. DNA sequence variation and chloroplast haplotypes revealed in this study.

Species	Population Code	Haplotype	Frequency	Variable Site						
				psbA-trnH			trnM-trnV			
				100	258	284	478	491	672	892
Castanopsis sclerophylla	KZ-XS	H1	8	C	T	A	C	C	G	C
	KZ-YLS	H2	8	C	T	A	C	A	G	C
Castanopsis × *kuchugouzhui*	KZGK-YLS	H2	1	C	T	A	C	A	G	C
Castanopsis tibetana	GK-LYS	H3	8	C	T	A	C	C	T	A
	GK-XS	H4	7	C	T	A	A	C	T	A
		H5	1	C	C	A	C	C	T	A
Castanopsis fabri	outgroup	H6	1	T	T	T	C	C	G	C

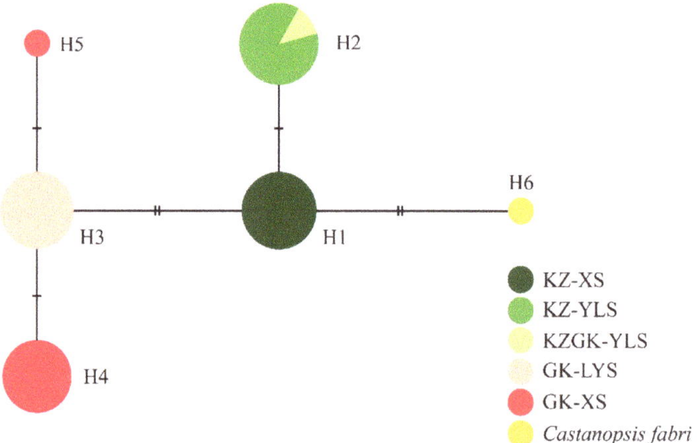

Figure 1. Haplotype relationships shown on median-joining network. The size of each circle is proportional to the haplotype frequency. Mutational steps between the haplotypes are indicated on the line. Population codes are the same as in Table 1.

3.2. Genetic Diversity at Nuclear Microsatellite Loci

In *C. sclerophylla*, a total of 195 alleles were revealed at 29 nuclear microsatellite loci (Table 3). Per locus, the number of alleles ranged from two to 15, and the observed heterozygosity (H_O) varied from 0.143 to 0.932. The within-population gene diversity (H_S) and the overall gene diversity (H_T) ranged from 0.286–0.900 and from 0.312–0.907, respectively. Over the 29 loci, the values of F_{IS} and F_{ST} were 0.080 and 0.053, respectively. Five of 29 loci significantly deviated from the Hardy–Weinberg equilibrium ($p < 0.01$).

Table 3. Genetic diversity at 29 nuclear microsatellite loci in *Castanopsis sclerophylla*.

Locus	Allele Size	A	H_O	H_S	H_T	F_{IS}	F_{ST}
CC-30080	151–163	5	0.473	0.549	0.545	0.132	−0.013
CC-33079-1	189–209	7	0.259	0.468	0.546	0.455 *	0.244
CC-935	131–149	6	0.668	0.662	0.669	−0.011	0.022
CsCAT34	149–161	5	0.475	0.486	0.521	0.026	0.128
CC-14826	228–244	9	0.423	0.461	0.462	0.079	0.006
CC-2994	269–272	2	0.150	0.498	0.499	0.714 *	0.004
CC-3722	131–145	5	0.548	0.570	0.598	0.042	0.087
CC-6538	169–185	6	0.734	0.713	0.727	−0.029	0.037
CC-4950	247–283	15	0.693	0.900	0.907	0.233 *	0.016
CcC02022	351–375	11	0.764	0.795	0.807	0.040	0.029
CFA22	156–162	3	0.186	0.286	0.312	0.354	0.150
CC-25032	176–194	5	0.143	0.454	0.455	0.688 *	0.002
CC-41284	271–289	10	0.759	0.825	0.862	0.073	0.081
CS05	156–178	7	0.764	0.741	0.744	−0.025	0.008
CC-43042	186–210	6	0.464	0.582	0.622	0.215	0.120
CFA71	174–204	9	0.809	0.840	0.845	0.037	0.011
CC-42621	288–297	3	0.366	0.308	0.319	−0.183	0.064
CsCAT14	124–154	15	0.932	0.856	0.865	−0.087	0.020
CC-4562	144–156	3	0.782	0.657	0.661	−0.194	0.011
CC-26213	122–152	4	0.505	0.619	0.637	0.190	0.055
CFA61	184–212	11	0.761	0.601	0.601	−0.268	0.001
CS24	123–144	7	0.784	0.730	0.725	−0.077	−0.015
CS43	148–168	7	0.755	0.702	0.718	−0.079	0.042
CT161	148–157	4	0.395	0.362	0.388	−0.094	0.124
CT128	163–202	4	0.455	0.456	0.480	−0.003	0.099
CS20	249–273	8	0.741	0.728	0.780	−0.024	0.128
CS44	135–147	7	0.748	0.797	0.807	0.050	0.025
CC-33079	243–259	6	0.310	0.370	0.387	0.171	0.081
CC-4323	231–249	5	0.309	0.521	0.543	0.411 *	0.077
Mean		6.7	0.557	0.605	0.622	0.080	0.053

A: total number of alleles detected, H_O: observed heterozygosity, H_S: gene diversity within populations, H_T: gene diversity in total population, F_{IS}: fixation index (a coefficient based on the difference among observed and expected heterozygosity), F_{ST}: genetic differentiation among populations (a coefficient based on the difference among expected heterozygosity within populations and expected heterozygosity in the species). * Deviated from Hardy–Weinberg equilibrium significantly ($p < 0.01$).

In *C. tibetana*, a total of 115 alleles were revealed at 29 nuclear microsatellite markers (Table 4). Per locus, the number of alleles ranged from two to nine, and the observed heterozygosity (H_O) varied from 0.021 to 0.921. The within-population gene diversity (H_S) and the overall gene diversity (H_T) ranged from 0.155–0.746 and from 0.171–0.786, respectively. Over the 29 loci, the values of F_{IS} and F_{ST} were −0.061 and 0.204, respectively. Five of 29 loci significantly deviated from the Hardy–Weinberg equilibrium ($p < 0.01$).

Table 4. Genetic diversity at 29 nuclear microsatellite loci in *Castanopsis tibetana*.

Locus	Allele Size	A	H_O	H_S	H_T	F_{IS}	F_{ST}
CC-30080	148–166	7	0.465	0.680	0.722	0.268 *	0.113
CC-33079-1	197–205	3	0.644	0.495	0.547	−0.293	0.173
CC-935	131–133	2	0.372	0.356	0.356	−0.030	−0.002
CsCAT34	147–153	3	0.530	0.417	0.419	−0.246	0.008
CC-14826	226–244	6	0.702	0.614	0.734	−0.135	0.281
CC-2994	266–275	2	0.404	0.353	0.350	−0.126	−0.019
CC-3722	137–141	3	0.721	0.573	0.598	−0.275	0.079
CC-6538	161–187	7	0.627	0.547	0.711	−0.150	0.366
CC-4950	247–257	2	0.503	0.469	0.470	−0.049	0.003
CcC02022	353–361	5	0.753	0.697	0.731	−0.088	0.086
CFA22	156–158	2	0.342	0.228	0.285	−0.502	0.363
CC-25032	182–194	2	0.921	0.495	0.498	−0.876	0.010
CC-41284	271–273	2	0.419	0.361	0.503	−0.152	0.439
CS05	156–160	3	0.021	0.163	0.175	0.872 *	0.115
CC-43042	204–240	3	0.219	0.196	0.199	−0.125	0.032
CFA71	184–214	8	0.702	0.663	0.786	−0.045	0.269
CC-42621	291–294	2	0.354	0.231	0.439	−0.532	0.616
CsCAT14	130–138	2	0.167	0.194	0.198	0.089	0.037
CC-4562	144–168	6	0.659	0.575	0.772	−0.132 *	0.398
CC-26213	137–157	3	0.346	0.680	0.673	0.521 *	−0.021
CFA61	184–196	3	0.246	0.221	0.222	−0.113	0.004
CS24	117–144	9	0.602	0.640	0.683	0.042	0.119
CS43	144–164	7	0.791	0.746	0.765	−0.066	0.049
CT161	145–148	2	0.414	0.320	0.331	−0.279	0.066
CT128	181–199	6	0.451	0.468	0.722	0.027 *	0.528
CS20	239–253	7	0.555	0.672	0.707	0.171	0.095
CS44	137–141	3	0.325	0.508	0.513	0.363	0.020
CC-33079	247–257	3	0.539	0.448	0.591	−0.183	0.387
CC-4323	237–240	2	0.188	0.155	0.171	−0.208	0.156
Mean		4.0	0.482	0.454	0.513	−0.061	0.204

A: total number of alleles detected, H_O: observed heterozygosity, H_S: gene diversity within populations, H_T: gene diversity in total population, F_{IS}: fixation index (a coefficient based on the difference among observed and expected heterozygosity), F_{ST}: genetic differentiation among populations (a coefficient based on the difference among expected heterozygosity within populations and expected heterozygosity in the species). * Deviated from Hardy–Weinberg equilibrium significantly ($p < 0.01$).

Genetic diversity within populations is summarized in Table 1. *Castanopsis sclerophylla* exhibited a higher level of genetic diversity within population when compared to *C. tibetana*. At the population level, the average values of the number of alleles (A), allelic richness (A_R), and gene diversity (H) were 5.6, 5.428, and 0.605 in *C. sclerophylla*, but 3.1, 3.041, and 0.454 in *C. tibetana*. Population KZ-XS harbored the highest genetic diversity (A = 5.6, A_R = 5.426, and H = 0.629), while population GK-XS showed the lowest genetic diversity (A = 2.9, A_R = 2.862, and H = 0.427). The fixation index (F_{IS}) was 0.078 in *C. sclerophylla*, but −0.063 in *C. tibetana*. The population KZ-XS significantly ($p < 0.01$) deviated from the Hardy–Weinberg equilibrium across 29 loci.

3.3. STRUCTURE and NewHybrids Analyses Based on Nuclear Microsatellite Markers

In the STRUCTURE analysis (Figure 2A), the optimal K-value was found to be 2, indicating that all individuals sampled were assigned to two genetic clusters. One cluster corresponded to *C. sclerophylla* (42 individuals with cluster membership >0.993) and the other corresponded to *C. tibetana* (43 individuals with cluster membership >0.996). *Castanopsis* × *kuchugouzhui* showed genetic admixture with proportions of 0.606 and 0.394 for each cluster that appeared to be "heterozygous" with alleles inherited from *C. sclerophylla* and *C. tibetana*, suggesting that it could be a hybrid offspring of the two species. In the NewHybrids analysis (Figure 2B), 86 sampled individuals were clearly assigned to three genotype categories with high posterior probabilities (>0.999). Forty-two individuals of

C. sclerophylla and 43 individuals of *C. tibetana* were assigned to Parent 1 and Parent 2, respectively. *Castanopsis* × *kuchugouzhui* was classified as an F2 hybrid between *C. sclerophylla* and *C. tibetana*.

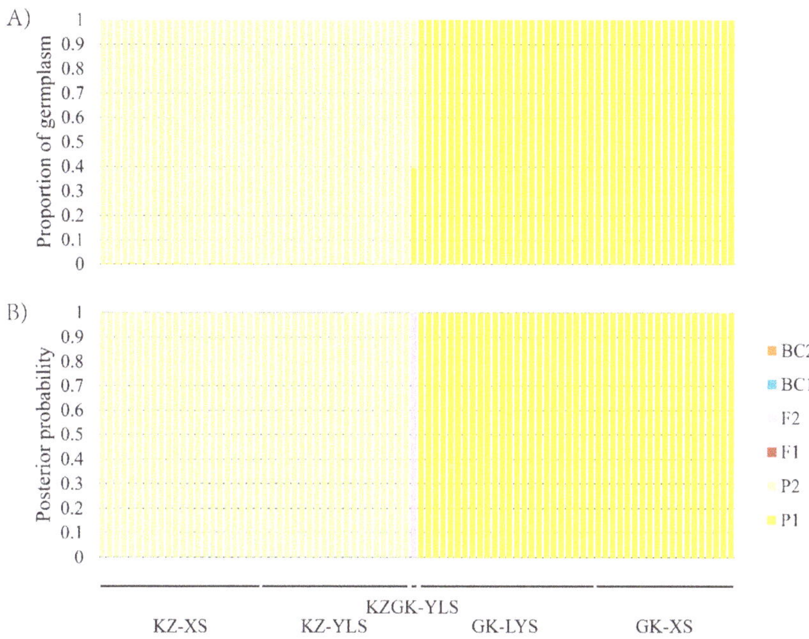

Figure 2. (**A**) Genetic admixture coefficient of the samples; (**B**) posterior probability of genotype category evaluated based on nuclear microsatellite markers. Population codes are the same as in Table 1.

4. Discussion

4.1. SSR Transferability among Castanopsis Species

The ability to transfer SSRs among species is related to the level of divergence between the species; a closer relationship, denotes higher transferability of the primers [40]. Ye et al. [23] screened 51 microsatellite markers originally developed from four *Castanopsis* species (*C. sclerophylla*, *Castanopsis chinensis*, *Castanopsis cuspidata*, and *C. sieboldii*) and found that 68.6% of SSR primer pairs successfully cross-amplified and 31.4% were polymorphic in *C. tibetana*. Li et al. [41] tested 124 EST(expressed sequence tags)-SSRs originally developed from *Castanea mollissima* and found that 42.7% of *C. mollissima* EST-SSR primers successfully cross-amplified and 56.6% showed polymorphism in *Castanopsis fargesii*. In this study, we screened 31 SSRs originally developed from *Castanopsis* species and 142 SSRs originally developed from *Castanea* species. We found that 80.6% of *Castanopsis* SSRs successfully cross-amplified in both *C. tibetana* and *C. sclerophylla*, and 52% showed polymorphism. In contrast, 35.2% of *Castanea* SSRs cross-amplified in both *C. tibetana* and *C. sclerophylla*, and 38% were polymorphic. These results were consistent with the expectation that successful cross-species amplification among closely related genera appears to be much lower than that within genera [40]. The moderate to very high cross-species transferability of SSRs among *Castanopsis* species implies that the species in this genus have more similar genetic makeup and may not be completely reproductively isolated; thus, there is a chance of natural hybridization between species in this genus.

4.2. Genetic Diversity of C. sclerophylla and C. tibetana

Genetic diversity is essential for populations to adapt to environmental change. Large populations of naturally outbreeding species usually have extensive genetic diversity, but it is generally reduced in small populations and endangered species. Habitat fragmentation caused by human interference would reduce the population size and increase the spatial isolation. Such changes will be accompanied by an erosion of genetic variation and an increase of inter-population genetic divergence due to increased genetic drift, elevated inbreeding, and reduced gene flow [42].

In the present study, moderate genetic variation was found in *C. sclerophylla*; the average number of alleles per locus was 6.7, and the mean observed heterozygosity was 0.557. The level of genetic diversity of *C. sclerophylla* was similar to that reported in other closely related species such as *C. fargesii* ($A = 6.7$, $H_O = 0.690$) [24], *Castanopsis acuminatissima* ($A = 10.8$, $H_O = 0.517$) [43], and *C. sieboldii* ($A = 5.2$, $H_O = 0.563$) [44]. Compared with the species above, a lower level of genetic diversity was observed in *C. tibetana* ($A = 4.0$, $H_O = 0.482$). The genetic diversity of *C. tibetana* may be greatly affected by habitat fragmentation and human interference given that *C. tibetana* was destroyed out on Yuelushan mountain and *C. tibetana* exhibited higher genetic differentiation than *C. sclerophylla*. Expanding the sampling of *C. tibetana* is required to examine how habitat fragmentation impacted the genetic diversity of this species.

4.3. Molecular Evidence of Natural Hybrid and Taxonomic Status for C. × kuchugouzhui

The genetic structure analyses based on nuclear microsatellite markers show a clear genetic differentiation between *C. sclerophylla* and *C. tibetana*, and two genetic clusters correspond well to the two species. The fact that *C. × kuchugouzhui* shows genetic admixture with proportions of 0.606 and 0.394 for each cluster indicated that it is a hybrid between *C. sclerophylla* and *C. tibetana*. Natural hybridization between *C. sclerophylla* and *C. tibetana* could be attributed to a full overlap in the flowering phenology of the two species that usually flower from April to May. *Castanopsis sclerophylla* is supposed to be the maternal parent of *C. × kuchugouzhui* since it shared with *C. sclerophylla* a common haplotype (H2) that is maternally inherited. *Castanopsis sclerophylla* inhabits lower elevation and would have a great chance of receiving pollen from *C. tibetana* that occupies higher elevation. *C. × kuchugouzhui* was assigned to an F2 hybrid between *C. sclerophylla* and *C. tibetana* with very high posterior probabilities in the NewHybrids analysis. However, the possible hybrid category that could occur after up to four generations of crossing between the two parent species was allowed [39]. The last specimen of *C. tibetana* on Yuelushan mountain was collected in 1977. Given that *C. × kuchugouzhui* is more than 100 years old, we can be sure that the hybridization event to form this hybrid occurred before *C. tibetana* disappeared from Yuelushan mountain. At the time of the presumable F1 hybrid formation, no other *Castanopsis* species was present in the region but *C. fargesii* and a preliminary test by four SSRs did not mark any gene flow from this species to *C. × kuchugouzhui* (data not shown). Because the dispersal distance of seeds and pollens of *Castanopsis* species is very limited [45–47], it is almost impossible that the formation of *C. × kuchugouzhui* was due to a hybrid seed dispersed from elsewhere or was contributed to by pollen of *C. tibetana* elsewhere.

The leaf and cupule morphologies of *C. × kuchugouzhui* were intermediate between *C. sclerophylla* and *C. tibetana* [18], and *C. × kuchugouzhui* showed mixed genetic characteristics between *C. sclerophylla* and *C. tibetana*. However, there is only one record of *C. × kuchugouzhui* up to now, and we did not identify natural hybridization elsewhere, such as in Xiushui, where the two species coexisted together. These facts suggest that natural hybridization between *C. sclerophylla* and *C. tibetana* is a very rare event. This is consistent with our expectation since the hybrid offspring between two species will suffer deleterious consequences termed outbreeding depression. The hybrid offspring in the F1 and subsequent generation will be rapidly eliminated by natural selection due to their minor fitness [48]. Therefore, instead of listing as a separate species, *C. × kuchugouzhui* should be comprehensively treated as a natural hybrid between *C. sclerophylla* and *C. tibetana*.

In recent years, *C. tibetana* on Yuelushan mountain disappeared due to serious disturbance from human beings such as tourism development. According to our investigation, *C. sclerophylla* inhabits lower elevations of approximately 200–1000 m and favors plenty sunshine, while *C. tibetana* generally prefers to grow in humid conditions at slightly higher elevations. These facts indicate that *C. sclerophylla* and *C. tibetana* have some differentiation in ecological niche occupation. The individuals of the two species seldom grow together just like they do on Yuelushan mountain and Xiushui county, although they could be found in the same forest communities. In this study, only one hybrid individual was corroborated. The very rare natural hybrid between *C. sclerophylla* and *C. tibetana* may imply that the two species have strong but not complete reproductive isolation, which may be caused by their ecological differentiation. Natural hybridization will generate new genotypes and increase genetic diversity, which is important for trees to adapt new environments in the face of rapid global change; thus, this new genotype of *C.* × *kuchugouzhui* is worth conserving as an important germplasm.

5. Conclusions

In this study, we provided compelling evidence for the natural hybrid of *C.* × *kuchugouzhui* using chloroplast DNA sequences and 29 nuclear microsatellite markers. *Castanopsis* × *kuchugouzhui* is a very rare event of natural hybridization, where introgression occurred between *C. sclerophylla* and *C. tibetana*. The genetic analysis of this rare natural hybrid is very helpful for us to understand the genetic differentiation and gene exchange between *Castanopsis* species.

Supplementary Materials: The following are available online at http://www.mdpi.com/1999-4907/11/8/873/s1: Table S1. Alleles at 29 loci used for STRUCTURE and NewHybrids analysis.

Author Contributions: Conceptualization, Y.S., X.Q., and Z.Z.; methodology, X.Z., R.C., Y.B., and Y.S.; software, X.Z.; validation, X.Z., R.C., and Y.S.; formal analysis, X.Z.; investigation, X.Z., R.C., Y.B., and Y.S.; resources, Y.S., X.Q., and Z.Z.; data curation, X.Z.; writing—original draft preparation, X.Z.; writing—review and editing, X.Z., R.C., and Y.S.; visualization, X.Z.; supervision, Y.S.; project administration, Y.S.; funding acquisition, Y.S. All authors read and agree to the published version of the manuscript.

Funding: This research was funded by the National Natural Science Foundation of China, grant number 31770698.

Acknowledgments: We are grateful to the editor and three anonymous reviewers for their insightful comments and suggestions that greatly helped improve our manuscript.

Conflicts of Interest: The authors declare no conflicts of interest.

References

1. Harrison, R.G.; Larson, E.L. Hybridization, Introgression, and the Nature of Species Boundaries. *J. Hered.* **2014**, *105*, 795–809. [CrossRef] [PubMed]
2. Mallet, J. Hybrid speciation. *Nature* **2007**, *446*, 279–283. [CrossRef] [PubMed]
3. Ungerer, M.C.; Baird, S.J.E.; Pan, J.; Rieseberg, L.H. Rapid hybrid speciation in wild sunflowers. *Proc. Natl. Acad. Sci. USA* **1998**, *95*, 11757–11762. [CrossRef] [PubMed]
4. Zhang, J.L.; Zhang, C.Q.; Gao, L.M.; Yang, J.B.; Li, H.T. Natural hybridization origin of *Rhododendron agastum* (Ericaceae) in Yunnan, China: Inferred from morphological and molecular evidence. *J. Plant Res.* **2007**, *120*, 457–463. [CrossRef] [PubMed]
5. Dai, S.; Wei, W.; Zhang, R.; Liu, T.; Chen, Y.; Shi, S.; Zhou, R. Molecular evidence for hybrid origin of *Melastoma intermedium*. *Biochem. Syst. Ecol.* **2012**, *41*, 136–141. [CrossRef]
6. Liu, T.; Chen, Y.; Chao, L.; Wang, S.; Wu, W.; Dai, S.; Wang, F.; Fan, Q.; Zhou, R. Extensive Hybridization and Introgression between *Melastoma candidum* and *M. sanguineum*. *PLoS ONE* **2014**, *9*, e96680. [CrossRef]
7. Rieseberg, L.H. Hybrid Origins of Plant Species. *Annu. Rev. Ecol. Syst.* **1997**, *28*, 359–389. [CrossRef]
8. Rieseberg, L.H.; Carney, S.E. Plant hybridization. *New Phytol.* **1998**, *140*, 599–624. [CrossRef]
9. Barton, N.H. The role of hybridization in evolution. *Mol. Ecol.* **2001**, *10*, 551–568. [CrossRef]
10. Zalapa, J.E.; Brunet, J.; Guries, R.P. The extent of hybridization and its impact on the genetic diversity and population structure of an invasive tree, *Ulmus pumila* (Ulmaceae). *Evol. Appl.* **2010**, *3*, 157–168. [CrossRef]
11. Ellstrand, N.C.; Whitkus, R.; Rieseberg, L.H. Distribution of Spontaneous Plant Hybrids. *Proc. Natl. Acad. Sci. USA* **1996**, *93*, 5090–5093. [CrossRef] [PubMed]

12. Rieseberg, L.H.; Raymond, O.; Rosenthal, D.M.; Lai, Z.; Livingstone, K.; Nakazato, T.; Durphy, J.L.; Schwarzbach, A.E.; Donovan, L.A.; Lexer, C. Major Ecological Transitions in Wild Sunflowers Facilitated by Hybridization. *Science* **2003**, *301*, 1211–1216. [CrossRef] [PubMed]
13. Stukenbrock, E.H. The Role of Hybridization in the Evolution and Emergence of New Fungal Plant Pathogens. *Phytopathology* **2016**, *106*, 104–112. [CrossRef] [PubMed]
14. Stebbins, G.L. The Role of Hybridization in Evolution. *Proc. Am. Philos. Soc.* **1959**, *103*, 231–251.
15. Isoda, K.; Shiraishi, S.; Watanabe, S.; Kitamura, K. Molecular evidence of natural hybridization between *Abies veitchii* and *A. homolepis* (Pinaceae) revealed by chloroplast, mitochondrial and nuclear DNA markers. *Mol. Ecol.* **2000**, *9*, 1965–1974. [CrossRef]
16. Ning, H.; Yu, J.; Gong, X. Bidirectional natural hybridization between sympatric *Ligularia vellerea* and *L. subspicata*. *Plant Divers.* **2017**, *39*, 214–220. [CrossRef]
17. Wu, R.; Zou, P.; Tan, G.; Hu, Z.; Wang, Y.; Ning, Z.; Wu, W.; Liu, Y.; He, S.; Zhou, R. Molecular identification of natural hybridization between *Melastoma malabathricum* and *Melastoama beccarianum* in Sarawak, Malaysia. *Ecol. Evol.* **2019**, *9*, 5766–5776. [CrossRef]
18. Huang, C.J.; Chang, Y.T. Flora Reipublicae Popularis Sinicae. In *Castanopsis*; Chun, W.Y., Huang, C.J., Eds.; Science Press: Beijing, China, 1998; pp. 13–80.
19. Huang, C.J.; Zhang, Y.T.; Bruce, B. Flora of China. In *Fagaceae*; Wu, Z.Y., Raven, P.H., Hong, D.Y., Eds.; Science Press and Missouri Botanical Garden Press: Beijing, China, 1999; pp. 315–333.
20. Kress, W.J.; Wurdack, K.J.; Zimmer, E.A.; Weigt, L.A.; Janzen, D.H. Use of DNA barcodes to identify flowering plants. *Proc. Natl. Acad. Sci. USA* **2005**, *102*, 8369–8374. [CrossRef]
21. Cheng, Y.P.; Hwang, S.Y.; Lin, T.P. Potential refugia in Taiwan revealed by the phylogeographic study of *Castanopsis carlesii* Hayata (Fagaceae). *Mol. Ecol.* **2005**, *14*, 2075–2085. [CrossRef]
22. Shi, Y.S.; Zhang, J.; Jiang, K.; Cui, M.Y.; Li, Y.Y. Development and characterization of polymorphic microsatellite markers in *Castanopsis sclerophylla* (Fagaceae). *Am. J. Bot.* **2011**, *98*, e19–e21. [CrossRef]
23. Ye, L.J.; Wang, J.; Sun, P.; Dong, S.P.; Zhang, Z.Y. The Transferability of Nuclear Microsatellite Markers in Four *Castanopsis* Species to *Castanopsis tibetana* (Fagaceae). *Plant Divers. Resour.* **2014**, *36*, 443–448.
24. Fu, D.D.; Wang, J.; Liu, Y.F.; Huang, H.W. Isolation of Microsatellite Markers for *Castanopsis fargesii* (Fagaceae). *J. Trop. Subtrop. Botany* **2010**, *18*, 541–546.
25. Ueno, S.; Aoki, K.; Tsumura, Y. Generation of Expressed Sequence Tags and development of microsatellite markers for *Castanopsis sieboldii* var. sieboldii (Fagaceae). *Ann. For. Sci.* **2009**, *66*, 509. [CrossRef]
26. Waikham, P.; Thongkumkoon, P.; Chomdej, S.; Liu, A.; Wangpakapattanawong, P. Development of 13 microsatellite markers for *Castanopsis tribuloides* (Fagaceae) using next-generation sequencing. *Mol. Biol. Rep.* **2018**, *45*, 27–30. [CrossRef] [PubMed]
27. Marinoni, D.; Akkak, A.; Bounous, G.; Edwards, K.J.; Botta, R. Development and characterization of microsatellite markers in *Castanea sativa* (Mill.). *Mol. Breed.* **2003**, *11*, 127–136. [CrossRef]
28. Li, C.; Sun, Y.; Huang, H.W.; Cannon, C.H. Footprints of divergent selection in natural populations of *Castanopsis fargesii* (Fagaceae). *Heredity* **2014**, *113*, 533–541. [CrossRef]
29. Holland, M.M.; Parson, W. GeneMarker® HID: A Reliable Software Tool for the Analysis of Forensic STR Data. *J. Forensic Sci.* **2011**, *56*, 29–35. [CrossRef]
30. Alzohairy, A.M. BioEdit: An important software for molecular biology. *GERF Bull. Biosci.* **2011**, *2*, 60–61.
31. Kumar, S.; Stecher, G.; Tamura, K. MEGA7: Molecular Evolutionary Genetics Analysis Version 7.0 for Bigger Datasets. *Mol. Biol. Evol.* **2016**, *33*, 1870–1874. [CrossRef]
32. Rozas, J.; Ferrer-Mata, A.; Sánchez-DelBarrio, J.C.; Guirao-Rico, S.; Librado, P.; Ramos-Onsins, S.E.; Sánchez-Gracia, A. DnaSP 6: DNA Sequence Polymorphism Analysis of Large Data Sets. *Mol. Biol. Evol.* **2017**, *34*, 3299–3302. [CrossRef]
33. Bandelt, H.J.; Forster, P.; Röhl, A. Median-joining networks for inferring intraspecific phylogenies. *Mol. Biol. Evol.* **1999**, *16*, 37–48. [CrossRef] [PubMed]
34. Goudet, J. *FSTAT (Version 2.9.4), A Program (for Windows 95 and Above) to Estimate and Test Population Genetics Parameters*; University of Lausanne: Lausanne, Switzerland, 2003.
35. Pritchard, J.K.; Wen, W. *Documentation for STRUCTURE Software: Version 2*; University of Chicago: Chicago, IL, USA, 2003.
36. Evanno, G.; Regnaut, S.; Goudet, J. Detecting the number of clusters of individuals using the software STRUCTURE: A simulation study. *Mol. Ecol.* **2005**, *14*, 2611–2620. [CrossRef] [PubMed]

37. Jakobsson, M.; Rosenberg, N.A. CLUMPP: A cluster matching and permutation program for dealing with label switching and multimodality in analysis of population structure. *Bioinformatics* **2007**, *23*, 1801–1806. [CrossRef] [PubMed]
38. Anderson, E.C.; Thompson, E.A. A Model-Based Method for Identifying Species Hybrids Using Multilocus Genetic Data. *Genetics* **2002**, *160*, 1217–1229.
39. Milne, R.I.; Abbott, R.J. Reproductive isolation among two interfertile *Rhododendron* species: Low frequency of post-F_1 hybrid genotypes in alpine hybrid zones. *Mol. Ecol.* **2008**, *17*, 1108–1121. [CrossRef]
40. Peakall, R.; Gilmore, S.; Keys, W.; Morgante, M.; Rafalski, A. Cross-Species Amplification of Soybean (*Glycine max*) Simple Sequence Repeats (SSRs) Within the Genus and Other Legume Genera: Implications for the Transferability of SSRs in Plants. *Mol. Biol. Evol.* **1998**, *15*, 1275–1287. [CrossRef]
41. Li, C.; Sun, Y. Transferability analysis of EST-SSR markers of *Castanea mollissima* to *Castanopsis fargesii*. *Guihaia* **2012**, *32*, 293–297.
42. Young, A.; Boyle, T.; Brown, T. The population genetic consequences of habitat fragmentation for plants. *Trends Ecol. Evol. (Amst.)* **1996**, *11*, 413–418. [CrossRef]
43. Blakesley, D.; Pakkad, G.; James, C.; Torre, F.; Elliott, S. Genetic diversity of *Castanopsis acuminatissima* (Bl.) A. DC. in northern Thailand and the selection of seed trees for forest restoration. *New For. (Dordr.)* **2004**, *27*, 89–100.
44. Aoki, K.; Tamaki, I.; Nakao, K.; Ueno, S.; Kamijo, T.; Setoguchi, H.; Murakami, N.; Kato, M.; Tsumura, Y. Approximate Bayesian computation analysis of Est-associated microsatellites indicates that the broadleaved evergreen tree *Castanopsis sieboldii* survived the Last Glacial Maximum in multiple refugia in Japan. *Heredity* **2019**, *122*, 326–340. [CrossRef]
45. Liu, B.; Wang, R.; Liu, Y.L.; Xu, G.F.; Chen, X.Y. Seed dispersal and predation of *Castanopsis sclerophylla* by small rodents in habits with different disturbance intensity. *Chin. J. Ecol.* **2011**, *30*, 1668–1673.
46. Xu, Y.; Shen, Z.; Li, D.; Guo, Q. Pre-dispersal seed predation in a species-rich forest community: Patterns and the interplay with determinants. *PLoS ONE* **2015**, *10*, e0143040. [CrossRef] [PubMed]
47. Nakanish, A.; Yoshimaru, H.; Tomaru, N.; Miura, M.; Manabe, T.; Yamamoto, S. Patterns of pollen flow in a dense population of the insect-pollinated canopy tree species *Castanopsis sieboldii*. *J. Hered.* **2012**, *103*, 547–556. [CrossRef] [PubMed]
48. Frankham, R.; Ballou, J.D.; Briscoe, D.A. *Introduction to Conservation Genetics*; Cambridge University Press: Cambridge, UK, 2002; pp. 366–394.

Article

Genome Cytosine Methylation May Affect Growth and Wood Property Traits in Populations of *Populus tomentosa*

Kaifeng Ma [1,2,3], Yuepeng Song [1,2], Dong Ci [1,2], Daling Zhou [1,2], Min Tian [1,2] and Deqiang Zhang [1,2,4,*]

1 National Engineering Laboratory for Tree Breeding, Beijing Forestry University, Beijing 100083, China; makaifeng@bjfu.edu.cn (K.M.); yuepengsong@bjfu.edu.cn (Y.S.); cdshanx@163.com (D.C.); zag08108042@163.com (D.Z.); minna_tian@163.com (M.T.)

2 Key Laboratory of Genetics and Breeding in Forest Trees and Ornamental Plants, Ministry of Education, College of Biological Sciences and Technology, Beijing Forestry University, Beijing 100083, China

3 Beijing Key Laboratory of Ornamental Plants Germplasm Innovation & Molecular Breeding, National Engineering Research Center for Floriculture, Beijing Forestry University, Beijing 100083, China

4 Xining Forestry Science Research Institute, Xining 810003, China

* Correspondence: DeqiangZhang@bjfu.edu.cn; Tel.: +86-10-6233-6007

Received: 18 June 2020; Accepted: 23 July 2020; Published: 29 July 2020

Abstract: Growth and wood formation are crucial and complex biological processes during tree development. These biological regulatory processes are presumed to be controlled by DNA methylation. However, there is little direct evidence to show that genes taking part in wood regulation are affected by cytosine methylation, resulting in phenotypic variations. Here, we detected epimarkers using a methylation-sensitive amplification polymorphism (MSAP) method and performed epimarker–trait association analysis on the basis of nine growth and wood property traits within populations of 432 genotypes of *Populus tomentosa*. Tree height was positively correlated with relative full-methylation level, and 1101 out of 2393 polymorphic epimarkers were associated with phenotypic traits, explaining 1.1–7.8% of the phenotypic variation. In total, 116 epimarkers were successfully sequenced, and 96 out of these sequences were linked to putative genes. Among them, 13 candidate genes were randomly selected for verification using quantitative real-time PCR (qRT-PCR), and it also showed the expression of nine putative genes of *PtCYP450*, *PtCpn60*, *PtPME*, *PtSCP*, *PtGH*, *PtMYB*, *PtWRKY*, *PtSTP*, and *PtABC* were negatively correlated with DNA methylation level. Therefore, it suggested that changes in DNA methylation might contribute to regulating tree growth and wood property traits.

Keywords: growth trait; wood property; cytosine methylation; epimarker; candidate gene; gene expression

1. Introduction

In plant species, phenotypic plasticity is a phenomenon that one genotype displays alternative phenotypes induced by changing environments and epigenetic variation [1–4]. Epigenetic regulation is not based on alterations in DNA sequence [5,6]; rather, it gives rise to 'epialleles', meaning alleles with identical DNA sequences yet diverse levels of gene expression resulting from differences in their epigenetic status [7,8]. Research shows that epigenetic changes are potentially reversible, often exist in metastable states [9,10], and can be passed from one generation to the next via mitosis or meiosis [11–13]. This inherited characteristic provides an opportunity to plot epigenetic linkage maps and perform intraspecific association analysis between epigenetic alleles and traits [14–17].

Cytosine methylation, primarily the addition of a methyl group to the C5 position of a cytosine residue, is one of the best known epigenetic modifications [18]. The modification participates in various important biological responses, e.g., repressing the expression of transposons and repetitive elements, responding to biotic/abiotic stress, and taking part in early embryogenesis, stem cell differentiation, X chromosome inactivation, and genomic imprinting [19–24]. Differences in the extent of DNA methylation can give rise to varied gene expression patterns, leading to phenotypic variation [25]; furthermore, specific sites showing different patterns of 5 mC (either hypermethylation or demethylation) between different individuals or tissues can affect genetic transcription, resulting in morphological changes [26–28].

For some time, researchers focused attention on selected methylated genes and their function in individual plants, such as *Arabidopsis SUPERMAN* genes with cytosine methylation affecting flower development [29], *PHOSPHORIBOSYLANTHRANILATE ISOMERASE* (*PAI*) genes supply insufficient PAI activity with cytosine methylation [30], *DEMETER* genes imprint *MEDEA Polycomb* gene by excising 5-methylcytosine [31], and *FLOWERING WAGENINGEN* (*FWA*) transcription factor gene silenced by DNA methylation in vegetative tissue but it is demethylated in the central cell of the female ovule [13,32]. In recent years, epimarkers were developed to explore epigenetic diversity and structure, and associated genes, on the basis of epialleles that can be detected using the methylation-sensitive amplified polymorphism (MSAP) method in hybrids and wild populations [16,33,34]. Furthermore, these methods led to the notion that germplasm resources carrying epimutations associated with beneficial traits could be selected from plant populations [15–17].

Chinese white poplar (*Populus tomentosa* Carr.), a native tree species with $2n = 2x = 38$ chromosomes (diploid; n and x represent chromosome numbers of gametophyte and haploid, respectively) and widely distributed in the Yellow River basin, plays an important role in ecological and environmental protection. The tree has also been cultivated for commercial timber production and pulpwood because of its low contents of fermentation-inhibiting extractives, as well as high biomass conversion efficiency [35–37]. It has also been used as a model species for perennial woody plants in physiological and biochemical research, genetic diversity and structure analysis, and the investigation of functional genes that participate in lignocellulose biosynthesis and growth processes [38]. Previously, we have explored the genetic/epigenetic diversity and structure of *P. tomentosa* and performed marker–trait association analysis [39,40], as well as showing that candidate wood-related genes with unique methylation patterns play critical roles in xylem biosynthesis [41]. However, epimarker–trait association analysis and seeking functional genes regulated by methylation status have not yet been performed in natural populations of *P. tomentosa*. Therefore, here, we investigate the relationship between global DNA cytosine methylation markers (using MSAP) and variations in tree growth and wood property parameters in order to elucidate the possible epigenetic regulation of xylem formation.

2. Materials and Methods

2.1. Plant Materials

Plant xylem materials were harvested from the germplasm bank of *P. tomentosa* located at Guanxian County (36°30′54″ N, 115°21′45″ E), Shandong Province, China. The germplasm bank, including 1047 genotypes collected from the region of China in which the species is naturally distributed, was constructed between 1982 and 1984 using root segment propagation techniques [42]. Each genotype was planted with at least three replicates with row spacing 4 m and plant spacing 4 m. In the present study, 432 unrelated genotypes (each genotype has three clones) originating from Beijing (20), Hebei (114), Shandong (19), Henan (114), Shaanxi (64), Shanxi (80), Gansu (6), Anhui (10), and Jiangsu (5) were selected randomly for phenotype measurements and genotyping. To deal with each of the 1296 trees, we first stripped off the bark (approximately 5 cm × 5 cm) at breast height and cut out xylem samples with a sharp blade; each of the xylem samples was divided into two parts equally, one part of the xylem material was frozen rapidly in liquid nitrogen for nucleic acid extraction and the other part was stored in plastic bags for wood property determination at normal temperature status, respectively.

2.2. Phenotypic Data Collection

Phenotypic data, including growth traits and wood properties, were collected and then used in statistical analysis. The growth phenotype traits of tree height (H) and diameter at breast height (DBH) were measured by using the enclosed ruler method and hypsometer, respectively. The volume of timber (V) was estimated according to the following equation [43]: $V = \pi \times (DBH/2)^2 \cdot H \cdot f$ with $f = 0.488$.

The xylem materials that were stored in plastic bags were taken back to the laboratory and dried naturally for determining wood properties, including fiber length, fiber width, microfibril angle (MFA), and contents of lignin, holocellulose, and α-cellulose. Firstly, the MFA was measured using an X-ray diffractometer (PanalyticalX'Pert Pro, Philips, Eindhoven, Netherlands) with the angle between the incident ray and the receiving optical path set to 22.4°. The main parameter settings were: tube voltage 40 kV; tube current 40 mA; scanning step 0.5°; and angle of rotation 0–360°. The MFA was calculated using the method of 0.6 T [44,45]. Secondly, we cut a small amount of each xylem sample into small pieces and then macerated the tissue in a solution of hydrogen peroxide (30%) and glacial acetic acid (1:1, *v/v*) at 70 °C for 48 h. For fiber length and fiber width determination, the fibrous material was washed with deionized water, processed with safranin staining, and placed uniformly over a slide, then measured using a computer image analysis system with a color television video camera (VM-60N, Olympus, Japan) according to Hart et al. [46]. Then, the xylem samples were ground into fine powder and passed through a 40–60 mesh screen; later, the contents of lignin, holocellulose, and α-cellulose were determined and measured using wet chemistry analysis techniques described by Porth et al. [47]. At the same time, near-infrared reflectance (NIR) spectroscopy was used to detect the NIR absorption spectra of the powder samples, and calibration models were developed to predict the contents of holocellulose, α-cellulose, and lignin [48].

The xylem materials that were stored in plastic bags were taken back to the laboratory and dried naturally for determining wood properties, including fiber length, fiber width, microfibril angle (MFA), and contents of lignin, holocellulose, and α-cellulose. Firstly, the MFA was measured using an X-ray diffractometer (PanalyticalX'Pert Pro, Philips, Eindhoven, Netherlands) with the angle between the incident ray and the receiving optical path set to 22.4°. The main parameter settings were: tube voltage 40 kV; tube current 40 mA; scanning step 0.5°; and angle of rotation 0–360°. The MFA was calculated using the method of 0.6 T [44,45]. Secondly, we cut a small amount of each xylem sample into small pieces and then macerated the tissue in a solution of hydrogen peroxide (30%) and glacial acetic acid (1:1, *v/v*) at 70 °C for 48 h. For fiber length and fiber width determination, the fibrous material was washed with deionized water, processed with safranin staining, and placed uniformly over a slide, then measured using a computer image analysis system with a color television video camera (VM-60N, Olympus, Japan) according to Hart et al. [46]. Then, the xylem samples were ground into fine powder and passed through a 40–60 mesh screen; later, the contents of lignin, holocellulose, and α-cellulose were determined and measured using wet chemistry analysis techniques described by Porth et al. [47]. At the same time, near-infrared reflectance (NIR) spectroscopy was used to detect the NIR absorption spectra of the powder samples, and calibration models were developed to predict the contents of holocellulose, α-cellulose, and lignin [48].

The plain vanilla ANOVA (analysis of variance) was used to see the significant difference ($p < 0.05$) in the phenotypic variables among the nine populations. Additionally, we used Duncan's multiple range test to detect the significance difference ($p < 0.05$) within the pairs of means of each phenotypic parameter for the populations [49].

2.3. Genotyping by Methylation-Sensitive Amplified Polymorphism

The frozen xylem materials were ground into fine powder in porcelain mortars with liquid nitrogen for the extraction of nucleic acid DNA and RNA. Half of each powder sample was used for DNA isolation by the Cetyltrimethyl Ammonium Bromide (CTAB) method [50], and DNA was quantified using a NanoVue UV/visible spectrophotometer (GE Healthcare Company). We processed the DNA for methylation-sensitive amplified polymorphism (MSAP) detection, including double digestion, ligation, pre- and selective amplification, and capillary electrophoresis with fluorescence detection, as described by Ma et al. [40]. Each DNA sample was digested separately by restricted endonuclease combinations of *Eco*RI–*Hpa*II and *Eco*RI–*Msp*I, and the electrophoresis results for the digested mixtures were analyzed using GeneMarker V1.7.1: four patterns of genotyping were generated (denoted '1,1'; '1,0'; '0,1'; and '0,0') according to the presence ('1') or absence ('0') of the relevant fragment in the electrophoresis lanes of the final amplification products pre-digested by *Eco*RI–*Hpa*II and *Eco*RI–*Msp*I, respectively.

2.4. Linear Correlation Analysis and Association Analysis

Four patterns of genotyping were generated from the procedure for MSAP detection, and the percentage of each pattern relative to the total number of bands was defined as relative hemi-methylation ('1,0'), full-methylation ('0,1'), non-methylation ('1,1'), or uninformative site ('0,0'). The total relative methylation level was defined as the sum of hemi-methylation and full-methylation percentages. Pearson correlation analysis was carried out between relative methylation levels and traits (tree height, diameter at breast height, volume of timber, fiber length, fiber width, MFA, and contents of lignin, holocellulose, and α-cellulose) to investigate the relationship between methylation and phenotype. Meanwhile, the relationship between gene expression level and DNA methylation level was also estimated by using linear correlation analysis. Primary percentage data were transformed as $x_{ij}' = \arcsin\sqrt{x_{ij}}$ before calculation (x_{ij} is the j observed value in the i group).

In order to identify correlations between phenotypic traits and the MSAP markers with four patterns at an epigenetic locus, single marker association analysis was carried out using single-factor ANOVA with a significance threshold of $p = 0.05$ or $p = 0.01$ in general. The contribution to trait variation of each MSAP marker was evaluated as the percentage of the square deviation among groups and population variance [39]. The structural network of associated epimarkers and the traits was constructed using a Cytoscape V3.5.1 software (https://cytoscape.org/what_is_cytoscape.html) according to the instructions.

2.5. Candidate Gene Screening and Gene Expression

The associated MSAP markers were separated from the selective amplification products using electrophoresis on 6% denaturing polyacrylamide gels and detected with silver staining [51]. Then, the candidate marker fragments were extracted from the gels with the Wizard SV and PCR Clean-Up System (Promega, Madison, WI, USA), and the short PCR fragments were sequenced by Biomed Company (Beijing, China) after transformation and cloning processes [38]. Sequence homology analysis and function prediction were performed using web databases, including National Center of Biotechnology Information (NCBI) and Joint Genome Institute (JGI).

We performed total RNA isolation using the RNeasy plant mini kit (Qiagen, Shanghai, China) from powdered xylem (mentioned in Section 2.3), and used the first-strand cDNA (synthesized using Reverse Transcription System, Promega) for quantitative real-time PCR (qRT-PCR) processes following the description by Song et al. and Ma et al. [52,53]. Primers were designed according to the functional annotation of the linked genes obtained from the homology sequence alignment. The specificity of each primer set was checked by sequencing the PCR products, and 13 candidate genes were selected for verification using *PtACTIN* as the internal control gene [54]. Primers for qRT-PCR were designed using Primer Express 3.0 software (Applied Biosystems) and were listed in Table S1. All reactions were performed in triplicate as technical and biological repetitions in 45 genotypes (135 trees or clones).

3. Results

3.1. Variation in Growth and Wood Characteristics

The tree phenotype, including growth traits (diameter at breast height, height of tree and volume of timber) and wood characteristics (fiber length, fiber width, microfibril angle and contents of lignin, holocellulose, and α-cellulose) were quantified in natural populations planted in the germplasm bank in Guanxian County. In total, the phenotypic variation was analyzed in 432 genotypes (1296 clones) collected from nine different geographic provenances: it was considered that this range of genotypes should provide substantial materials for selective breeding and genetic improvement. The contents of lignin, holocellulose, and α-cellulose were 20.87 ± 2.67% (mean ± SD; SD, standard deviation), 72.59 ± 10.59%, and 40.08 ± 8.78%, respectively. The values of fiber length and width were distributed within the range 0.866–1.512 mm and 16.984–29.850 μm with mean values 1.169 ± 0.085 mm and 23.161 ± 1.973 μm, respectively. The mean microfibril angle was 17.815° ± 4.526°. The commercially important parameters of diameter at breast height (21.38 ± 5.67 cm), the height of the tree (14.57 ± 2.88 m), and the volume of timber (0.29 ± 0.19 m^3) were also measured.

We performed one-way ANOVA to investigate the genetic variation in the growth and wood property traits among the nine natural populations (Figure 1). Three parameters of wood properties (contents of holocellulose and α-cellulose, and fiber width) showed significant differences among the populations; similarly, the growth traits of diameter at breast height, the height of the tree, and the volume of timber displayed statistically significant significances ($p < 0.05$).

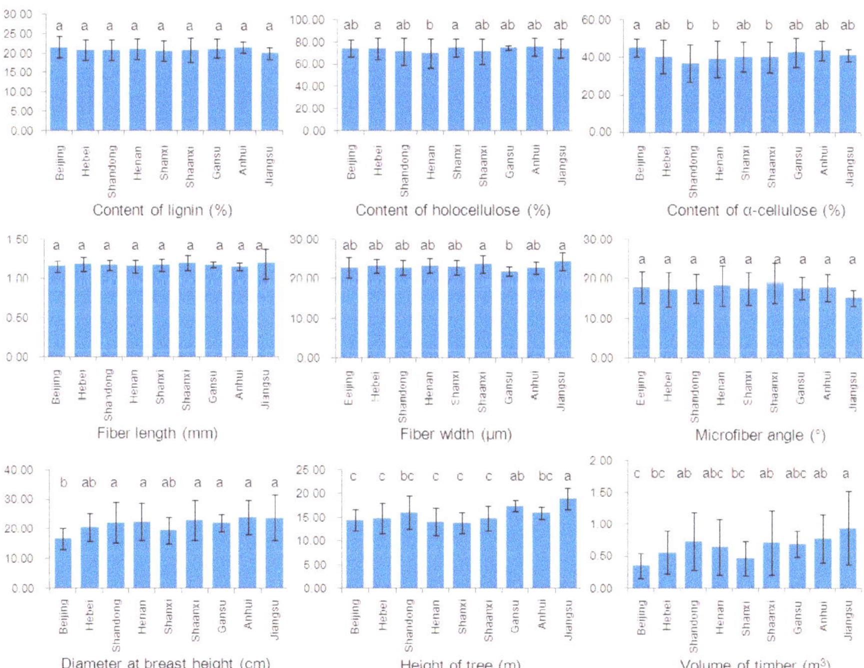

Figure 1. Differences in phenotypic traits among nine natural populations of *Populus tomentosa*. The x- and y-axes indicate provenance and phenotypic value, respectively. Mean ± SD (standard deviation); different letters above the bars (SD) indicate significant differences ($p < 0.05$) according to Duncan's multiple range test (within each sub-graph, it shows a significant difference between each two means if the pair of data sets have no common letter above the bars).

3.2. Linear Correlation between Phenotype and DNA Methylation Levels

Linear correlation analysis was performed to determine trait–trait phenotypic correlations and trait—methylation level correlations among *P. tomentosa* genotypes (Table 1). Lignin content was negatively correlated with cellulose (holocellulose, α-cellulose) content. Three wood and growth parameters (fiber length, timber volume, diameter at breast height) were significantly correlated with the other traits. For instance, fiber length was positively correlated with fiber width, diameter at breast height, tree height and timber volume; in contrast, fiber length was negatively correlated with microfibril angle ($p < 0.01$). Timber volume was positively correlated with lignin content, diameter at breast height, as well as tree height, but was negatively correlated with α-cellulose content ($p < 0.01$). These results suggest that it was possible to select individuals for long fibers and large quantities of stem volume according to measurements of diameter at breast height and tree height.

We also analyzed the relationships between phenotypic traits and relative gene methylation levels estimated from 2393 polymorphic MSAP markers (epimarkers) out of 2408 bands according to our previous data files [40]. The relative total methylation and non-methylation levels were 26.55% and 42.71%, respectively; the relative hemi-methylation level (13.47%) was larger than that of full-methylation (13.10%) ($p < 0.001$) [40]. Here, it showed that tree height was significantly negatively correlated with relative non-methylation level but positively correlated with relative full-methylation level ($p < 0.05$) (Table 1).

3.3. MSAP Markers Associated with Phenotypic Traits within the Populations

Though we investigated the linear correlations between phenotypic traits and relative gene methylation levels, the relationship was unclear between traits and each of the polymorphic epimarker. On the basis of the single marker ANOVA, we considered that 1101 (630 trait-specific and 471 multi-function epimarkers) out of 2393 polymorphic MSAP markers were associated with the nine phenotypic traits ($p < 0.05$). We constructed structural networks (Figure 2) to illustrate their relationships and they showed that 125 (77 trait-specific), 127 (66 trait-specific), and 72 (30 trait-specific) epimarkers, explaining 1.19% ($p = 0.034$) to 4.69% ($p < 0.001$) of variation, were associated with contents of lignin, holocellulose, and α-cellulose, respectively (Table S2). Similarly, 1.06% ($p = 0.043$) to 7.15% ($p < 0.001$) of variation in fiber length, fiber width and microfibril angle was explained by 190 (91 trait-specific), 134 (75 trait-specific), and 140 (64 trait-specific) associated epimarkers, respectively. Finally, 312 (281 trait-specific), 414 (254 trait-specific) and 345 (309 trait-specific) epimarkers were associated with diameter at breast height, the height of the tree, and the volume of timber, respectively, explaining 1.13% ($p = 0.035$) to 7.78% ($p < 0.001$) of variation in those traits (Figure 2, Table S2). The analysis suggested that these epimarkers might be associated with important functions in the regulation of complex quantitative traits, including wood property and tree growth.

Table 1. Correlations among phenotype parameters and between phenotype and genome methylation/non-methylation levels in the xylem tissue of *Populus tomentosa* [a,b].

	L (%)	HC (%)	α- (%)	FL (mm)	FW (μm)	MFA (°)	DBH (cm)	H (m)	V (m^3)	Non- (%)	CNG (%)	CG (%)	Uninformative (%)
HC (%)	−0.229 **												
α- (%)	−0.248 **	0.611 **											
FL (mm)	−0.000	−0.012	−0.034										
FW (μm)	−0.004	−0.017	−0.041	0.206 **									
MFA (°)	−0.008	0.043	0.037	−0.383 **	−0.093								
DBH (cm)	0.159 **	−0.120 *	−0.181 **	0.280 **	0.157 **	−0.114 *							
H (m)	0.083	−0.001	−0.095	0.246 **	−0.001	−0.143 **	0.634 **						
V (m^3)	0.134 **	−0.080	−0.161 **	0.273 **	0.095	−0.085	0.945 **	0.728 **					
Non- (%)	0.010	−0.011	−0.049	−0.039	0.037	0.015	0.010	−0.109 *	−0.049				
CNG (%)	0.035	0.028	0.032	0.016	−0.002	0.041	−0.064	0.002	−0.021	−0.741 **			
CG (%)	−0.039	0.011	0.017	0.056	0.071	−0.005	0.077	0.122 *	0.096	−0.413 **	0.356 **		
CG/CNG (%)	−0.032	−0.017	0.032	0.014	−0.088	−0.060	0.009	0.097	0.044	−0.518 **	−0.103 *	−0.254 **	
Total (%)	0.013	0.026	0.032	0.035	0.027	0.030	−0.019	0.051	0.022	−0.748 **	0.931 **	0.671 **	−0.181 **

[a] Abbreviations: L, lignin; HC, holocellulose; α-, α-cellulose; FL, fiber length; FW, fiber width; MFA, microfibril angle; DBH, diameter at breast height; H, height of tree; V, volume of timber; Non-, relative non-methylation level; CNG, relative hemi-methylation level; CG, relative full-methylation level; Uninformative, relative level of uninformative site; [b] The primary percentage data were transformed as $x_{ij}' = \arcsin\sqrt{x_{ij}}$ before calculation; * $p < 0.05$ (2-tailed), ** $p < 0.01$ (2-tailed).

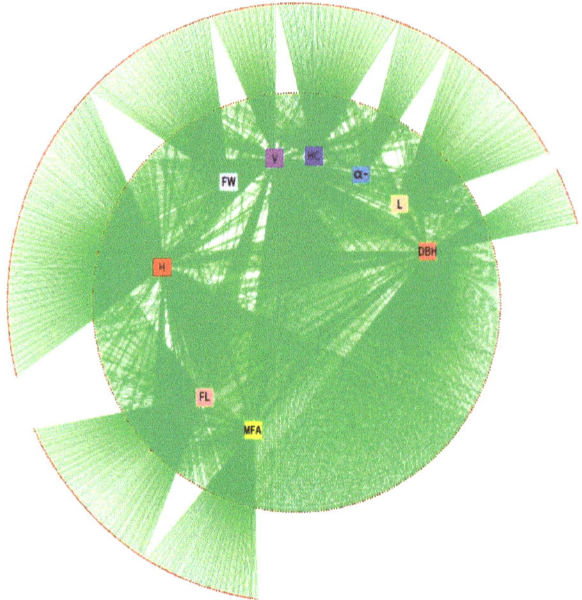

Figure 2. Relationships between all of the significant associated epimarkers and phenotypic traits, as represented by a structural network. The square nodes in the innermost circle represent fiber length (FL), fiber width (FW), microfibril angle (MFA), diameter at breast height (DBH), the height of the tree (H), the volume of timber (V), and the contents of lignin (L), holocellulose (HC) and α-cellulose (α-). Nodes with red around the central circle represent multi-function associated epimarkers. Nodes with red around the outermost circle represent trait-specific associated epimarkers. The green lines connecting traits with epimarkers represent the phenotype variants explained by the associated epimarkers ($p < 0.05$).

3.4. Sequencing and Functional Prediction by Homology Alignment

In order to investigate the function of genes linked with candidate epimarkers that were sequenced by denaturing polyacrylamide gel electrophoresis and silver staining, recovery from the gels, transformation and cloning processes must take place. In total, out of 180 randomly selected MSAP markers, we successfully sequenced 116 epimarker fragments (NCBI, Nos. MN757649–MN757764; Data S1). We aligned the sequences to the reference genome of *P. trichocarpa* and identified 96 markers that were linked to putative functional genes. The linked genes were predicted to take part in, e.g., encoding cytochrome P450 family proteins (MSAP-602), regulatory MYB and WRKY transcription factor families (MSAP-1222, MSAP-2105), ATPase activity regulation (MSAP-803, MSAP-811), regulating transmembrane transport (MSAP-1250), and glycosyl hydrolase family 1 protein (MSAP-2313), as well as other important functions (Table S3). The evidence indicated that the genes linked to epimarkers were essential for plant development, gene regulation via transcription factors, and energy metabolism in *P. tomentosa*.

3.5. Quantitative Expression of Candidate Genes

To examine whether the expression of putative genes was influenced by DNA methylation, we investigated the relationship between different patterns of gene exSpression and the relative DNA methylation levels. Thirteen candidate genes were randomly selected for verification using qRT-PCR within 45 genotypes of *P. tomentosa*: We found that the genes of *Potri.002G228400*, *Potri.010G254400*, and *Potri.018G070900* were expressed at high levels

(Figure 3a). We then observed that the genes of *Potri.001G223800*, *Potri.002G228400*, *Potri.004G133800*, *Potri.005G091700*, and *Potri.018G070900* were expressed at high levels and showed relatively high non-methylation levels (Figure 3b, in red); however, the genes *Potri.001G223800*, *Potri.005G091700*, and *Potri.018G070900* were expressed at lower levels and showed relatively high hemi-methylation, full-methylation, and total-methylation levels, respectively. Negative correlations were also detected between some of the other genes and DNA methylation levels in different patterns (Figure 3b, Table S4). Homology analysis showed that the nine correlated putative genes were *PtCYTOCHROME P450* (*PtCYP450*, *Potri.018G070900*), *PtCHAPERONES* (*PtCpn60*, *Potri.004G133800*), *PtPECTINMETHYLESTERASE* (*PtPME*, *Potri.015G127500*), *PtSERINECARBOXYPEPTIDASE* (*PtSCP*, *Potri.005G091700*), *PtGLYCOSIDE HYDROLASE* (*PtGH*, *Potri.001G223800*), *PtMYB* (*Potri.015G075800*), *PtWRKY* (*Potri.002G228400*), *PtSUGAR TRANSPORT PROTEIN* (*PtSTP*, *Potri.002G096000*), and *PtATP-BINDING CASSETTE TRANSPORTER* (*PtABC*, *Potri.010G254400*) (Table S3).

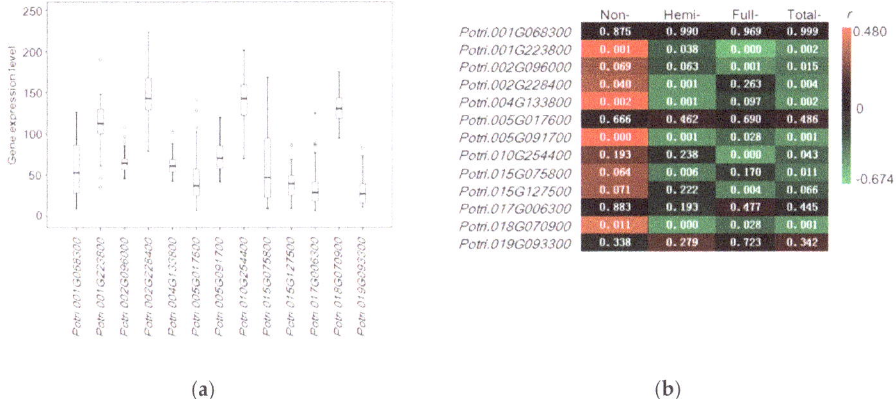

Figure 3. Expression levels of candidate genes and their correlations with relative DNA methylation levels. (**a**) Gene expression level determined by quantitative real-time polymerase chain reaction (qRT-PCR). The x- and y-axes indicate, respectively, gene name and expression level relative to *PtACTIN*. (**b**) Linear correlations between gene expression levels and relative DNA methylation levels. Non-, Hemi-, Full-, and Total- indicate relative non-methylation, hemi-methylation, full-methylation, and total-methylation levels, respectively, estimated by the Pearson correlation coefficient (r, red-black-green color scale from positive to negative correlation). The white number (p-value) in each cell indicates the significance of difference.

On the basis of previous research, the *Arabidopsis* CYP450 Reductase 2 (ATR2), providing electrons from NADPH to a large number of CYP450 [55], appears to be induced during lignin biosynthesis and under stress [56]. Cpn60 are large double-ring assemblies that assist in the folding of the key proteins in plant growth and development [57–59]. The roles of PME have not been fully elucidated; however, this enzyme may be involved in cell elongation by modifying cell wall pectin [60]. It has long been proposed that *PMEs* have roles in growth regulation, including stem growth [61,62], and *PttPME1* was demonstrated involving in mechanisms determining fiber width and length in the wood cells of aspen trees [63]. The serine carboxypeptidase (SCP) family, also called SCP-like (SCPL) family, plays a key part in plant growth, development and stress responses. Transgenic plants of *Nicotiana tabacum* over-expressing *NtSCP1* show reduced cell elongation [64]. In *Triticum aestivum*, SCP regulates cell death during vascular tissue development [65]. Glycosyl hydrolase (GH) proteins are broadly distributed in organisms, and β-glucosidases belonging to GH family 1 have been implicated in several fundamental processes, including lignification [66,67].

MYB and WRKY are two transcription factor families involved in most biological processes. It was shown that the MYB61, regulated by NAC29/31, binds with *CELLULOSE SYNTHASE*, which in turn activates gene expression in secondary wall cellulose synthesis in *Oryzasativa* [68]. Similarly, *PtrMYB152* is over-expressed in secondary wall-forming cells, resulting in the specific activation of lignin biosynthetic genes in *P. trichocarpa* [69]. It was also reported that PtrWRKY19 and VvWRKY2 may function as regulators of pith secondary wall formation in poplar and lignification in grapevine, respectively [70,71].

Sugar transport protein (STPs) and ATP-binding cassette transporter (ABC) transporters are important for transmembrane transport, an essential biological process in cells. STPs are high-affinity hexose transporters with sink-specific tissue expression [72–74]; constitutive over-expression of *STP13* resulted in seedlings with increased biomass when grown on media supplemented with sugar [75]. It was suggested that *MeSTPs* may play a role in early tuber growth, the period when these genes were mainly expressed in *Manihot esculenta* [76]. A breakthrough study demonstrated that a transporter of the ABC family pumps the monolignol *p*-coumaryl alcohol, one of the three main monolignols synthesized in the cytosol, across the plasma membrane [77,78]. It was also reported that the stem vascular morphology was slightly disorganized in *abcb14-1* mutants, with decreased phloem area in the vascular bundle and decreased xylem vessel lumen diameter [79]. Therefore, it seemed that cytosine methyaltion might take part in regulating the expression of the putative genes and affect the growth and wood property traits.

4. Discussion

Growth traits and wood properties, determining the business value and potential for energy production, are the most important phenotypes for commercial timber. *P. tomentosa* is a native species mainly used for timber in vast regions of North China; thus, it is imperative to investigate the mechanisms of growth regulation and wood formation to underpin the selection of germplasm resources and breeding for genetic improvement. In this paper, we measured and calculated values of nine parameters of phenotypic traits (diameter at breast height, height of tree, and volume of timber; content of lignin, content of holocellulose, content of α-cellulose, fiber length, fiber width, and microfibril angle) among 1296 plants (432 genotypes) from natural tree populations. Then, we explored the relationships between those traits and variation in DNA methylation detected using MSAP and qRT-PCR methods.

On the basis of the statistics of the different genotypes, we found that the coefficient of variation in traits was 7.24–66.91%, and the growth and wood property traits showed significant variation among the nine natural populations, with the exception of lignin content, fiber length and microfibril angle. This phenotypic variation, or plasticity, might correlate with genome DNA methylation, the level of which plays an important role in genome stabilization and transposable element repression [19,22]. Previous evidence showed that the methylation level is correlated with phenotype [80]. For instance, a negative correlation was detected between genomic methylation level and gene expression in maize; similarly, negative correlations were reported between methylation level and energy use efficiency and crop yield in *Brassica napus* [80,81]. However, in the woody ornamental plant mei, the leaf length, width, and area were positively correlated with relative full and total methylation levels [17]. In hybrids of poplar, a positive correlation was demonstrated between DNA methylation percentage and productivity [82]. We previously showed that the net photosynthetic rate, tree height and diameter at breast height were positively correlated with relative total methylation and hemi-methylation levels in an intraspecific hybrid population of *P. tomentosa* [39]; here, we found a similar result among the natural populations—the relative full-methylation level was positively correlated with tree height.

Previous studies have showed that methylation is essential for gene exploration and epimarker-assisted analysis in plant populations [15,17,34]. For instance, *PtHT1.1*, *PtHT1.2*, *PtPsbK*, *PtPIN1.2*, *PtMYB60*, and *PtMYB61* were found to be modified by DNA methylation and regarded as playing roles in leaf formation and regulation of photosynthesis in *P. simonii* [34]. In our research, we detected 1101 epimarkers associated with growth and wood property traits. From these, we identified 96 epimarker-linked putative genes, 13 of which were randomly selected for qRT-PCR analysis and it was found that *PtCYP450, PtCpn60, PtPME, PtSCP, PtGH, PtMYB, PtWRKY, PtSTP*, and *PtABC* were negatively correlated with relative DNA methylation level. Meanwhile, it was demonstrated that *PttPME1* involved in mechanisms determining fiber width and length in the wood cells of aspen trees [63]. The transcription factor gene *PtrMYB152* is over-expressed in secondary wall-forming cells, resulting in the specific activation of lignin biosynthetic genes, and PtrWRKY19 may function as regulators of pith secondary wall formation in *P. trichocarpa* [69,71]. The *ABCB14-1* plays a role in decreasing the phloem area in the vascular bundle and xylem vessel lumen diameter in *Arabidopsis* [79]. The above evidence seems to suggest that genes involved in regulating growth and wood development in *P. tomentosa* might be affected by cytosine methylation modifications. The epimarkers that needed to be verified may also provide a new sight for breeding programs in this commercially important tree species.

5. Conclusions

The quantitative traits of the growth and wood property are important for commercial timber tree species. The objective of the present study was to reveal the relationship between tree growth, wood property traits and cytosine methylation, respectively. It was found that the tree height was positively correlated with relative full methylation level. And 1101 single- and multi-function MSAP markers explaining 1.1–7.8% of phenotypic variation were associated with growth traits (diameter at breast height, the height of the tree and the volume of timber) and wood characteristics (fiber length, fiber width, microfibril angle, lignin, holocellulose, and α-cellulose). It was demonstrated that 96 sequences, out of 116 successfully sequenced epimarkers, were linked to putative genes. The expression levels of nine putative genes (*PtCYP450, PtCpn60, PtPME, PtSCP, PtGH, PtMYB, PtWRKY, PtSTP*, and *PtABC*) were negatively correlated with DNA methylation. Our results imply that the widespread natural variation of DNA methylation might contribute to regulating tree growth and wood formation, and the findings will enhance our understanding of epigenetics in tree growth and xylem formation.

Supplementary Materials: The following are available online at http://www.mdpi.com/1999-4907/11/8/828/s1, Table S1: Primer sequences for Real-time PCR, Table S2: MSAP markers associated with phenotypic traits and the variation explanation in natural population of *Populus tomentosa*, Table S3: Homologous alignment and gene function analysis, Table S4: The gene expression levels detected using qRT-PCR and relative cytosine methylation levels within 45 genotypes of *Populus tomentosa*. Data S1: Sequences for the 116 MSAP markers.

Author Contributions: D.Z. (Deqiang Zhang) designed the experiments. K.M., Y.S., and D.C. collected the plant materials and phenotypic data. K.M., D.Z. (Daling Zhou), and M.T. did the experiment in gene cloning and quantification of gene expression. K.M. detected the molecular markers and analyzed all of the data profiles. K.M. and D.Z. (Deqiang Zhang) wrote the manuscript. Y.S., D.C., D.Z. (Daling Zhou), and M.T. provided suggestions for manuscript revision. All authors have read and agreed to the published version of the manuscript.

Funding: This work was supported by the National Natural Science Foundation of China (Nos. 31872671 and 31670333), and the 111 Project (No. B20050).

Conflicts of Interest: The authors declare no conflict of interest.

Sequencing Data: All of the sequences were submitted to NCBI (https://www.ncbi.nlm.nih.gov/) with accession numbers: MN757649–MN757764.

References

1. Labra, M.; Ghiani, A.; Citterio, S.; Sgorbati, S.; Sala, F.; Vannini, C.; Vannini, C.; Ruffini-Castiglione, M.; Bracale, M. Analysis of cytosine methylation pattern in response to water deficit in pea root tips. *Plant Biol.* **2002**, *4*, 694–699. [CrossRef]
2. Nicotra, A.B.; Atkin, O.K.; Bonser, S.P.; Davidson, A.M.; Finnegan, E.J.; Mathesius, U.; Poot, P.; Purugganan, M.D.; Richards, C.L.; Valladares, F.; et al. Plant phenotypic plasticity in a changing climate. *Trends Plant Sci.* **2010**, *15*, 684–692. [CrossRef]

3. Niederhuth, C.E.; Bewick, A.J.; Ji, L.; Alabady, M.S.; Kim, K.D.; Li, Q.; Rohr, N.A.; Rambani, A.; Burke, J.M.; Udall, J.A.; et al. Widespread natural variation of DNA methylation within angiosperms. *Genome Biol.* **2016**, *17*, 194. [CrossRef]
4. Richards, E.J. Natural epigenetic variation in plant species: A view from the field. *Curr. Opin. Plant Biol.* **2011**, *14*, 204–209. [CrossRef]
5. Teixeira, F.K.; Colot, V. Repeat elements and the *Arabidopsis* DNA methylation landscape. *Heredity* **2010**, *105*, 14–23. [CrossRef]
6. Herrera, C.M.; Bazaga, P. Epigenetic differentiation and relationship to adaptive genetic divergence in discrete populations of the violet *Viola cazorlensis*. *New Phytol.* **2010**, *187*, 867–876. [CrossRef]
7. Rakyan, V.K.; Preis, J.J.; Morgan, H.; Whitelaw, E. The marks, mechanisms and memory of epigenetic statesin mammals. *Biochem. J.* **2001**, *356*, 1–10. [CrossRef] [PubMed]
8. Rakyan, V.K.; Blewitt, M.E.; Druker, R.; Preis, J.I.; Whitelaw, E. Metastable epialleles in mammals. *Trends Genet.* **2002**, *18*, 348–351. [CrossRef]
9. Foerster, A.M.; Dinh, H.Q.; Sedman, L.; Wohlrab, B.; Scheid, O.M. Genetic rearrangements can modify chromatin features at epialleles. *PLoS Genet.* **2011**, *7*, e1002331. [CrossRef]
10. Vaughn, M.W.; Tanurdžic, M.; Lippman, Z.; Jiang, H.; Carrasquillo, R.; Rabinowicz, P.D.; Dedhia, N.; McCombie, W.R.; Agier, N.; Bulski, A.; et al. Epigenetic natural variation in *Arabidopsis thaliana*. *PLoS Biol.* **2007**, *5*, e174. [CrossRef] [PubMed]
11. Angers, B.; Castonguay, E.; Massicotte, R. Environmentally induced phenotypes and DNA methylation: How to deal with unpredictable conditions until the next generation and after. *Mol. Ecol.* **2010**, *19*, 1283–1295. [CrossRef] [PubMed]
12. Dowen, R.H.; Pelizzola, M.; Schmitz, R.J.; Lister, R.; Dowen, J.M.; Nery, J.R.; Dixon, J.E.; Ecker, J.R. Widespread dynamic DNA methylation in response to biotic stress. *Proc. Natl. Acad. Sci. USA* **2012**, *109*, E2183–E2191. [CrossRef] [PubMed]
13. Fujimoto, R.; Sasaki, T.; Kudoh, H.; Taylor, J.M.; Kakutani, T.; Dennis, E.S. Epigenetic variation in the *FWA* gene within the genus *Arabidopsis*. *Plant J.* **2011**, *66*, 831–843. [CrossRef] [PubMed]
14. Zhang, X.; Shiu, S.; Cal, A.; Borevitz, J.O. Global analysis of genetic, epigenetic and transcriptional polymorphisms in *Arabidopsis thaliana* using whole genome tiling arrays. *PLoS Genet.* **2008**, *4*, e1000032. [CrossRef]
15. Long, Y.; Xia, W.; Li, R.; Wang, J.; Shao, M.; Feng, J.; King, G.J.; Meng, J. Epigenetic QTL mapping in *Brassica napus*. *Genetics* **2011**, *189*, 1093–1102. [CrossRef] [PubMed]
16. Foust, C.; Preite, V.; Schrey, A.W.; Alvarez, M.; Robertson, M.; Verhoeven, K.J.F.; Richards, C.L. Genetic and epigenetic differences associated with environmental gradients in replicate populations of two salt marsh perennials. *Mol. Ecol.* **2016**, *25*, 1639–1652. [CrossRef] [PubMed]
17. Ma, K.; Sun, L.; Cheng, T.; Pan, H.; Wang, J.; Zhang, Q. Epigenetic variance, performing cooperative structure with genetics, is associated with leaf shape traits in widely distributed populations of ornamental tree *Prunus mume*. *Front. Plant Sci.* **2018**, *9*, 41. [CrossRef]
18. Cervera, M.T.; Ruiz-Garcia, L.; Martinez-Zapater, J.M. Analysis of DNA methylation in *Arabidopsis thaliana* based on methylation-sensitive AFLP markers. *Mol. Genet. Genom.* **2002**, *268*, 543–552. [CrossRef]
19. Lisch, D. Epigenetic regulation of transposable elements in plants. *Ann. Rev. Plant Biol.* **2009**, *60*, 43–66. [CrossRef]
20. Hsieh, T.; Shin, J.; Uzawa, R.; Silva, P.; Cohen, S.; Bauer, M.J.; Hashimoto, M.; Kirkbride, R.C.; Harada, J.J.; Zilberman, D.; et al. Regulation of imprinted gene expression in *Arabidopsis* endosperm. *Proc. Natl. Acad. Sci. USA* **2011**, *108*, 1755–1762. [CrossRef]
21. Uthup, T.K.; Ravindran, M.; Bini, K.; Thakurdas, S. Divergent DNA methylation patterns associated with abiotic stress in *Hevea brasiliensis*. *Mol. Plant* **2011**, *4*, 996–1013.
22. Diez, C.M.; Roessler, K.; Gaut, B.S. Epigenetics and plant genome evolution. *Curr. Opin. Plant Biol.* **2014**, *18*, 1–8. [CrossRef] [PubMed]
23. Rico, L.; Ogaya, R.; Barbeta, A.; Pnuelas, J. Changes in DNA methylation fingerprint of *Quercus ilex* trees in response to experimental field drought simulating projected climate change. *Plant Biol.* **2014**, *16*, 419–427. [CrossRef] [PubMed]

24. Tang, X.; Tao, X.; Wang, Y.; Ma, D.; Li, D.; Yang, H.; Ma, X. Analysis of DNA methylation of perennial ryegrass under drought using the methylation-sensitive amplification polymorphism (MSAP) technique. *Mol. Genet. Genom.* **2014**, *289*, 1075–1084. [CrossRef] [PubMed]
25. Kakutani, T.; Munakata, K.; Richards, E.; Hirohiko, H. Meiotically and mitotically stable inheritance of DNA hypomethylation induced by *ddm1* mutation of *Arabidopsis thaliana*. *Genetics* **1999**, *151*, 831–838. [PubMed]
26. Cubas, P.; Vincent, C.; Coen, E. An epigenetic mutation responsible for natural variation in floral symmetry. *Nature* **1999**, *401*, 157–161. [CrossRef] [PubMed]
27. Kashkush, K.; Khasdan, V. Large-scale survey of cytosine methylation of retrotransposons and the impact of read out transcription from long terminal repeatson expression of adjacent rice genes. *Genetics* **2007**, *177*, 1975–1985. [CrossRef]
28. Vining, K.J.; Pomraning, K.R.; Wilhelm, L.J.; Priest, H.D.; Pellegrini, M.; Mockler, T.C.; Freitag, M.; Strauss, S.H. Dynamic DNA cytosine methylation in the *Populus trichocarpa* genome: Tissue-level variation and relationship to gene expression. *BMC Genom.* **2012**, *13*, 27. [CrossRef]
29. Jacobsen, S.E.; Meyerowitz, E.M. Hypermethylated *SUPERMAN* epigenetic alleles in *Arabidopsis*. *Science* **1997**, *277*, 1100–1103. [CrossRef]
30. Bender, J.; Fink, G.R. Epigenetic control of an endogenous gene family is revealed by a novel blue fluorescent mutant of *Arabidopsis*. *Cell* **1995**, *83*, 725–734. [CrossRef]
31. Choi, Y.; Gehring, M.; Johnson, L.; Hannon, M.; Harada, J.J.; Goldberg, R.B.; Jacobsen, S.E.; Fischer, R.L. DEMETER, a DNA glycosylase domain protein, is required for endosperm gene imprinting and seed viability in *Arabidopsis*. *Cell* **2002**, *110*, 33–42. [CrossRef]
32. Kinoshita, T.; Miura, A.; Choi, Y.; Kinoshita, Y.; Cao, X.; Jacobsen, S.E.; Fischer, R.L.; Kakutani, T. One-way control of *FWA* imprinting in *Arabidopsis* endosperm by DNA methylation. *Science* **2004**, *303*, 521–523. [CrossRef] [PubMed]
33. Schmitz, R.J.; He, Y.; Valdés-López, O.; Khan, S.M.; Joshi, T.; Urich, M.A.; Nery, J.R.; Diers, B.; Xu, D.; Stacey, G.; et al. Epigenome-wide inheritance of cytosine methylation variants in a recombinant inbred population. *Genome Res.* **2013**, *23*, 1663–1674. [CrossRef] [PubMed]
34. Ci, D.; Song, Y.; Du, Q.; Tian, M.; Han, S.; Zhang, D. Variation in genomic methylation in natural populations of *Populus simonii* is associated with leaf shape and photosynthetic traits. *J. Exp. Bot.* **2016**, *67*, 723–737. [CrossRef]
35. Zhu, Z. Collection, conservation and utilization of plus tree resources of *Populus tomentosa* in China. *J. Beijing For. Univ.* **1992**, *14*, 1–25.
36. Zhu, Z.; Zhang, Z. The status and advavces of genetic improvement of *Populus tomentosa* Carr. *J. Beijing For. Univ.* **1997**, *6*, 1–7.
37. Zhang, D.; Zhang, Z.; Yang, K.; Li, B. Genetic mapping in (*Populus tomentosa* × *Populus bolleana*) and *P. tomentosa* Carr. Using AFLP markers. *Theor. Appl. Genet.* **2004**, *108*, 657–662. [CrossRef] [PubMed]
38. Du, Q.; Pan, W.; Xu, B.; Li, B.; Zhang, D. Polymorphic simple sequence repeat (SSR) loci within cellulose synthase (*PtoCesA*) genes are associated with growth and wood properties in *Populus tomentosa*. *New Phytol.* **2013**, *197*, 763–776. [CrossRef]
39. Ma, K.; Song, Y.; Jiang, X.; Zhang, Z.; Li, B.; Zhang, D. Photosynthetic response to genome methylation affects the growth of Chinese white poplar. *Tree Genet. Genomes* **2012**, *8*, 1407–1421. [CrossRef]
40. Ma, K.; Song, Y.; Yang, X.; Zhang, Z.; Zhang, D. Variation in genomic methylation in natural populations of Chinese white poplar. *PLoS ONE* **2013**, *8*, e63977. [CrossRef]
41. Wang, Q.; Ci, D.; Li, T.; Li, P.; Song, Y.; Chen, J.; Quan, M.; Zhou, D.; Zhang, D. The role of DNA methylation in xylogenesis in different tissues of poplar. *Front. Plant Sci.* **2016**, *7*, 1003. [CrossRef] [PubMed]
42. Du, Q.; Wang, B.; Wei, Z.; Zhang, D.; Li, B. Genetic diversity and population structure of Chinese white poplar (*Populus tomentosa*) revealed by SSR markers. *J. Hered.* **2012**, *103*, 853–862. [CrossRef] [PubMed]
43. Yang, H.; Meng, X.; Cheng, J.; Zhao, L.; Wang, X. Study on estimating the volume of a single tree using normal form factor. *Forest Res. Manag.* **2005**, *1*, 39–41.
44. Andersson, S.; Serimaa, R.; Torkkeli, M.; Paakkari, T.; Saranpaa, P.; Pesonen, E. Microfibril angle of Norway spruce [*Piceaabies* (L.) Karst.] compression wood: Comparison of measuring techniques. *J. Wood Sci.* **2000**, *46*, 343–349. [CrossRef]

45. Wang, Y.; Liu, C.; Zhao, R.; Mccord, J.; Rials, T.G.; Wang, S. Anatomical characteristics, microfibril angle and micromechanical properties of cottonwood (*Populus deltoides*) and its hybrids. *Biomass Bioenergy* **2016**, *93*, 72–77. [CrossRef]
46. Hart, J.F.; De-Araujo, F.; Thomas, B.R.; Mansfield, S.D. Wood quality and growth characterization across intra- and inter-specific hybrid aspen clones. *Forests* **2013**, *4*, 786–807. [CrossRef]
47. Porth, I.; Klápště, J.; Skyba, O.; Lai, B.S.K.; Geraldes, A.; Muchero, W.; Tuskan, G.A.; Douglas, C.J.; El-Kassaby, Y.A.; Mansfield, S.D. *Populus trichocarpa* cell wall chemistry and ultrastructure trait variation, genetic control and genetic correlations. *New Phytol.* **2013**, *197*, 777–790. [CrossRef]
48. Schimleck, L.R.; Kube, P.D.; Raymond, C.A. Genetic improvement of kraft pulp yield in *Eucalyptus nitens* using cellulose content determined by near infrared spectroscopy. *Can. J. Forest Res.* **2004**, *34*, 2363–2370. [CrossRef]
49. Aleong, J.; Howard, D. Extensions of the Duncan's Multiple Range Test for unbalanced data. *J. Appl. Stat.* **1985**, *12*, 83–90. [CrossRef]
50. Murray, M.G.; Thompson, W.F. Rapid isolation of high molecular weight plant DNA. *Nucl. Acids Res.* **1980**, *8*, 4321–4325. [CrossRef]
51. Tixier, M.H.; Sourdille, P.; Röder, M.S.; Leroy, P.; Bernard, M. Detection of wheat microsatellites using a non-radioactive silver-nitrate staining method. *J. Genet. Breed.* **1997**, *51*, 175–177.
52. Song, Y.; Ma, K.; Bo, W.; Zhang, Z.; Zhang, D. Sex-specific DNA methylation and gene expression in andromonoecious poplar. *Plant Cell Rep.* **2012**, *31*, 1393–1405. [CrossRef] [PubMed]
53. Ma, K.; Song, Y.; Huang, Z.; Lin, L.; Zhang, Z.; Zhang, D. The low fertility of Chinese white poplar: Dynamic changes in anatomical structure, endogenous hormone concentrations, and key gene expression in the reproduction of a naturally occurring hybrid. *Plant Cell Rep.* **2013**, *32*, 401–414. [CrossRef] [PubMed]
54. Zhang, D.; Du, Q.; Xu, B.; Zhang, Z.; Li, B. The actin multigene family in *Populus*: Organization, expression and phylogenetic analysis. *Mol. Genet. Genom.* **2010**, *284*, 105–119. [CrossRef] [PubMed]
55. Niu, G.; Zhao, S.; Wang, L.; Dong, W.; Liu, L.; He, Y. Structure of the *Arabidopsis thaliana* NADPH-cytochrome P450 reductase 2 (ATR2) provides insight into its function. *FEBS J.* **2017**, *284*, 754–765. [CrossRef] [PubMed]
56. Sundin, L.; Vanholme, R.; Geerinck, J.; Goeminne, G.; Hofer, R.; Kim, H.; Ralph, J.; Boerjan, W. Mutation of the inducible *ARABIDOPSIS THALIANA CYTOCHROME P450 REDUCTASE2* alters lignin composition and improves saccharification. *Plant Physiol.* **2014**, *166*, 1956–1971. [CrossRef] [PubMed]
57. Hill, J.E.; Hemmingsen, S.M. *Arabidopsis thaliana* type I and II chaperonins. *Cell Stress Chaperon* **2001**, *6*, 190–200. [CrossRef]
58. Peng, L.; Fukao, Y.; Myouga, F.; Motohashi, R.; Shinozaki, K.; Shikanai, T. A chaperonin subunit with unique structures is essential for folding of a specific substrate. *PLoS Biol.* **2011**, *9*, e1001040. [CrossRef]
59. Tiwari, L.D.; Grover, A. Cpn60β4 protein regulates growth and developmental cycling and has bearing on flowering time in *Arabidopsis thaliana* plants. *Plant Sci.* **2019**, *286*, 78–88. [CrossRef]
60. Micheli, F. Pectin methylesterases: Cell wall enzymes with important roles in plant physiology. *Trends Plant Sci.* **2001**, *6*, 414–419. [CrossRef]
61. Wolf, S.I.; Mouille, G.; Pelloux, J. Homogalacturonan methyl-esterification and plant development. *Mol. Plant* **2009**, *2*, 851–860. [CrossRef] [PubMed]
62. Hongo, S.; Sato, K.; Yokoyama, R.; Nishitani, K. Demethylesterification of the primary wall by *PECTIN METHYLESTERASE35* provides mechanical support to the *Arabidopsis* stem. *Plant Cell* **2012**, *24*, 2624–2634. [CrossRef] [PubMed]
63. Siedlecka, A.; Wiklund, S.; Peronne, M.; Micheli, F.; Lesniewska, J.; Sethson, I.; Edlund, U.; Richard, L.; Sundberg, B.; Mellerowicz, E.J. Pectin methyl esterase inhibits intrusive and symplastic cell growth in developing wood cells of *Populus*. *Plant Physiol.* **2007**, *146*, 554–565. [CrossRef] [PubMed]
64. Bienert, M.D.; Delannoy, M.; Navarre, C.; Boutry, M. *NtSCP1* from tobacco is an extracellular serine carboxypeptidase III that has an impact on cell elongation. *Plant Physiol.* **2012**, *158*, 1220–1229. [CrossRef]
65. Dominguez, F.; Gonzalez, M.D.; Cejudo, F.J. A germination-related gene encoding a serine carboxypeptidase is expressed during the differentiation of the vascular tissue in wheat grains and seedlings. *Planta* **2002**, *215*, 727–734. [CrossRef]
66. Escamillatrevino, L.L.; Chen, W.; Card, M.L.; Shih, M.; Cheng, C.; Poulton, J.E. *Arabidopsis thalianaβ*-glucosidases BGLU45 and BGLU46 hydrolyse monolignolglucosides. *Phytochemistry* **2006**, *67*, 1651–1660. [CrossRef]

67. Chapelle, A.; Morreel, K.; Vanholme, R.; Le-Bris, P.; Morin, H.; Lapierre, C.; Boerjan, W.; Jouanin, L.; Demont-Caulet, N. Impact of the absence of stem-specific β-glucosidases on lignin and monolignols. *Plant Physiol.* **2012**, *160*, 1204–1217. [CrossRef]
68. Huang, D.; Wang, S.; Zhang, B.; Shang-guan, K.; Shi, Y.; Zhang, D.; Liu, X.; Wu, K.; Xu, Z.; Fu, X.; et al. A gibberellin-mediated DELLA-NAC signaling cascade regulates cellulose synthesis in rice. *Plant Cell* **2015**, *27*, 1681–1696. [CrossRef]
69. Li, C.; Wang, X.; Lu, W.; Liu, R.; Tian, Q.; Sun, Y.; Luo, K. A poplar R2R3-MYB transcription factor, PtrMYB152, is involved in regulation of lignin biosynthesis during secondary cell wall formation. *Plant Cell Tiss. Org.* **2014**, *119*, 553–563. [CrossRef]
70. Guillaumie, S.; Mzid, R.; Méchin, V.; Léon, C.; Hichri, I.; Destrac-Irvine, A.; Trossat-Magnin, C.; Delrot, S.; Lauvergeat, V. The grapevine transcription factor WRKY2 influences the lignin pathway and xylem development in tobacco. *Plant Mol. Biol.* **2010**, *72*, 215–234. [CrossRef]
71. Yang, L.; Zhao, X.; Yang, F.; Fan, D.; Jiang, Y.; Luo, K. PtrWRKY19, a novel WRKY transcription factor, contributes to the regulation of pith secondary wall formation in *Populus trichocarpa*. *Sci. Rep.* **2016**, *6*, 18643. [CrossRef] [PubMed]
72. Büttner, M.; Truernit, E.; Baier, K.; Scholz-starke, J.; Sontheim, M.; Lauterbach, C.; Huss, V.A.R.; Sauer, N. AtSTP3, a green leaf-specific, low affinity monosaccharide-H^+symporter of *Arabidopsis thaliana*. *Plant Cell Environ.* **2000**, *23*, 175–184. [CrossRef]
73. Schneidereit, A.; Scholzstarke, J.; Buttner, M. Functional characterization and expression analyses of the glucose-specific AtSTP9 monosaccharide transporter in pollen of *Arabidopsis*. *Plant Physiol.* **2003**, *133*, 182–190. [CrossRef]
74. Scholz-Starke, J.; Büttner, M.; Sauer, N. AtSTP6, a new, pollen-specific H^+-monosaccharide symporter from *Arabidopsis*. *Plant Physiol.* **2003**, *131*, 70–77. [CrossRef]
75. Schofield, R.A.; Bi, Y.; Kant, S.; Rothstein, S.J. Over-expression of *STP13*, a hexose transporter, improves plant growth and nitrogen use in *Arabidopsis thaliana* seedlings. *Plant Cell Environ.* **2009**, *32*, 271–285. [CrossRef]
76. Liu, Q.; Dang, H.; Chen, Z.; Wu, J.; Chen, Y.; Chen, S.; Luo, L. Genome-wide identification, expression, and functional analysis of the sugar transporter gene family in cassava (*Manihotesculenta*). *Int. J. Mol. Sci.* **2018**, *19*, 987. [CrossRef]
77. Alejandro, S.; Lee, Y.; Tohge, T.; Sudre, D.; Osorio, S.; Park, J.; Bovet, L.; Lee, Y.; Geldner, N.; Fernie, A.R.; et al. AtABCG29 is a monolignol transporter involved in lignin biosynthesis. *Curr. Biol.* **2012**, *22*, 1207–1212. [CrossRef]
78. Takeuchi, M.; Watanabe, A.; Tamura, M.; Tsutsumi, Y. The gene expression analysis of *Arabidopsis thaliana* ABC transporters by real-time PCR for screening monolignol-transporter candidates. *J. Wood Sci.* **2018**, *64*, 477–484. [CrossRef]
79. Kaneda, M.; Schuetz, M.; Lin, B.S.P.; Chanis, C.; Hamberger, B.; Western, T.L.; Ehlting, J.; Samuels, A.L. ABC transporters coordinately expressed during lignification of *Arabidopsis* stems include a set of ABCBs associated with auxin transport. *J. Exp. Bot.* **2011**, *62*, 2063–2077. [CrossRef]
80. Hauben, M.; Haesendonckx, B.; Standaert, E.; Kelen, K.V.D.; Azmi, A.; Akpo, H.; Breusegem, F.V.; Guisez, Y.; Bots, M.; Lambert, B.; et al. Energy use efficiency is characterized by an epigenetic component that can be directed through artificial selection to increase yield. *Proc. Natl. Acad. Sci. USA* **2009**, *106*, 20109–20114. [CrossRef]
81. Tsaftaris, A.S.; Kafka, M.; Polidoros, A.; Tani, W. Epigenetic Changes in Maize DNA and Heterosis. In *The Genetics and Exploitation of Heterosis in Crops*; CIMMYT: Mexico City, Mexico, 1997; pp. 112–113.
82. Gourcilleau, D.; Bogeat-Triboulot, M.; Thiec, D.L.; Lafon-Placette, C.; Delaunay, A.; El-Soud, W.A.; Brignolas, F.; Maury, S. DNA methylation and histone acetylation: Genotypic variations in hybrid poplars, impact of water deficit and relationships with productivity. *Ann. For. Sci.* **2010**, *67*, 208. [CrossRef]

 *forests* 

Article

Preliminary Evidence for Domestication Effects on the Genetic Diversity of *Guazuma crinita* in the Peruvian Amazon

Lady Laura Tuisima-Coral [1,2,*], Petra Hlásná Čepková [3], John C. Weber [4] and Bohdan Lojka [1,*]

[1] Department of Crop Sciences and Agroforestry, Faculty of Tropical AgriSciences, Czech University of Life Sciences Prague, Kamycka 129, 16500 Prague, Czech Republic

[2] Estación Experimental Agraria El Chira, Dirección de Recursos Genéticos y Biotecnología, Instituto Nacional de Innovación Agraria, Carretera Sullana–Talara, Km. 1027, Sullana, Piura 20120, Peru

[3] Crop Research Institute, Department of Gene Bank, Drnovska 507/73 Praha 6-Ruzyne, 16521 Prague, Czech Republic; hlasna@vurv.cz

[4] World Agroforestry Centre, Lima 15074, Peru; johncrweber@aol.com

* Correspondence: ct_elchira@inia.gob.pe (L.L.T.-C.); lojka@ftz.czu.cz (B.L.);

Received: 30 June 2020; Accepted: 20 July 2020; Published: 23 July 2020

Abstract: *Guazuma crinita*, a fast-growing timber tree species, was chosen for domestication in the Peruvian Amazon because it can be harvested at an early age and it contributes to the livelihood of local farmers. Although it is in an early stage of domestication, we do not know the impact of the domestication process on its genetic resources. Amplified fragment length polymorphic (AFLP) fingerprints were used to estimate the genetic diversity of *G. crinita* populations in different stages of domestication. Our objectives were (i) to estimate the level of genetic diversity in *G. crinita* using AFLP markers, (ii) to describe how the genetic diversity is distributed within and among populations and provenances, and (iii) to assess the genetic diversity in naturally regenerated, cultivated and semi-domesticated populations. We generated fingerprints for 58 leaf samples representing eight provenances and the three population types. We used seven selective primer combinations. A total of 171 fragments were amplified with 99.4% polymorphism at the species level. Nei's genetic diversity and Shannon information index were slightly higher in the naturally regenerated population than in the cultivated and semi-domesticated populations (He = 0.10, 0.09 and 0.09; I = 0.19, 0.15 and 0.16, respectively). The analysis of molecular variation showed higher genetic diversity within rather than among provenances (84% and 4%, respectively). Cluster analysis (unweighted pair group method with arithmetic mean) and principal coordinate analysis did not show correspondence between genetic and geographic distance. There was significant genetic differentiation among population types ($F_{st} = 0.12$ at $p < 0.001$). The sample size was small, so the results are considered as preliminary, pending further research with larger sample sizes. Nevertheless, these results suggest that domestication has a slight but significant effect on the diversity levels of *G. crinita* and this should be considered when planning a domestication program.

Keywords: genetic diversity; genetic differentiation; natural regeneration; cultivated population; semi-domesticated population

---

1. Introduction

Tropical forests provide many valuable products, including rubber, fruits and nuts, medicinal herbs, lumber, firewood, and charcoal [1]. Natural forest populations typically possess considerable genetic variation [2]. However, deforestation due to slash-and-burn agriculture [3], over-harvesting and other unsustainable forestry practices are reducing tree genetic diversity in many areas in the tropics [1]. Tree domestication has been promoted as a strategy to conserve genetic resources for tropical species [1,4]. Domestication involves the selection and propagation of desirable trees, so we expect that genetic diversity is lower in domesticated populations compared with natural populations. However, very few studies have assessed the difference in genetic diversity between domesticated and natural forest tree populations in the tropics [5,6]. This information is necessary for planning tree domestication strategies that maintain high levels of genetic diversity in the domesticated population.

Guazuma crinita Mart. (Malvaceae) was identified as a priority timber species for tree domestication in the Peruvian Amazon [7]. It is a pioneer species in the Amazon basin of Peru, Ecuador and Brazil [8]. It can be inter-cultivated with food crops because it has a small crown with thin branches and the older branches naturally self-prune. It provides wood products at an early age, can be coppiced for successive harvests and contributes significantly to farmers' income [9,10]. Due to its initial fast growth (up to 3 m per year), it has been promoted in reforestation programs and agroforestry systems [11,12]. In addition, it can be vegetative propagated for commercial purposes [10]. It has promising national and international markets for lumber products [10,13].

G. crinita is a cross pollinated species that produces fruit at an early age, and can potentially disperse seeds over a long distance by both wind and water, allowing it to colonize forest gaps and potentially form dense stands of natural regeneration [14]. These dispersal characteristics should produce extensive gene flow among populations, resulting in high levels of genetic diversity within population and relatively low genetic differentiation among populations [15–17]. Local farmers manage *G. crinita* in its natural ecological niche for timber, while also engaging in other agricultural activities.

The objectives of this study were (i) to estimate the level of genetic diversity in *G. crinita* using AFLP (amplified fragment length polymorphism) markers, (ii) to describe how the genetic diversity is distributed within and among populations and provenances, and (iii) to assess the genetic diversity in naturally regenerated, cultivated and semi-domesticated populations. Based on the reproductive characteristics of this species, we hypothesized that there would be (i) a high level of polymorphism within populations and provenances, (ii) relatively little genetic differentiation among populations, and (iii) greater genetic diversity in the naturally regenerated population compared with the cultivated and semi-domesticated populations.

2. Materials and Methods

2.1. Sampling

In this study, we analyzed the genetic diversity of *G. crinita* populations in three stages of tree domestication. A natural population included wild, naturally regenerated trees from one provenance that farmers retained in their fields. In the cultivated population, we sampled trees from one provenance that farmers planted in a home garden using seedlings produced in a home nursery. The semi-domesticated population included trees from six provenances in a clonal garden. Genotypes in the clonal garden were selected over a period of years from progeny trees originating from an extensive collection of 200 mother trees. The natural, cultivated and semi-domesticated populations are in the second, fourth and sixth stages, respectively, of the seven stages of domestication proposed by Vodouhe and Dansi [18].

A total of 84 individuals from the three different population types (natural, cultivated and semi-domesticated) were sampled from eight *G. crinita* provenances in the Peruvian Amazon (Figure 1). Thirty individuals from the village of Nuevo Piura were randomly sampled from a population of natural regeneration located in Campo Verde district, Ucayali region (150 m.a.s.l). In the city of Tingo Maria, Huanuco region, 30 cultivated individuals were sampled in a home garden (564 m.a.s.l). In addition, 24 vegetative propagated trees were sampled from a clonal multiplication garden at the Peruvian Amazon Research Institute (IIAP), located 12.4 km from Pucallpa, Ucayali Region (154 m.a.s.l). They represent selected genotypes from six provenances in two watersheds in the Peruvian Amazon. Young leaf tissues were collected from individual plants and then dried in silica gel for DNA extraction.

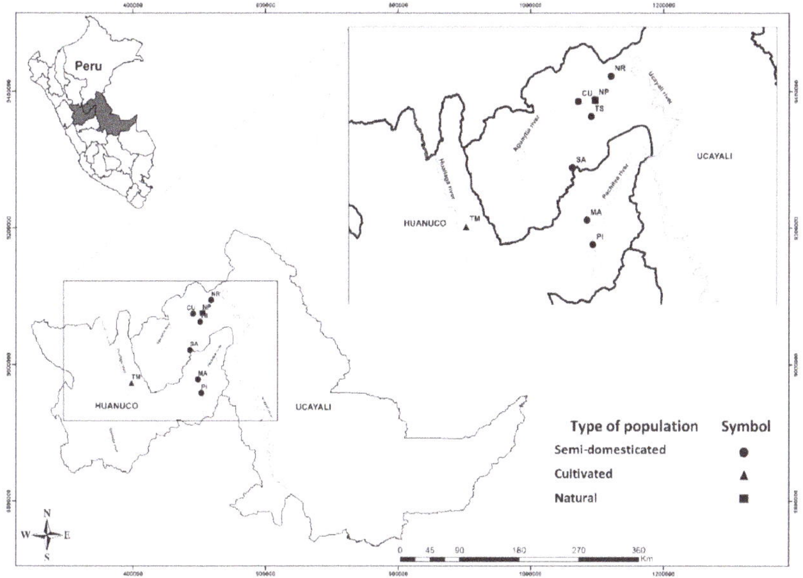

Figure 1. Map of the geographic distribution of eight *Guazuma crinita* provenances sampled in Peru. NP = Nuevo Piura—naturally regenerated population. TM = Tingo Maria—cultivated population. NR = Nueva Requena, SA = San Alejandro, PI = Puerto Inca, CU = Curimana, MA = Macuya and TS = Tahuayo—semi-domesticated populations.

The sample size in this study was small so we consider the results as preliminary. Other studies of genetic diversity in tropical tree species have also used small sample sizes [17,19–22] and reported genetic diversity patterns consistent with studies based on large sample sizes.

2.2. DNA Extraction

DNA from the 84 leaf samples was extracted using the CTAB (cetyltrimethylammonium bromide) method [23] with a slight modification (adding a trace of polyvinylpyrrolidone (PVP) and 5 µL of RNase). The DNA quality was determined by 0.8 % agarose gel electrophoresis using a Nanodrop Spectrophotometer (Thermo Scientific, Delaware, USA). We only obtained genomic DNA of sufficient quality for amplification from 58 of the 84 samples (Table 1). It was diluted to 50 ng/µL and stored at −20 °C.

Table 1. Origin of the 58 *G. crinita* individuals analysed by amplified fragment length polymorphic (AFLP) markers.

Provenance	Region	No. of Samples (Code)	Population Type
Nuevo Piura (NP)	Ucayali	19 (NP 1, 2, 3, ... 19)	Natural regeneration
Tingo Maria (TM)	Huanuco	15 (TM 1, 2, 3, ... 15)	Cultivated [1]
Nueva Requena (NR)	Ucayali	5 (NR 7, 8, 11, 12, 13)	Semi-domesticated [2]
Tahuayo Stream (TS)	Ucayali	3 (TS 1, 2,9)	Semi-domesticated
San Alejandro (SA)	Ucayali	5 (SA 1, 10, 11, 14, 15)	Semi-domesticated
Curimana (CU)	Ucayali	4 (CU 3, 4, 6, 7)	Semi-domesticated
Puerto Inca (PI)	Huanuco	3 (PI 1, 3, 13)	Semi-domesticated
Macuya (MA)	Huanuco	5 (MA11, 33, 36, 42, 46)	Semi-domesticated

[1] Samples cultivated in a home garden; [2] genotypes established in a clonal multiplication garden.

2.3. AFLP Amplification

Molecular AFLP markers were used because no previous genome information is required and a large number of polymorphic loci can be analysed simultaneously [24]. With the use of AFLP, we expected to successfully assess the genetic relationships between *G. crinita* populations in the Peruvian Amazon.

Techniques for the AFLP analysis of *G. crinita* were adapted from those described by Vos et al. [25]. Commercial AFLP kits (Stratec Molecular, Berlin, Germany) were used for the restriction, ligation and pre-amplification steps.

An AFLP Core Plant Reagent Kit I (Stratec Molecular, Berlin, Germany) was used for restriction and ligation. The restriction reaction volume was 5 µL and included the following: 1 uL of 5 × Reaction Buffer (50 mM Tris-HCl (pH 7.5), 50 mM Mg-acetate, 250 mM k-acetate); 0.4 µL of enzyme mixture EcoRI/MseI (1.25 U/µL each in 10 mM Tris-HCl (pH 7.4), 50 mM NaCl, 0.1 mM EDTA, 1 mM DTT, 0.1 mg/mL BSA, 50% glycerol (*v/v*), 0.1% Triton®X-100); 1.1 µL of sterile water and 2.5 µL of DNA (50 ng/µL). After mixing the reaction we incubated it in a thermocycler at 37 °C for 2 h. Ligation of the adapters included the following: 4.8 µL of Adapter/Ligation Solution (EcoRI/MseI adapters, 0.4 mM ATP, 10 mM Tris-HCl (pH 7.5), 10 mM Mg-acetate, 50 mM K-acetate); and 0.2 µL of T4 DNA Ligase (1 U/µL in 10 mM Tris-HCl (pH 7.5), 1 mM DTT, 50 mM KCl, 50% (v/v) glycerol). This volume was added into a microtube with the restriction products from previous reactions. The reaction was left at 37 °C for 2 h.

For pre-amplification, we used AFLP Pre-Amp Mix I (Stratec Molecular, Berlin, Germany). The cycle profile for pre-amplification PCR was as follows: an initial step at 72 °C for 2 min, followed by 20 cycles of 94 °C for 10 s, at 56 °C for 30 s and at 72 °C for 2 min and final elongation at 60 °C for 30 min; containing 4.0 µL of pre-amplification mix, 0.5 µL of 10 × Buffer for RedTaq Polymerase (100 mM Tris-HCl (pH 8.3), 500 mM KCl, 11 nM $MgCl_2$ and 0.1% gelatin) (Sigma-Aldrich, Saint Louis, USA), 0.1 µL RedTaq Polymerase (Sigma-Aldrich, Saint Louis, USA) and 0.5 µL of DNA after restriction and ligation. The product was visualized on 1.8% TBE agarose gel. After amplification, the product was diluted by the addition of 15 µL of ddH_2O.

The selective amplification reactions with slight modifications were performed following the protocol described in Mikulášková et al. [26], with a total volume of 9.8 µL, comprising 2.3 µL of preamplified DNA, 5.1 µL ddH2O, 1 µL 10 × polymerase buffer (100 mM Tris-HCl (pH 8.3), 500 mM KCl, 11 nM $MgCl_2$ and 0.1% gelatin) (Sigma-Aldrich, Saint Louis, USA), 0.2 mM dNTP (Thermo Scientific, USA), 0.5 pmol fluorescent dye-labelled EcoRI primer (Applied Biosystems, Foster city, California, USA), 0.5 pmol MseI primer (Generi Biotech, Hradec Králové, Czech Republic) and 0.2 U RedTaq DNA polymerase (Sigma Aldrich, Saint Louis, USA). Selective PCR amplifications were carried out using the following cycle profile: 92 °C for 2 min, 65 °C for 30 s and 72 °C for 2 min. A touchdown protocol was applied in the following eight cycles at 94 °C for 1 s, at 64 °C (1 °C decrease each cycle) for 30 s, and at 72 °C for 60 s. This was followed by 23 cycles of 94 °C for 1 s, at 56 °C for 30 s and at 72 °C for 2 min. Final elongation was carried out at 60 °C for 30 min.

Eleven primer combinations were tested but only seven were selected for final analysis because they produced distinct polymorphic bands. For all PCR amplifications T100TM Thermal Cycler (Bio-Rad Laboratories, California, USA) was used. The final products after selective amplification were visualized on 1.8% agarose gels buffered in 1 × TBE. Following a successful amplification, the AFLP products were prepared for analysis on 3500 Genetic Analyser, automated sequencer (Applied Biosystems, Foster city, California, USA). Ten percent of the samples were analyzed twice for error rate estimation.

2.4. Data Analysis

AFLP fragments were analyzed using GeneMarker v 2.0.2 (SoftGenetics, USA). Polymorphic and strong peaks were scored as present or absent and then converted into a binary matrix. The data were used to calculate the percentage of polymorphic fragments, gene diversity (He) and Shannon's information index (*I*) using POPGENE v1.32 [27].

Analysis of molecular variance (AMOVA) was carried out to evaluate genetic diversity within and among samples, as well as to estimate genetic differentiation indexes, using GenAlEx v6 [28]. Principal coordinate analysis (PCoA) was carried out to assess genetic relationship among samples, also using GenAlEx v6.

Patterns of genetic relationships among samples was also investigated using cluster analysis. A dendrogram was constructed based on Jaccard's dissimilarity index with UPGMA using DARwin5 [29]. The software STRUCTURE v2.3.2.1 [30] was used to identify the number of similar population clusters (K) and the proportion of membership of each population in each of the *K* clusters. The analysis of the number of clusters was performed using the recessive allele model with a burn-in and run lengths of 100,000 and 1,000,000 interactions, respectively. The number of clusters was determined following the guidelines of Pritchard and Wen [31] and Evano et al. [32] using the online software Structure Harvester [33], and subsequently visualized using DISTRUCT 1.1 [34]. AFLP percentage of reproducibility was calculated following Bonin et al. [35].

3. Results

3.1. AFLP Fingerprint

The seven primer combinations selected for the analysis revealed 10 to 35 fragments in the 58 *G. crinita* samples, with the mean of 24 fragments. Of the 171 total fragments, 99.7% were polymorphic. The fragments were in a size range of 52 to 336 bp (Table 2). The first primer combination (EcoRI-ACG/*Mse*I- CTT) was the most successful with a polymorphic rate of 20.5%. The least successful was EcoRI-ACG/*Mse*I- CAG (5.8%).

Table 2. Description of the primer combinations.

No.	Primer Combination	No. of Fragments	Fragment Range Size (pb)	Polymorphic Bands (%)
1	EcoRI-ACG/*Mse*I- CTT	35	58–231	20.6
2	EcoRI-ACG/*Mse*I- CTG	23	60–215	13.5
3	EcoRI-ACG/*Mse*I- CTA	19	69–198	11.1
4	EcoRI-ACG/*Mse*I- CAT	27	52–176	15.9
5	EcoRI-ACG/*Mse*I- CAG	10	71–140	5.9
6	EcoRI-ACG/*Mse*I- CAC	29	64–336	17.0
7	EcoRI-ACG/*Mse*I- CAA	28	52–228	16.4
	Total	171		

Ten percent of the sample size was independently replicated with the same primer combinations, resulting in 85% of the fragment reproducibility among replicated samples.

3.2. Genetic Diversity and Population Structure

In a single measurement of intra-population diversity, e.g., the percentage of polymorphic fragments, samples from Nuevo Piura provenance (natural population) exhibited the highest diversity (72.5%), followed by the samples from Tingo Maria (cultivated population, 42.2%). There was less diversity in the provenances in the semi-domesticated population (20.7% on average) (Table 3). However, in the semi-domesticated provenances considered as one population, the percentage of polymorphic fragments was 54.4%, Nei's genetic diversity was 0.09 and the Shannon index was 0.16.

Table 3. Measurements of genetic diversity in three types of populations of *G crinita:* natural regeneration, cultivated and semi domesticated.

Samples	N° of Samples	N° of PF	PPF (%)	Nei's Gene Diversity (*He*)	Shannon Information Index (*I*)
Types of population					
Natural regeneration	19	124	72.5	0.10	0.19
Cultivated	15	84	49.1	0.09	0.15
Semi-domesticated	24	93	54.4	0.09	0.16
Provenances					
Nuevo Piura (NP)	19	124	72.5	0.10	0.19
Tingo Maria (CP)	15	84	49.1	0.09	0.15
Nueva Requena (SDP)	4	35	20.5	0.06	0.10
Tahuayo Stream (SDP)	3	41	23.9	0.09	0.13
San Alejandro (SDP)	5	43	25.2	0.07	0.12
Curimana (SDP)	4	30	17.5	0.06	0.09
Puerto Inca (SDP)	3	31	18.1	0.07	0.10
Macuya (SDP)	5	32	18.7	0.07	0.10
Species level	58	170	99.4	0.11	0.20

PF = polymorphic fragments; PPF = percentage of polymorphic fragments; NP = natural population, CP = cultivated population, SDP = semi-domesticated population.

Based on 170 polymorphic fragments from the 58 *G. crinita* samples, Nei's genetic diversity values ranged from 0.06 to 0.10 and the Shannon information index (*I*) ranged from 0.09 to 0.19 (Table 3). Comparing the three population types, all measure (polymorphic fragments (PF), percentage of polymorphic fragments (PPF), *He*, and *I*) were slightly higher in the population of natural regeneration.

The coefficient of genetic differentiation (G_{st}) among the three population types was 0.10. This indicates that 10% of the genetic diversity was distributed among the population types. Nei's genetic identity comparison between population types indicated that the highest identity (0.011) was between natural and cultivated populations, and the lowest identity (0.022) was between cultivated and semi-domesticated populations.

Pairwise genetic distance between provenances ranged from 0.011 to 0.063 (Table 4). Nueva Piura and Tingo Maria were the most similar with the minimum distance value of 0.011, while the highest value of genetic distance (0.063) was between Nuevo Piura, Puerto Inca, and Tahuayo, San Alejandro.

Table 4. Nei's genetic identity (above diagonal) and genetic distance (below diagonal) among eight *G. crinita* provenances analysed by AFLP.

Provenances	NR	TS	SA	CR	NP	TM	PI	MA
NR	****	0.960	0.960	0.972	0.967	0.962	0.959	0.972
TS	0.040	****	0.939	0.948	0.950	0.951	0.943	0.951
SA	0.041	0.063	****	0.978	0.947	0.952	0.972	0.971
CR	0.028	0.054	0.023	****	0.971	0.975	0.971	0.979
NP	0.033	0.051	0.054	0.029	****	0.989	0.939	0.963
TM	0.038	0.050	0.049	0.026	0.011	****	0.942	0.968
PI	0.042	0.059	0.029	0.030	0.063	0.059	****	0.972
MA	0.029	0.050	0.030	0.022	0.037	0.032	0.029	****

NR = Nueva Requena, TS = Tahuayo Stream, SA = San Alejandro, CR = Curimana River, NP = Nuevo Piura, TM = Tingo Maria, PI = Puerto Inca, MA = Macuya.

Analysis of molecular variation (AMOVA) showed that 12% of the variation was among population types, 4% was among provenances and 84% was within provenances (Table 5). The level of differentiation among provenances was higher ($F_{st} = 0.16$) than among population types ($F_{st} = 0.12$) at $p < 0.001$.

Table 5. Results of the analysis of molecular variation (AMOVA) of 58 *G. crinita* individuals representing three population types and eight provenances.

Source of Variance	Degree of Freedom (*df*)	Sum of Square (SS)	Variance Component	Variance (%)	*p* Value [a]
Among population type	2	99.43	1.60	12	<0.001
Among provenances	5	69.95	0.58	4	<0.001
Within provenances	50	583.56	11.67	84	<0.001
Total	57	752.95	13.85	100	

[a] Significance tests after 999 permutations.

Patterns of a genetic relationship were visualized using principal coordinate analysis (PCoA) (Figure S1) and a dendrogram based on Jaccard's dissimilarity, which grouped the 58 samples into two main clusters with seven sub-clusters (Figure S2). The number of clusters (K value) assessed by STRUCTURE analysis suggested two was the optimal K because it had the largest delta K value. Under this K = 2 model, provenance from the semi-domesticated population (NR, TS, PI, MA, SA and CU) had some individuals with mixed assignment membership in cluster 1 (black bar) and cluster 2 (white bar, Figure 2). The analysis also provided membership assignment, with the higher membership ranging from 56.6% (CU provenance in cluster 1) to 89.3% (SA provenance in cluster 1). In natural and cultivated populations (NP and TM, respectively), the membership was 73.2% and 80.4%, respectively in cluster 2.

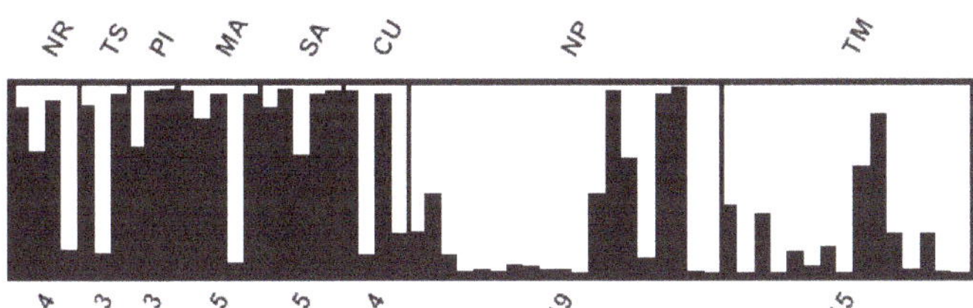

Figure 2. Population structure for the 58 *G. crinita* samples from eight provenances at two *K* value. The provenance codes are given above, and the numbers of samples per provenance are given below.

4. Discussion

Studies of genetic variation in growth and wood traits of *Guazuma crinita* have been published [14,36,37], but genetic variation in morphological traits represents a small part of a total genetic variation in a species [38]. This research assesses genetic diversity in *G. crinita* based on amplified fragment length polymorphism (AFLP) markers, and we found 99.4% polymorphism. In another study involving eleven provenances of *G. crinita* in the Peruvian Amazon, there was 93.8% polymorphism based on Inter Simple Sequence Repeat (ISSR) [17]. Although the methods were different, both studies confirm high levels of genetic diversity in *G. crinita*. The high levels of diversity are probably related to the fact that *G. crinita* is a pioneer species and has long-distance seed dispersal, which results in extensive gene flow [39,40].

We analyzed the genetic diversity of *G. crinita* from three different population types (natural, cultivated and semi-domesticated). Comparing the genetic diversity parameters, such as PPF, *He,* and *I*, the naturally regenerated population had slightly greater genetic diversity than the cultivated and semi-domesticated populations. This suggests that artificial selection in the domestication process has reduced the levels of *G. crinita* genetic diversity. Other studies also confirmed that wild populations usually maintain higher levels of genetic diversity compared with cultivated populations [5,6,41,42].

Higher diversity in natural populations is expected because they are not affected by artificial selection. Maintaining high genetic diversity in natural populations is important because it reduces the risk of local extinction under natural conditions [1,43]. The conservation of cultivated populations is also important to conserve genetic diversity, particularly for those cultivated populations with superior individuals. Genetic diversity parameters were slightly higher in the semi-domesticated population compared with the cultivated population. This probably is due to the larger genetic base of the semi-domesticated population. The semi-domesticated population included six provenances with individuals selected from offspring of 200 mother trees (details of the initial collection were reported by Rochon et al. [14]), while the cultivated population represented only one provenance and a few mother trees. The number of mother trees used to establish a population is a key factor that affects inbreeding and genetic diversity: a low number will cause inbreeding among progeny, while a large number will increase genetic diversity and reduce differentiation among plantations [44].

In this study, three parameters were used to assess genetic differentiation among the three population types, and they gave similar results (AMOVA = 12%, G_{st} coefficient = 0.10 and F_{st} = 0.12). This indicates that about 12% of the variation was due to the domestication stage. In contrast, genetic differentiation among naturally regenerated and managed stands of *Picea abies* (L.) Karst in Europe are much lower (F_{st} = 0.012), suggesting that tree breeding activities have not greatly altered gene frequencies compared with natural populations of this species [45]. Geographical and climatic factors can also affect genetic differentiation, and their effects should be assessed in future studies of *G. crinita* [46,47].

The relatively low level of genetic differentiation (4%) among the eight *G. crinita* provenances can be explained by the high gene flow value (*Nm* = 12.9) reported by Tuisima et al. [17]. The high gene flow probably reflects the long-distance dispersal of its small seed by wind and water [37]. Lower genetic differentiation is also expected for cross-pollinated species [48]. The high level of genetic diversity within provenances in this study was consistent with reports based on phenotypic traits [14,37].

Other studies have also reported relatively low genetic differentiation among tree populations in the Amazon Basin. Russell et al. [49] reported 9% variation among populations of *Calycophylum spruceanum* Benth from several watersheds in the Peruvian Amazon Basin, using seven AFLP primer combinations. This species also produces small seeds that are dispersed over long distances by both wind and water. Nassar et al. [50] found low levels of diversity among populations of three native species from the Amazon Basin based on allozymes (8%, 6% and 7%, respectively, among populations of *Samanea saman* (Jack.) Merr. (Fabaceae), *Guazuma ulmifolia* (Malvaceae) and *Hura crepitans* L. (Euphorbiaceae)).

Many trees species have adaptations that allow long-distance seed dispersal [48,51]. One may expect that trees sampled over a relatively small geographical range (as in our study) would show low levels of variation among populations [15]. However, in some studies in the tropics, trees were sampled over extensive geographical ranges and still showed low differentiation among population (e.g., *Swietenia macrophylla* King [52]; *Vitellaria paradoxa* Gaertn [53]; *Inga edulis* Mart [22]).

According to STRUCTURE analysis, individuals within provenances were assigned mixed membership in the two clusters, so provenances were not distinctly separated. This is consistent with the AMOVA, which showed much greater genetic diversity within than among provenances. As a result, it was not possible to correctly identify groups [54]. Nevertheless, we notice similarity among provenances from the semi-domesticated population (cluster 1), and similarity between provenances from the naturally regenerated and cultivated populations (cluster 2).

5. Conclusions

AFLP markers were successful and effective for the assessment of the genetic diversity and structure of *G. crinita* populations in different stages of the domestication process. A high level of genetic diversity was observed at the species level, and this probably reflects extensive gene flow due to long-distance seed dispersal.

Genetic diversity appears to be slightly greater in the natural population compared to the cultivated and semi-domesticated populations, while significant genetic differentiation was detected among the three population types. These results are preliminary, given the small sample size, and suggest the presence of a slight, but significant genetic bottleneck in the cultivated and semi-domesticated populations. The semi-domesticated population appears to have a slightly higher genetic diversity than the cultivated population.

There appears to be significant differentiation among the natural, cultivated and semi-domesticated populations, a result presented with caution given the small sample size employed. Future studies should include larger sample sizes in different domestication stages to confirm the results reported in this paper.

The in situ and *circa situ* conservation and sustainable management of naturally regenerated populations are recommended to maintain *G. crinita* genetic resources in order to cope with potential inbreeding depression and environmental changes.

To increase genetic variation in planted populations, we recommend further sampling, the collection of *G. crinita* seeds over an extensive geographic range (including various natural stands), and the establishment of seedlings and clonal seed orchards.

Supplementary Materials: The following are available online at http://www.mdpi.com/1999-4907/11/8/795/s1, Figure S1: Dendrogram based on 171 AFLP loci for 58 samples of *G. crinita*. Samples in green, black and blue are from natural, cultivated and semi-domesticated populations, respectively. Figure S2: Principal coordinate analysis of the 58 samples belonging to eight provenances of *G. crinita*, based on AFLP analysis.

Author Contributions: Conceptualization, L.L.T.-C. and P.H.Č.; methodology, L.L.T.-C. and P.H.Č.; formal analysis, L.L.T.-C. and P.H.Č. writing—original draft preparation, L.L.T.-C. and B.L.; writing—review and editing, P.H.Č., B.L., J.C.W.; Supervision of the research, B.L., P.H.Č.; project administration, B.L.; funding acquisition, B.L. All authors have read and agreed to the published version of the manuscript.

Funding: The research was funded by the Internal Grant Agency of Czech University of Life Science Prague CIGA (Project No. 20205003).

Acknowledgments: The authors thank the collaboration of the Peruvian Amazon Research Institution (IIAP) and to Ing. Walter Rios Perez and Gabriel Morales Alejo in the collection of samples. We would like to thank Bohumil Mandák for support on Structure analysis and Antony Del Aguila Heller for help with Figure 1.

Conflicts of Interest: The authors declare no conflict of interest.

References

1. O'Neill, G.; Dawson, I.; Sotelo-Montes, C.; Guarino, L.; Guariguata, M.; Current, D.; Weber, J.C. Strategies for genetic conservation of trees in the Peruvian Amazon. *Biodivers Conserv.* **2001**, *10*, 837–850. [CrossRef]
2. Schnabel, A.; Hamrick, J.L. Organization of genetic diversity within and among populations of *Gleditsia triacanthos* (Leguminosae). *Am. J. Bot.* **1990**, *77*, 1060–1069. [CrossRef]
3. Mertz, O.; Wadley, R.L.; Nielsen, U.; Bruun, B.T.; Colfer, C.J.P.; de Neergaard, A.; Jepsen, M.R.; Martinussen, T.; Zhao, Q.; Noweg, G.T.; et al. A fresh look at shifting cultivation: Fallow length and uncertain indicator of productivity. *Agric. Syst.* **2008**, *96*, 75–84. [CrossRef]
4. Weber, J.C.; Sotelo-Montes, C.; Vidaurre, H.; Dawson, I.K.; Simons, A.J. Participatory domestication of agroforestry trees: An example from the Peruvian Amazon. *Dev. Pract.* **2001**, *11*, 425–433. [CrossRef]
5. Lengkeek, A.G.; Muchugi-Mwangi, A.; Agufa, C.A.C.; Ahenda, J.O.; Dawson, I.K. Comparing genetic diversity in agroforestry systems with natural forest: A case study of the important timber tree *Vitex fischeri* in central Kenya. *Agrofor. Syst.* **2006**, *67*, 293–300. [CrossRef]
6. Hollingsworth, P.; Dawson, I.; Goodall-Copestake, W.; Richardson, J.; Weber, J.C.; Sotelo-Montes, C.; Pennington, T. Do farmers reduce genetic diversity when they domesticate tropical trees? A case study from Amazonia. *Mol. Ecol.* **2005**, *14*, 497–501. [CrossRef]
7. Sotelo-Montes, C.; Weber, J.C. Priorización de especies arbóreas para sistemas agroforestales en la selva baja del Perú. *Agrofor. Am.* **1997**, *4*, 12–17.
8. Encarnación, F. *Nomenclatura de las Especies Forestales Comunes en el Perú*; Ministerio de Agricultura, Instituto Nacional Forestal y de Fauna: Lima, Peru, 1983; p. 149.
9. Labarta, R.A.; Weber, J.C. Valorización económica de bienes tangibles de cinco especies arbóreas agroforestales en la Cuenca Amazónica Peruana. *Rev. For. Centroam.* **1998**, *23*, 12–21.

10. Putzel, L.; Cronkleton, P.; Larson, A.; Pinedo-Vasquez, M.; Salazar, O.; Sears, R. Peruvian Smallholder Production and Marketing of Bolaina (Guazuma Crinita), a Fast-Growing Amazonian Timber Species: Call for a Pro-Livelihoods Policy Environment. Available online: http://www.cifor.org/library/4257 (accessed on 15 April 2017).
11. Sotelo-Montes, C.; Vidaurre, H.; Weber, J.C.; Simons, A.J.; Dawson, I. Producción de semillas a partir de la domesticación participativa de árboles agroforestales en la amazonia peruana. In *Proceedings of the Memorias del Segundo Symposio sobre Avances en la Producción de Semillas Forestales en América Latina. Proyecto de Semillas Forestales (PROSEFOR), Centro de Agricultura Tropical y de Enseñanza (CATIE) y International Union of Forest Research Organizations (IUFRO): Santo Domingo, República Dominica, 18–22 October 1999*; Salazar, R., Ed.; CATIE: Turrialba, Costa Rica; pp. 65–72.
12. IIAP (Instituto de Investigación de la Amazonía Peruana, PE). *Evaluación Económica de Parcelas de Regeneración Natural y Plantaciones de Bolaina Blanca, Guazuma Crinita, en el Departamento de Ucayali*; Instituto de Investigación de la Amazonía Peruana: Loreto, Peru, 2009; p. 54.
13. Toledo, E.; Rincon, C. *Utilización Industrial de Nuevas Especies Forestales en el Perú*; Cámara Nacional Forestal, Instituto Nacional de Recursos Naturales, Organización Internacional de las Maderas Tropicales: Lima, Peru, 1996.
14. Rochon, C.; Margolis, H.A.; Weber, J.C. Genetic variation in growth of *Guazuma crinita* (Mart.) trees at an early age in the Peruvian Amazon. *Forest Ecol. Manag.* **2007**, *243*, 291–298. [CrossRef]
15. Hamrick, J.L.; Godt, M.J.; Sherman Broyles, S.L. Factors influencing levels of genetic diversity in woody plant species. *New Forest* **1992**, *6*, 95–124. [CrossRef]
16. Ouborg, N.J.; Piquot, Y.; Van Groenendael, M. Population genetics, molecular markers and the study of dispersal in plants. *J. Ecol.* **1990**, *87*, 551–568. [CrossRef]
17. Tuisima-Coral, L.; Hlásná-Čepková, P.; Lojka, B.; Weber, J.C.; Filomeno-Alves-Milo, S. Genetic diversity in *Guazuma crinita* from eleven provenances in the Peruvian Amazon revealed by ISSR markers. *Bosque* **2016**, *37*, 63–70. [CrossRef]
18. Vodouhe, R.; Dansi, A. The "Bringing into Cultivation" Phase of the Plant Domestication Process and its Contributions to In Situ Conservation of Genetic Resources in Benin. *Sci. World J.* **2012**, 176939. [CrossRef]
19. Hernández, R.C.A.; Kafuri, L.A.; Isaza, R.A.; Arias, M.L. Analysis of genetic variation in clones of rubber (*Hevea brasiliensis*) from Asian, South and Central American origin using RAPDs markers. *Rev. Col. Biotecnol.* **2006**, *8*, 29–34.
20. Thangjam, R. Inter-Simple Sequence Repeat (ISSR) Marker Analysis in *Parkia timoriana* (DC.) Merr. Populations from Northeast India. *Appl. Biochem. Biotech.* **2014**, *172*, 1727–1734. [CrossRef] [PubMed]
21. Nazareno, A.G.; Bemmels, J.B.; Dick, C.W.; Lohmann, L.I.G. Minimum sample sizes for population genomics: An empirical study from an Amazonian plant species. *Mol. Ecol. Resour.* **2017**, *17*, 1136–1147. [CrossRef] [PubMed]
22. Rollo, A.; Ribeiro, M.M.; Costa, R.L.; Santos, C.; Clavo, P.Z.M.; Mandák, B.; Kalousová, M.; Vebrová, H.; Chuquilin, E.; Torres, S.G.; et al. Genetic Structure and Pod Morphology of *Inga edulis* Cultivated vs. Wild Populations from the Peruvian Amazon. *Forests* **2020**, *11*, 655. [CrossRef]
23. Doyle, J.J.; Doyle, J.L. A rapid DNA isolation procedure for small quantities of fresh tissue. *Phytochem. Bull.* **1987**, *19*, 11–15.
24. Russell, J.R.; Fuller, J.D.; Macaulay, M.; Hatz, B.G.; Jahoor, A.; Powell, W.; Waugh, R. Direct comparison of levels of genetic variation among barley accessions detected by RFLPs, AFLPs, SSRs and RAPDs. *Theor. Appl. Genet.* **1997**, *95*, 714–722. [CrossRef]
25. Vos, P.; Hogers, R.; Bleeker, M.; Reijans, M.; Lee van de, T.; Hornes, M.; Frijters, A.; Pot, J.; Peleman, J.; Kuiper, M.; et al. AFLP: A new technique for DNA fingerprinting. *Nucleic Acids Res.* **1995**, *23*, 4407–4414. [CrossRef]
26. Mikulášková, E.; Fér, T.; Kučabová, V. The effect of different DNA isolation protocols and AFLP fingerprinting optimizations on error rate estimates in the bryophyte *Campylopus introflexus*. *Lindbergia* **2012**, *35*, 7–17.
27. Yeh, F.C.; Yang, R.C.; Boyle, T.B.; Ye, Z.H.; Mao, J.X. POPGENE ver. 1.32 the User-Friendly Shareware for Population Genetic Analysis. Molecular Biology and Biotechnology Center, University of Alberta: Edmonton, AB, Canada, 1997. Available online: http://www.ualberta.ca/$\sim$fyeh/ (accessed on 10 March 2020).
28. Peakall, R.; Smouse, P.E. GenAlEx 6.5: Genetic analysis in Excel. Population genetic software for teaching and research-an update. *Bioinformatics* **2012**, *28*, 2537–2539. [CrossRef] [PubMed]

29. Perrier, X.; Jacquemoud-Collet, J.P. Darwin software. Available online: http://darwin.cirad.fr/darwin (accessed on 12 March 2020).
30. Pritchard, J.; Stephens, M. Inference of population structure using multilocus genotype data. *Genetics* **2000**, *155*, 945–959. [PubMed]
31. Pritchard, J.K.; Wen, W. Documentation for structure 2.0b. Available online: http://pritch.bsd.uchicago.edu (accessed on 1 April 2020).
32. Evanno, G.; Regnaut, S.; Goudet, J. Detecting the number of clusters of individuals using the software structure: A simulation study. *Mol. Ecol.* **2005**, *14*, 2611–2620. [CrossRef]
33. Earl, D.A.; von Holdt, B.M. STRUCTURE HARVESTER: A website and program for visualizing STRUCTURE output and implementing the Evanno method. *Conserv. Genet. Resour.* **2012**, *4*, 359–361. [CrossRef]
34. Rosenberg, N.A. DISTRUCT: A program for the graphical display of population structure. *Mol. Ecol. Notes* **2004**, *4*, 137–138. [CrossRef]
35. Bonin, A.; Bellemain, E.; Bronken-Eidesen, P.; Pompanon, F.; Brochmann, C.; Taberlet, P. How to track and assess genotyping errors in population genetic studies. *Mol. Ecol.* **2004**, *13*, 3261–3273. [CrossRef]
36. Weber, J.C.; Sotelo-Montes, C. Geographic variation in tree growth and wood density of *Guazuma crinita* Mart. in the Peruvian Amazon. *New Forest* **2008**, *36*, 29–52. [CrossRef]
37. Weber, J.C.; Montes, C.S.; Cornelius, J.; Ugarte, J. Genetic variation in tree growth, stem form and mortality of *Guazuma crinita* in slower- and faster-growing plantations in the Peruvian Amazon. *Silvae Genet.* **2011**, *30*, 70–78. [CrossRef]
38. White, T.L.; Adams, W.T.; Neale, D.B. *Within Population Variation, Genetic Diversity, Mating Systems and Stand Structures*; CABI Publishing, Ed.; Forest Genetics: Cambridge, UK, 2007; pp. 149–283.
39. Hamrick, J.L.; Godt, M.J.W. Allozyme diversity in plant species. In *Plant Population Genetics, Breeding and Genetic Resources*; Brown, H.D., Clegg, M.T., Kahler, A.L., Weir, B.S., Eds.; Sinauer Associates: Sunderland, MA, USA, 1989; pp. 43–64.
40. Reynel, C.; Pennington, R.; Pennington, T.; Flores, D.; Daza, C.A. *Árboles útiles de la Amazonía Peruana, Manual de Identificación Ecológica y Propagación de las Especies*; Universidad Nacional Agraria La Molina: Lima, Peru, 2003; p. 509.
41. Abo-elwafa, A.K.; Shimada, M.T. Intra and inter specific variations in Lens revealed by RAPD markers. *Theor. Appl. Genet.* **1995**, *90*, 335–340. [CrossRef]
42. Lu, Y.; Chen, C.; Wang, R.; Egan, A.; Fu, C. The effects of domestication on genetic diversity in *Chimonanthus praecox*: Evidence from cpDNA and AFLP data. *J. Syst. Evol.* **2015**, *53*, 239–251. [CrossRef]
43. Simons, A.J.; MacQueen, D.J.; Stewart, J.L. Strategic concepts in the breeding of non-industrial trees. In *Tropical Trees: The Potential for Domestication and the Rebuilding of Forest Resources, Proceedings of a conference organized by the Edinburgh Centre for Tropical Forests, Heriot-Watt University, Edinburgh, UK, 23-28 August 1992*; Leakey, R.R.B., Newton, A.C., Eds.; HMSO: Edinburgh, UK, 1994; pp. 91–102.
44. Aravanopoulos, F.A.P. Do Silviculture and Forest Management Affect the Genetic Diversity and Structure of Long-Impacted Forest Tree Populations? *Forests* **2018**, *9*, 355. [CrossRef]
45. Rungis, D.; Luguza, S.; Bāders, E.; Šķipars, V.; Jansons, Ā. Comparison of Genetic Diversity in Naturally Regenerated Norway Spruce Stands and Seed Orchard Progeny Trials. *Forests* **2019**, *10*, 926. [CrossRef]
46. Wang, B.; Mao, J.F.; Zhao, W.; Wang, X.R. Impact of geography and climate on the genetic differentiation of the subtropical pine *Pinus yunnanensis*. *PLoS ONE* **2013**, *8*, e67345. [CrossRef] [PubMed]
47. Tong, Y.W.; Lewis, B.J.; Zhou, W.M.; Mao, C.R.; Wang, Y.; Zhou, L.; Yu, D.P.; Dai, L.M.; Qi, L. Genetic Diversity and Population Structure of Natural *Pinus koraiensis* Populations. *Forests* **2020**, *11*, 39. [CrossRef]
48. Russell, J.R.; Weber, J.C.; Booth, A.; Powell, W.; Sotelo-Montes, C.; Dawson, I.K. Genetic variation of *Calycophyllum spruceanum* in the Peruvian Amazon basin, revealed by AFLP analysis. *Mol. Ecol.* **1999**, *8*, 199–204. [CrossRef]
49. Nassar, J.M.; García-Rivas, A.E.; González, J.A. Patrones de diversidad genética en especies arbóreas de bosques fragmentados en Venezuela. *Interciencia* **2011**, *36*, 914–922.
50. Pär, K.I.; Dahlberg, H. The effects of clonal forestry on genetic diversity in wild and domesticated stands of forest trees. *Scand. J. For. Res.* **2019**, *34*, 370–379. [CrossRef]
51. Finkeldey, R.; Hattemer, H.H. *Tropical Forest Genetics*; Springer: Berlin/Heidelberg, Germany, 2007; p. 315. [CrossRef]

52. Lemes, M.R.; Gribel, R.; Procor, J.; Grattapaglia, D. Population genetic structure of mahogany (*Swietenia macrophylla* King, Meliaceae) across the Brazilian Amazon, based on microsatellite at the loci: Implication for conservation. *Mol. Ecol.* **2003**, *12*, 2875–2883. [CrossRef]
53. Kelly, B.A.; Hardy, O.J.; Bouver, J.M. Temporal and spatial genetic structure in *Vitellaria paradoxa* (shea tree) in an agroforestry system in southern Mali. *Mol. Ecol.* **2004**, *13*, 1231–1240. [CrossRef]
54. Nelson, M.F.; Anderson, N.O. How many marker loci are necessary? Analysis of dominant marker data sets using two popular population genetic algorithms. *Ecol. Evol.* **2013**, *3*, 3455–3470. [CrossRef] [PubMed]

Article

Plus Tree Selection of *Quercus salicina* Blume and *Q. glauca* Thunb. and Its Implications in Evergreen Oaks Breeding in Korea

In Sik Kim *, Kyung Mi Lee, Donghwan Shim, Jin Jung Kim and Hye-In Kang

Division of Forest Tree Improvement, Department of Forest Bio-Resources, National Institute of Forest Science, 39 Onjeong-ro, Gwonseon-gu, Suwon 16331, Korea; kmile@korea.kr (K.M.L.); shim104@korea.kr (D.S.); acgwnd88@korea.kr (J.J.K.); hyeinkang@korea.kr (H.-I.K.)

* Correspondence: kimis02@korea.kr

Received: 21 May 2020; Accepted: 2 July 2020; Published: 6 July 2020

Abstract: This study was conducted to select plus trees of two evergreen oaks, *Quercus salicina* and *Q. glauca*, in Korea. Evergreen oaks are distributed in subtropical region in Korea and have recently emerged as one of the alternative tree species against climate change. Accordingly, a tree breeding program is underway to foster evergreen oaks as a reforestation species for the future. Through intensive survey on the distribution range, 15 stands (8 for *Q. salicina*, 3 *for Q. glauca*, and 4 for both species) were selected as base populations. To select candidate trees, we developed a subjective grading system with six characteristics in three categories and introduced a weighted generalized value (GVI_w) to compare superiority of candidate trees. The candidate trees were screened using baseline value '0', i.e., if $GVI_w > 0$, then accepted and if $GVI_w < 0$, then rejected. After then, adjustment was conducted to avoid biasing the selection of plus trees for a particular location. Through this process, 44 candidate trees in *Q. salicina* and 41 candidate trees in *Q. glauca* were selected as plus trees. Finally, the results and implications were discussed in relation to evergreen oak breeding in Korea.

Keywords: tree improvement; evergreen oak; phenotypic selection; selection criteria; seed orchard; generalized value; conservation

---

1. Introduction

The genus *Quercus* is one of the most important angiosperms in the northern hemisphere in terms of species diversity, ecological dominance, and economic value. Oaks are dominant members of a wide variety of habitats including temperate deciduous forest, temperate and subtropical evergreen forest, subtropical and tropical savannah, and subtropical woodlands [1,2].

In South Korea, six deciduous oaks (*Q. accutissima* Carruth., *Q. variabilis* Blume, *Q. mongolica* Fisch. Ex Ledeb., *Q. serrata* Thunb. Ex Murray, *Q. dentata* Thunb. and *Q. aliena* Blume) and five evergreen oaks (*Q. myrsinifolia* Blume, *Q. acuta* Thunb., *Q. glauca* Thunb., *Q. salicina* Blume, and *Q. gilva* Blume) are naturally distributed. While the deciduous oaks are growing all over the country, from temperate to subtropical regions, evergreen oaks are restricted to subtropical regions, the southernmost part of Korea [3]. Traditionally, deciduous oaks have been used as a wood resource (timber, media for mushroom cultivation, charcoal, etc.) and their nuts are also used for starch production, i.e., acorn jelly. Evergreen oaks are also used as a timber production in subtropical region, but the portion is very small. Accordingly, deciduous oaks have more attention in tree breeding program than evergreen oaks [4].

As global warming progresses, the distribution and composition of forest tree species are expected to change. According to climate change scenario in Korea, it is predicted that the distribution range of conifers will decrease while the range of subtropical tree species will expand northward [5,6]. The assisted migration and development of alternative tree species were suggested as countermeasures against global warming [7,8]. In Korea, evergreen oaks have recently emerged as one of the alternative tree species against climate change. Moreover, there have been reports that the leaf, branch, bark, and acorn of evergreen oaks contain useful substances that can be used for food or medicine, which have attracted more attention [9,10]. Thus, as a preliminary study, plus tree selection and establishment of seedling seed orchard of *Q. glauca* had been conducted with small scale mainly in Jeju island [11]. After that, Korea Forest Service established a plan for additional plus tree selection and expansion of seed orchard of evergreen oaks, which was major motivation for conducting this study.

Although the utilization potential of evergreen oaks is highly evaluated, there is a problem that the amount of available resources is relatively limited compared to other common species. For example, they are restrictedly distributed in a narrow area of subtropical region with small population size and low appearance frequency. There is also a need to prepare measures to conserve genetic resource. Thus, in the process of reviewing the breeding plan, it is highly recommended to consider the utilization and genetic resources conservation together. Accordingly, unlike a previous study conducted by considering only growth characteristics [11], this study aimed to consider growth, adaptability, and seed production characteristics simultaneously.

Tree breeding is an integral part of modern silviculture used to increase the economic profit through enhanced wood production. In first-generation tree improvement programs, mass selection or phenotypic selection would be the first step in breeding cycle [12]. The philosophy underlying plus tree selection is that the favorable deviation of plus tree from the population mean is due at least partly to genetic rather than environmental or random effects [13]. There are several methods for plus tree selection, i.e., ocular or subjective method, comparison tree method, baseline method, and regression method [14,15]. The comparison tree method is generally preferred and widely applied in even-aged, pure-species stands. In uneven-aged and mixed-species stands, most common for hardwoods, a grading system is required than standard comparison tree method. The regression or baseline method aims to adjust tree scores for differences in ages, competition and/or environmental gradients. The ocular or subjective method is used when environmental noise is too large to be overcome by regression methods or visual inspection will be effective at locating superior trees. A determination of the best selection techniques depends on several factors, including species characteristics, past history, the present condition of the forest, variability and inheritance pattern of important characteristics, and objectives of the particular tree improvement program [12,16].

As deciduous oaks are dominant or co-dominant tree species with high density in natural stands, comparison tree method is usually used for plus tree selection. However, evergreen oaks have a characteristic of being scattered in a mixed forest with a low density, making it difficult to apply the comparison tree method. Regression method or baseline method would be considered as alternatives. However, the data needed to obtain a baseline or regression equation for the target traits was not accumulated at present. Thus, we have to consider ocular selection method for plus tree selection in evergreen oaks. Ocular method rely upon a subjective assessment where trees that appear to be better than average are chosen as plus trees without measuring the candidate or neighboring trees [14,15]. Actually, simple direct (ocular) selection method was used for plus tree selection in previous study on *Q. glauca* [11]. They were selected 35 plus trees only based on growth traits such as height, diameter, stem straightness, and timber height in lowlands and valleys of Mt. Halla in Jeju island. It seemed to achieve the selection objectives relatively easily using ocular selection method. However, they were not present the detailed selection criteria and methods. In addition, it was regrettable to overlook the characteristics such as flowering, fruiting, resistance to pests or diseases, and adaptability to environment for production of improved seed and reforestation. In other case, Stringer et al. [17] compared the growth and stem quality of 19-year-old progeny from superior and comparison trees in *Q. rubra*. They suggested that selecting several phenotypically above-average candidate trees may be more effective than rigorously selecting a smaller number of phenotypically superior trees. Combining these two cases, it seemed that development of some modified selection method was required for this study, because the selection work has to be operated at several stands with different environments in the entire distribution and the selection results have to be evaluated objectively by some standard.

Accordingly, we firstly tried to make selection criteria and method for evergreen oaks. Then, we applied the method to plus tree selection of *Q. salicina* and *Q. glauca* for case study. Finally, the results and implications were discussed in relation to evergreen oak breeding in Korea.

2. Materials and Methods

2.1. Selection of Base Populations

To screen candidate stands and select base populations of *Q. salicina* and *Q. glauca*, some relevant literatures had been searched [3,18–22]. However, there were limitations in understanding of population size, density and growth performance of target species, which were needed for the selection of base populations. To supplement this, we had interviewed with related researchers and experts to get more information. In addition, consideration was also given not to overlap with the stands selected in previous study [11].

Through this process, 25 stands (15 for *Q. salicina* and 10 for *Q. glauca*) were screened. After then, preliminary survey was conducted to determine if they were appropriate or not as a base population. In the fields, we examined stand size, frequency and density of target species, growth state, existence of candidate trees, and environmental condition. These findings were used to select base populations as well as set up the selection criteria and method.

For selection of base population, the priority was given to whether there were enough individuals to perform selection and whether there were candidate trees that met the selection criteria. For example, a stand with small size and no available candidate trees was excluded. Finally, 15 stands were selected as a base population, i.e., 8 for *Q. salicina*, 3 for *Q. glauca*, and 4 for both species (Figure 1, Table 1).

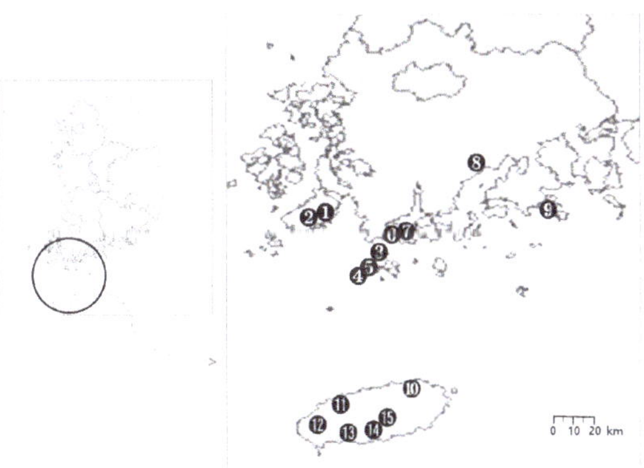

Figure 1. Location map of base populations for candidate tree selection in *Quercus salicina* and *Q. glauca*. The numbers in the black circle indicate the No. of stands shown in the Table 1.

2.2. Selection Criteria and Methods

In Korea, selection criteria and method for broad-leaved tree species are already established and have been used to plus tree selection of several tree species [23]. However, the method is based on comparison tree method, which is not suitable to evergreen oaks as mentioned above. Through literature review and in-depth discussion, a subjective grading system was developed and applied for plus tree selection of evergreen oaks.

Firstly, the dominant trees that occupy the upper crown canopy of the stand were selected as a candidate trees. The interval among adjacent candidate trees was kept at 30~50 m to avoid selection of relatives. In addition, selection of candidate trees was made to include as diverse location as possible to avoid biased selection from a particular site.

Table 1. List of 15 candidate stands of *Q. salicina* and *Q. glauca* stands used in this study.

Region	Stand	Latitude (°N)	Longitude (°E)	Altitude (m)	Target Species	Mean Annual Temperature (MAP, °C)	Mean Annual Precipitation (MAP, mm)	Relative Humidity (RH, %)
Jindo	1. Sacheon1	34.4701	126.3087	108	*Q. salincina*	139	1568.5	78.2
	2. Sacheon2	34.4618	126.3211	154	*Q. salincina*			
Wando	3. Yesong	34.1555	126.5744	58	*Q. salincina*	14.1	1532.7	72.5
	4. Jeongja	34.1612	126.5095	37	*Q. salincina* *Q. glauca*			
	5. Buhwang	34.1507	126.5348	72	*Q. salincina*			
	6. Jangjoa	34.3555	126.7177	86	*Q. glauca*			
	7. Daeya	34.3655	126.7160	80	*Q. salincina*			
Boseong	8. Yulpo	34.6791	127.0846	62	*Q. salincina*	13.6	1779.9	70.7
Goheung	9. Sayang	34.4627	127.4493	33	*Q. glauca*	13.6	1453.4	69.7
Jeju-si	10. Seonhul	33.5191	126.7129	113	*Q. salincina* *Q. glauca*	15.5	1456.9	73.3
	11. Napup	33.4339	126.3304	90	*Q. glauca*			
	12. Cheongsu	33.2969	126.2534	105	*Q. salincina* *Q. glauca*			
Seguipo-si	13. Gamsan	33.2573	126.3565	125	*Q. salincina* *Q. glauca*	16.2	1850.8	70.7
	14. Seoho	33.3018	126.5250	602	*Q. salincina*			
	15. Sanghyo	33.3294	126.5857	465	*Q. salincina*			

To select candidate trees, we developed subjective grading system with six characteristics in three categories (Table 2). Each characteristic was rated and scored as 5 grades, i.e., 5—very good, 4—good, 3—moderate, 2—poor, and 1—very poor. The grade was given by relative superiority or degree compared to the average of the stand.

Table 2. Selection criteria and weights of characteristics used for candidate tree selection in *Quercus salicina* and *Q. glauca*.

Category (Weight)	Characteristic (Weight)	Remark
Growth (0.5)	Superiority of Growth (SG) (0.30)	Determined by considering growth status of shoot, branch, crown, and stem.
	Superiority of Tree form (STF) (0.2)	Determined by considering number of stem, stem straightness, and timber height.
Adaptability (0.2)	Adaptability to Disturbance (AD) (0.08)	Determined by considering adaptability to disturbance or damage by abiotic and biotic stresses
	Adaptability to Environment (AE) (0.12)	Determined by considering vitality and adaptability to the environments of the site.
Seed production (0.3)	Superiority of seed production (SSP) (0.18)	Determined by considering current status and recent history of seed production including buried or stand floor seed.
	Potential of seed production (PSP) (0.12)	Determined by considering seed production capacity reflecting the advantages and disadvantages of current micro-environmental conditions affecting seed production.

Growth category was aimed to evaluate the superiority of growth (SG) and superiority of tree form (STF). The SG was determined by considering growth status of shoot, branches, crown, and stem. This is an indicator for selecting large and well-growing individuals. The STF was determined by considering number of stem, stem straightness, and timber height. This is an indicator for selecting individuals with straight stem and high timber height for long-log production.

Adaptability category was aimed to evaluate the adaptability to disturbance (AD) and adaptability to environment (AE). The AD was determined by considering adaptability to disturbance such as biotic (pests or diseases) and abiotic (artificial disturbance) stresses. The symptoms or damages by pests or diseases were immediately identified by ocular observation. Artificial disturbance usually appeared in the form of collecting leaves or small branches, even stem for medicine. It was also possible to intuitively determine whether there was an impact on tree growth by ocular observation. The grade was scored by considering these two factors together. Thus, this is an indicator for selecting individuals with superior adaptability or resilience to disturbance. The AE was determined by considering vitality and adaptability to a given environment. When a tree does not adapt well to a given environment, its vitality was decreased. The adaptability can be indirectly identified through observations of growth state of leaves, i.e., small and fewer leaves in crown and/or abnormal growth responses of leaves such as curling, bending, marginal browning, chlorosis, and shedding. The grade was scored considering all aspects mentioned above. Thus, this is an indicator for selecting individuals with superior adaptability to environment.

Seed production category was aimed to evaluate the superiority of seed production (SSP) and potential of seed production (PSP). The SSP was determined by considering current status and recent history of seed production. First, the average level of flowering or fruiting of the stand was determined and the grade was scored according to the relative level of flowering or fruiting of an individual. However, the amount of flowering or fruiting usually varied from year to year. To compensate for this, the grade was adjusted considering the amount of buried or floor seed under the crown of individual tree. Thus, this is an indicator for selecting individuals with superior seed production. The PSP was determined by considering seed production capacity reflecting the advantages and disadvantages of micro-environment affecting flowering or fruiting such as spacing, light intensity, soil moisture, and humidity. The exact values of these factors cannot be accurately known without measuring equipment, but experienced foresters can intuitively determine whether the place is favorable or unfavorable for flowering or fruiting. The grade of PSP was scored considering these judgement results and SSP grade together. Thus, PSP is an indicator for selecting individuals with high potential of seed production.

In a grading system, weights of the characteristics were generally determined by its heritability and economic worth [16]. However, since there is no information on the heritability of evergreen oaks, breeding goal and expected economic worth were only considered for determining the weights of each category and characteristic in this study. The primary objective of this breeding program is to improve growth rate and stem straightness for timber production. Thus, the highest weight was given to growth category. Seed production was considered the second most important in terms of supply of materials for reforestation as well as food and medicine. So, the second highest weight was given to seed production category. Usually, adaptation is not well placed in a grading system, because it is automatically taken into account through common garden test of selected trees [16]. However, since evergreen oaks have the urgency of genetic resources conservation as mentioned before, adaptability category was included in this study considering adaptability to various environments for assisted migration or reforestation outside of current habitats range. However, it was judged that its importance in current state was relatively lower than those of others, so the lowest weight was given to adaptability category. The detailed weights of each category and characteristic were inevitably determined through discussion of relevant experts and researchers (Table 2).

In principle, individuals above the moderate grade in all characteristics were selected as candidates. Exceptionally, however, if the number of candidates to be selected was small and/or the growth-related characteristics were excellent, an individual with lower than moderate grade in some characteristic included in the candidates.

This subjective grading system is frequently used for hardwoods but is successful only if the grader is experienced and dedicated to finding the best tree possible [16,24]. In order to minimize the problem of deviation caused by technical errors or prejudices when selecting candidates, the survey team composed of three experienced researchers and standardized the research method through simulation selection according to the selection criteria before starting work. Two investigation teams were operated for candidate tree selection. Each team consisted of 3 people, i.e., selector, evaluator/recorder, and GPS manager. Selector was primarily responsible for selection and marking of candidate trees. Evaluator/recorder was responsible for evaluation and recording the grade of each characteristic in each candidate tree. GPS manger had recorded and managed the positional information. In all cases, if there were some problem in selection process, i.e., whether to select or not, how to rank, and etc., three persons worked together to draw a conclusion. To reduce the deviation of candidate tree selection, all members of each team had a meeting to evaluate and adjust daily results after finishing work every day.

2.3. Data Analysis

Since the primary goal of this study was to secure as many candidate trees as possible from various candidate stands, candidate selection was made on each stand basis according to the selection criteria and methods. However, it was suggested that it would be appropriate to combine geographically adjacent stand with similar environment into a group. If each stand was treated separately in the process of data analysis for plus tree selection, candidate trees selected under similar environmental conditions were more likely to be included into breeding population. The reason could be understood by looking at the relevant analysis method presented later in this section. For *Q. salicina*, Sacheon1 and Scaheon2 were tied together because they were adjacent to each other with a ridge in between them. Gamsan and Cheongsu, Seoho and Sanghyo, and Buhwang and Jeongja were also tied together respectively, because they are geographically adjacent under similar environment. For *Q. glauca*, Cheongsu and Gamsan were also tied for the same reason. The linear distance between stands tied together was ranged 1.4~10.4 km. All subsequent analysis was performed based on this group (Table 3).

Table 3. The adjusted groups and the number of candidate trees selected in each group in *Quercus salicina* and *Q. glauca*.

Q. salicina				*Q. glauca*			
Stand	**No. of Candidates**	**Group**	**Remark**	**Stand**	**No. of Candidates**	**Group**	**Remark**
Seonhul	11	A		Seonhul	15	A	
Cheongsu	9	B	+Gamsan	Cheongsu	33	B	+Gamsan
Seoho	14	C	+Sanghyo	Napup	10	C	
Sacheon	26	D	Sacheon1 + Sacheon2	Jeongja	9	D	
Yesong	10	E		Jangjoa	12	E	
Buhwang	8	F	+Jeongja	Sayang	3	F	
Daeya	6	G					
Yulpo	3	H					
Total	87	8			82	6	

Since a grading system was applied to select candidate trees in this study, the scoring data of measured characteristics were highly skewed to 4 or 5 grades. Thus, prior to performing the analysis, the data was transformed using logarithm function to improve normality [25–27]. Even after data transformation, the normality was not improved significantly while the homoscedasticity was satisfied. In this case, the robustness of analysis of variance (ANOVA) would be weakened. So, a non-parametric method such as the Kruskal-Wallis (KW) test was generally recommended as an alternative. However, in case of small sample size like this study, the ANOVA test will be a better option than KW test, even for non-normal data [28]. Moreover, when using rank data, the result of KW test was very similar to that of ANOVA F test. Since the grade score measured in this study was a kind of rank in each characteristic, it was considered possible to use ANOVA and multiple comparison methods. Thus, one-way ANOVA with fixed effect model and post hoc multiple comparison analysis were performed to examine difference across the group means and specific difference between pairs of groups, respectively. For multiple comparison analysis, Fisher's least significant difference (LSD) with Bonferroni adjustment was applied. To understand the relationship among six characteristics as well as height and diameter at breast height of candidate trees, correlation analysis was performed with Spearman rank correlation. All statistical analyses were performed using R software version 4.0.0.

Although we obtained scores applying grade system in each characteristic, the mean and distribution of the measurements were different among characteristics as well as base populations. If we used original grade scores measured for candidate selection, it was possible that individuals from a specific stand or group with overall high grade are more likely to be selected as candidates. In other words, if such an erroneous biased selection was made, it does not meet the purpose of this study to select candidate trees evenly from various stands as possible. Therefore, normalization (rescaling) of the data was carried out before applying the weights for each characteristic according to the selection criteria.

There were several methods typically used for normalization, i.e., scaling to a range, clipping, log scaling, z-score, etc. Depending on the nature of the data and the purpose of the study, researchers usually select and use the appropriate method [25–27]. In this study, we normalized the data by dividing the deviation by mean. When applying this method, the mean of the new data was set to '0' and the each value was reversely adjusted according to the relative value of the original mean. This feature was thought to be more suitable to our purpose, because we could select candidates more evenly from various stands or locations than other methods.

Then, we could consider two kinds of mean value for normalization, i.e., each population level (GV_p) and total population level (GV_t) like the below formula.

$$GV_{p,\,i} = (X_i - \overline{X}p)/\overline{X}p \tag{1}$$

where, $GV_{p,i}$ is a generalized value of i_{th} individual, X_i is a grade score i_{th} individual and $\overline{X}p$ is a mean value of a corresponding characteristic in each population.

$$GV_{t,i} = (X_i - \overline{X}t)/\overline{X}t \tag{2}$$

where, $GV_{t,i}$ is a generalized value of i_{th} individual, X_i is a grade score i_{th} individual and $\overline{X}t$ is overall mean value of a corresponding characteristic in total population.

The next question was which mean value was more effective for candidate selection. GV_t was assumed that the environmental differences among stands was negligibly small. If we only considered GV_t, it would be easy to compare and rank overall selected candidates, but it may lead to biased selection from a specific stand with superior properties without consideration of environmental differences. However, environmental impact on fitness and growth of provenances or families is already well known in several cases [16,29]. Thus, it is necessary to combine GV_p in candidate tree selection, which was assumed environmental differences among stands. Moreover, since the genetic test has not been conducted, the GxE interaction effect cannot be estimated. To compensate for this problem, we decided to temporarily apply the weighted generalized value, i.e., $GV_w = 0.7 \times GV_p + 0.3 \times GV_t$. These weights are not from a scientific basis but from a conceptual one suggested by researchers in the sense that consideration of each population level was more important in achieving the purpose of this study.

In actual data analysis, we firstly calculated GV_p and GV_t of each characteristic in each individual, respectively. To obtain the summation index of generalized value of each characteristic for each individual, GVI_p and GVI_t were calculated with GV_p and GV_t multiplying by the weights of each characteristic, respectively (Table 2).

$$GVI_{p,\,i} = \sum (GV_{p,\,i} \times C_j) \tag{3}$$

where $GVI_{p,i}$ is a generalized value of individual based on GV_p value and GV_j is a weight of corresponding characteristics.

$$GVI_{t,\,i} = \sum (GV_{t,\,i} \times C_j) \tag{4}$$

where $GVI_{t,i}$ is a generalized value of individual based on GV_t value and C_j is a weight of corresponding characteristics.

Then, the weighted generalized value of each individual (GVI_w) was calculated as like $GVI_w = 0.7 \times GVI_p + 0.3 \times GVI_t$. The candidate trees were truncated using GVI_w of each individual with baseline value '0', i.e., if $GVI_w > 0$, then accepted and if $GVI_w < 0$, then rejected. Finally, plus trees were selected through the adjustment process to avoid biasing the selection of candidate trees for a particular stand or group.

3. Results and Discussion

3.1. Characteristics of Base Populations

The base populations were horizontally distributed in the inlands and islands region of southern coast of Jeonnam-do, and southern parts of Jeju-do. The latitudinal range was from 33.33° (Sanghyo) to 34.47° N (Sacheon) and the most of altitudinal range was below 200 m above sea level, except for Seoguipo (465 m) and Seoho (602 m). The mean annual temperature (MAT), mean annual precipitation (MAP), and relative humidity (RH) of base populations were ranged 13.6~16.2 °C, 1453.4~1850.8 mm and 69.7~78.2%, respectively. These areas belonged to warm temperate and subtropical region, which were characterized by high temperature, high precipitation, and high humidity in Korea (Figure 1, Table 1).

In overall range of both species, most stands were discontinuously distributed with relatively small size and low appearance frequency, which would lead to restriction of genetic exchange and promote of genetic differentiation. The flowering and fruiting are well done in most stands, however, seedlings for next generation were not well observed within a stand, except for some forest gaps. This phenomenon was commonly found in mature and old-growth forests with closed canopy [30,31]. In addition, both species were being pushed out of competition with other companion species, such as *Distylium racemosum*, *Acer palmatum*, *Ficus erecta*, *Machilus therbergii*, *Litsea japonica*, *Celtis sinensis*, etc.

In most stands, except for Seonhul and Cheongsu, it was estimated that inbreeding was likely to occur due to limitation of pollen-mediated gene flow by low dominance and frequency of target species within a stand. Moreover, the morphological variation of leaves was observed. For example, *Q. salicina* typically has leathery, narrow-lanceolate, taper-pointed leaves having mostly entire margins with a few marginal teeth near the apex. The front side of leaves is glossy and the reverse side is grayish green, which is easily distinguishable from other evergreen oaks. However, in Yesong, Daeya, and Sacheon, there were variations on leaf size, shape, and reverse side color different from the typical characteristics. In relation to this, some varieties like *Q. salicina* var. *latifolia* and *Q. glauca* var. *nudata* were described as an ecotype [4]. Thus, it was difficult to assert that there were some degree of genetic differentiation among stands in target species.

While well managed stands generally showed good growth performances, the stands experienced with illegal or destructive logging had many multi- or crooked stems due to reproduction by sprouts [19]. Thus, it was also necessary to consider the history of disturbance of stand for candidate tree selection. For example, even if an individual had multi-stem, it is possible to select as a candidate if the growth characteristics of each stem is good.

Considering the investigation results, it was highly suggested that more active measures are needed for sustainable use of evergreen oaks. Although the distribution of evergreen oaks is expected to expand due to climate change [5,6], it is possible that the distribution will not expand as expected due to competition with other broad-leaved tree species. Even if it is possible to create new habitat by natural migration, the genetic diversity may decrease due to the founder effects, i.e., limited number of individuals will contribute in the formation of next generation and genetic drift may result in the loss or fixation of some allele within stand [16]. Thus, it was suggested that assisted migration by supplying seeds from seed orchard with reasonable genetic diversity [7,8]. To this end, this study attempted to select plus trees from as many stands and/or individuals as possible to secure genetic diversity, if the selection criteria were met.

3.2. Selection of Candidate Trees

According to selection criteria and methods, intensive survey and selection were conducted to each base population. Since we tried to assess all trees appearing within a stand as possible, there were large differences in the number of assessed trees per site depending on the stand size as well as density and frequency of the target species. For example, in small stands such as Yulpo, Goheung, and Gamsan, it was ranged approximately 50~100. In large stands such as Seonhul, Seoho, and Cheongsu, it was ranged approximately 250~300. A total of 169 candidate trees were primarily selected in two target species (Table 3). While the selection intensity of *Pinus* species with large pure stands was generally ranged 0.5~1.0% [23], the range in this study was approximately 5~8%. As a result of sampling and analyzing of candidate trees using increment borer, most of them were belonged to age class IV (30~40 years old). It indicated that they were mature enough to be selected. Although optimum selection age may vary depending on heritability and genetic correlation of target trait, more than 50% of rotation age is thought as an appropriate selection age to reduce the risk of early selection [12,23].

In *Q. salicina*, 87 candidate trees were selected from 8 groups (12 stands). The highest number of candidate trees was selected in Sacheon (26 individuals) and the smallest one was Yulpo (3 individuals). In *Q. glauca*, 82 candidate trees were selected from 6 groups (7 stands). While the highest numbers of candidate trees were obtained in Cheongsu (33 individuals), Sayang had the smallest candidate trees (3 individuals).

Although there were some limitations to discuss the differences among groups because the values were just only measured from selected candidate trees, it was observable significant differences among groups in some characteristics (Table 4). In *Q. salicina*, the mean values of superiority of growth (SG), adaptability to disturbance (AD), superiority of seed production (SSP) and potential of seed production (PSP) were significantly different among groups, but superiority of tree form (STF) and adaptability to environment (AE) were not. In *Q. glauca*, all characteristics except for adaptability to environment (AE) were significantly different among groups. These differences might reflect the differences in local environment and growth situation among groups.

Table 4. Multiple comparisons of group mean among candidate trees in *Quercus salicina* and *Q. glauca* by Fisher's least significant difference (LSD) with Bonferrroni adjustment.

Species	Group	Characteristics Used for Candidate Tree Selection						Growth Arranged after Finishing of Selection	
		Superiority of Growth (SG)	Superiority of Tree Form (STF)	Adaptability to Disturbance (AD)	Adaptability to Environment (AE)	Superiority of Seed Production (SSP)	Potential of Seed Production (PSP)	Height (HT)	Diameter at Breast Height (DBH)
Q. salicina	A	4.45^{ab}	4.09^{a}	4.18^{ab}	4.45^{a}	3.36^{c}	3.73^{ab}	14.8^{bc}	30.1^{bc}
	B	4.67^{ab}	4.11^{a}	3.89^{b}	4.33^{a}	3.78^{abc}	4.00^{ab}	13.9^{c}	28.8^{cd}
	C	4.88^{a}	4.14^{a}	4.57^{a}	4.42^{a}	3.43^{bc}	3.36^{b}	16.9^{c}	37.4^{ab}
	D	4.58^{ab}	4.50^{a}	4.19^{ab}	4.42^{a}	3.92^{ab}	4.27^{ab}	13.9^{c}	29.7^{c}
	E	4.50^{ab}	4.40^{a}	4.10^{ab}	4.60^{a}	3.80^{abc}	3.70^{ab}	13.7^{c}	23.1^{cd}
	F	4.00^{b}	4.25^{a}	3.88^{b}	4.25^{a}	3.63^{abc}	4.00^{ab}	10.5^{d}	20.3^{d}
	G	4.33^{ab}	4.33^{a}	3.83^{b}	4.33^{a}	4.17^{ab}	4.67^{a}	11.3^{d}	24.7^{cd}
	H	4.33^{ab}	4.33^{a}	3.67^{b}	4.33^{a}	4.37^{a}	4.33^{ab}	16.7^{ab}	50.0^{a}
	Pr.(>F)	*	ns	**	ns	**	**	**	**
	Mean	4.53	4.30	4.14	4.41	3.76	3.97	14.1	29.6
Q. glauca	A	4.33^{a}	4.00^{ab}	3.80^{ab}	4.00^{a}	3.53^{b}	3.60^{b}	14.7^{a}	26.8^{bc}
	B	4.33^{ab}	4.24^{a}	4.03^{ab}	4.15^{a}	3.91^{ab}	3.91^{ab}	12.4^{b}	23.6^{c}
	C	4.20^{ab}	3.60^{ab}	4.30^{a}	4.20^{a}	4.30^{a}	4.50^{a}	15.2^{a}	40.7^{a}
	D	3.67^{b}	4.11^{ab}	3.56^{b}	3.89^{a}	3.89^{ab}	3.67^{ab}	8.8^{c}	17.3^{d}
	E	4.33^{a}	4.17^{ab}	3.92^{ab}	4.33^{a}	4.17^{ab}	4.58^{a}	10.8^{bc}	22.9^{cd}
	F	4.33^{ab}	4.33^{a}	3.67^{ab}	4.33^{a}	3.67^{ab}	3.67^{ab}	11.0^{bc}	33.7^{ab}
	Pr.(>F)	*	*	**	ns	*	**	**	**
	Mean	4.24	4.10	3.94	4.13	3.92	3.99	12.5	25.9

Significant differences ($p < 0.05$) are indicated by different letters among groups. ns indicate not significant. *, and ** represent a significant at 95% and 99% probability level.

Although mean values of height (HT) and diameter at breast height (DBH) of candidate trees in each group were presented together (Table 4), candidate tree selection was made by only six characteristics. In other words, HT and DBH had no effect on candidate tree selection because they were arranged after finishing of selection. To understand the relationships among these variables, correlation analysis was conducted in each species (Table 5). Although the explanatory power was low due to weak correlation, a rough tendency for significantly correlated variables was examined.

Table 5. Spearman rank correlations among variables measured from candidate trees of *Quercus salicina* (upper diagonal) and *Q. glauca* (lower diagonal).

Variable	Superiority of Growth (SG)	Superiority of Tree Form (STF)	Adaptability to Disturbance (AD)	Adaptability to Environment (AE)	Superiority of Seed Production (SSP)	Potential of Seed Production (PSP)	Height (HT)	Diameter at Breast Height (DBH)
SG	-	−0.087 ns	0.320 **	0.114 ns	−0.052 ns	−0.171 ns	0.330 **	0.350 **
STF	−0.050 ns	-	0.010 ns	0.332 **	−0.049 ns	0.094 ns	0.021 ns	0.043 ns
AD	0.257 *	0.083 ns	-	−0.099 ns	−0.354 **	−0.279 **	0.321 **	0.178 ns
AE	0.235 *	0.086 ns	0.140 ns	-	−0.035 ns	−0.076 ns	0.071 ns	0.088 ns
SSP	−0.057 ns	−0.132 ns	0.149 ns	0.250 *	-	0.434 **	0.170 ns	0.083 ns
PSP	−0.051 ns	−0.300 **	0.152 ns	0.300 **	0.441 **	-	0.273 **	0.040 ns
HT	0.353 **	−0.215 ns	0.227 *	0.069 ns	−0.122 ns	−0.080 ns	-	0.713 **
DBH	0.256 ns	−0.298 **	0.228 *	0.197 ns	0.225 ns	0.267 **	0.597 **	-

ns indicate not significant. *, and ** represent a significant at 95% and 99% probability level.

In both species, there were no correlations between SG and STF in growth category and between AD and AE in adaptability category. However, there was a significant correlation between SSP and PSP in seed production category. It meant that candidate trees with higher fruiting at present were also graded having higher potential of reproduction, although the evaluation criteria were different.

In *Q. salicina*, AD had significant positive correlation with SG and significant negative correlations with SSP and PSP. It indicated that the candidate trees with high adaptability to disturbance also showed good growth and development, but the seed production was not. AE showed positive correlation with STF, which indicated that the candidate trees with high adaptability to environment had good tree form. HT of candidate trees had significant positive correlation with SG, AD and PSP. DBH of candidate trees also had significant positive correlation with SG. It meant that the candidate trees with superior growth, high adaptability to disturbance and high potential of seed production had higher HT and/or DBH.

In *Q. glauca*, AD had significant positive correlation with SG like *Q. salicina*. AE had significant positive correlation with SG, SSP, and PSP. It indicated that the candidate trees with higher adaptability showed higher growth and seed production. Meanwhile, STF had negative correlation with PSP, which indicated that candidate trees with superior tree form showed low potential of seed production. HT had significant positive correlation with SG and AD. DBH had significant positive correlation with AD and PSP but significant negative correlation with STF. It meant that the candidate trees with superior growth and high adaptability to disturbance had higher HT and/or DBH. However, candidate trees with higher STF or DBH showed lower PSP unlike *Q. salicina*.

By analogy with the results of this survey and the relevant research [32], it seemed to be related to the characteristics of the species. Initial adaptation of *Q. glauca* is favorable as a pioneer species, but if the stand is stabilized, it is disadvantageous in competition with other species. *Q. salicina* is estimated to be advantageous in competition once it is adapted because it has higher adaptability in relatively broad environments as a generalist species.

3.3. Data Analysis and Plus Tree Selection

The weighted generalized value of each individual (GVI_w) was calculated as $GVI_w = 0.7 \times GVI_p + 0.3 \times GVI_t$. The candidate trees were truncated with baseline value '0', i.e., if $GVI_w > 0$, then accepted and if $GVI_w < 0$, then rejected (Figure 2).

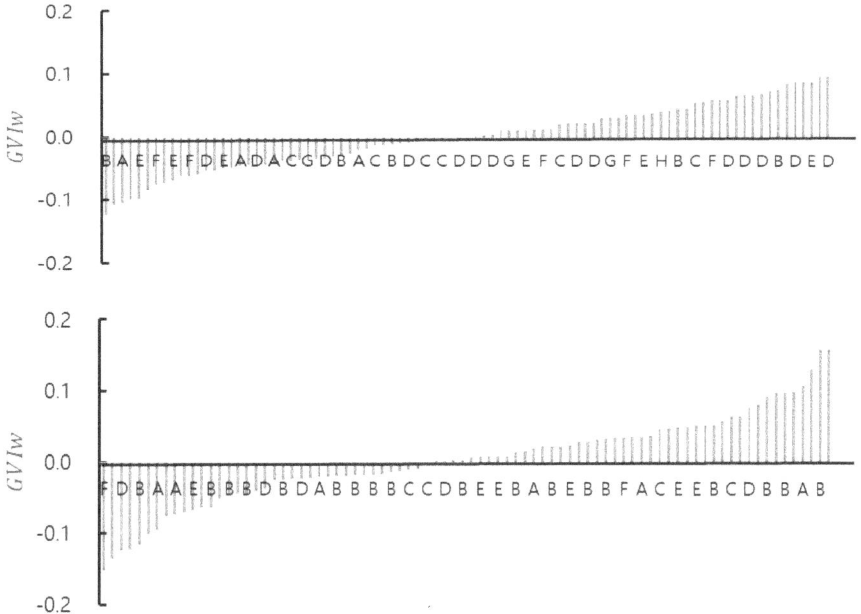

Figure 2. Truncation selection of candidate trees based on GVI_w values in *Quercus salicina* ***(top)*** *and Q. glauca* ***(bottom)***. If $GVI_w > 0$, then accepted and if $GVI_w < 0$, then rejected. Capital letters indicate the group to which each individual belongs.

In *Q. salicina*, 43 candidate trees were accepted and 44 candidate trees were rejected. In *Q. glauca*, 42 candidate trees were accepted and 40 candidate trees were rejected (Table 6). In group A of *Q. salicina*, for example, GVI_p and GVI_t of No. 4 candidate tree were 0.0242 and −0.0200, respectively. If we only considered GVI_p the candidate tree would be accepted (>0), but if GVI_t was only considered, the candidate tree would be rejected (<0). However, if we applied GVI_w (−0.0109) suggested in this study, it was classified as an accepted candidate (>0).

Table 6. Final list of plus trees selected by truncation ($GVI_w > 0$) and adjustment in *Quercus salicina* and *Q. glauca*.

	Q. salicina			*Q. glauca*	
Group	**Accepted by Truncation ($GVI_w > 0$)**	**Final Selection by Adjustment**	**Group**	**Accepted by Truncation ($GVI_w > 0$)**	**Final Selection by Adjustment**
A	1, 4, 11	1, 4, 6, 8, 11	A	1, 5, 7, 10, 11, 12, 15	1, 3, 5, 7, 10, 12, 15
B	1, 2, 3, 8	1, 2, 3, 8	B	2, 5, 8, 9, 10, 12, 13, 14, 15, 17, 20, 21, 22, 23, 24, 27, 29	2, 4, 5, 8, 9, 12, 13, 15, 17, 20, 21, 22, 23, 24, 27, 29
C	2, 3, 4, 9, 12	2, 3, 4, 8, 9, 12, 14	C	1, 4, 6, 7, 8	1, 4, 6, 7, 8
D	2, 3, 4, 5, 6, 7, 8, 9, 10, 11, 13, 19, 20, 22, 25	2, 3, 4, 6, 8, 9, 10, 11, 13, 19, 20, 22, 25	D	2, 6, 7, 9	2, 4, 6, 7, 9
E	1, 5, 6, 7, 8, 10	1, 5, 7, 8, 10	E	2, 3, 4, 7, 8, 9	2, 4, 7, 8, 9, 12
F	1, 3, 4, 8	1, 3, 4, 8	F	1, 3	1, 3
G	2, 3, 4, 5	2, 3, 4, 5	Total	42	41
H	1, 2	1, 2			
Total	43	44			

The final selection was decided through some adjustment process considering the imbalance of number of individuals selected, the specificity of stand identified at investigation, and biased selection according to the differences in micro-environment within the group.

The major adjustments performed in *Q. salicina* were as follows (Table 6). In group A, only 3 out of 11 candidate trees were included in accepted group. To balance the number of candidate trees among groups, No. 6 (−0.0054) and No. 8 (−0.0179) with higher GVI_w among rejected group were included in accepted group. In group C, Seoho (602 m) and Sanghyo (465 m) stand were located in highland area comparing other stands (less than 200 m). To consider specificity of stand and balance the number of candidate trees among groups, No. 8 (−0.0050) and No. 14 (−0.0050) with higher GVI_w among rejected group were included in accepted group. In group D, Sacheon1 and Sacheon2 stand were adjacent with a ridge in between them. However, the micro-environment was a little different, i.e., soil depth, direction of slope, and topology of Sacheon1 is better than Sacheon2. So, they were initially treated as a separate stand. Later, however, they were grouped into a group considering geographic proximity. Accordingly, 85.7% of candidate trees in Sacheon1 (6 out of 7) were accepted whereas only 47.4% of candidate trees of Sacheon2 (9 out of 19) were accepted. To balance the number of candidate trees within group, adjustment was made to include No. 5 (0.0033) and No. 7 (0.0047) with small GVI_w in the rejected group.

The following adjustments had been made in *Q. glauca* (Table 6). In Cheonsu stand in group B, No. 8~13, 15, and 20~24 candidate trees were selected from relatively good sites with deep soil depth and low gravel ratio. All except for No. 11 candidate tree were included in accepted group. To avoid biased selection to a specific site condition, two candidate trees with lower GVI_w, No. 10 (0.0172) and No. 14 (0.0188), were adjusted into rejected group. An opposite case was found in group D. No. 3~5 were selected in relatively poor site of Jeongja stand. Accordingly, all of them were rejected by GVI_w truncation. To broaden genetic variability, adjustment was made to include No. 4 candidate tree in accepted group. In the same sense, No. 3 candidate tree (0.0208) in Jangjoa stand of group E were adjusted into rejected group and No. 3 (−0.0228) and 11 (0.0067) in Seonhul of group A were adjusted into accepted group and rejected group, respectively.

Although the primary goal of this study was to select plus trees, the conservation aspect was also considered. Thus, we tried to include as many individuals as possible from stands of various environments. To this end, the selected stands were divided into groups as well as the adjustment was performed as above. Nevertheless, we cannot be sure that enough genetic diversity will be included in a breeding population. Thus, additional studies are required to evaluate genetic diversity, i.e., comparison of genetic diversity between natural stands and candidate trees including rejected trees. If genetic diversity is lower than expected, adding some rejected trees to breeding population may be an option.

Through adjustment process, 44 candidate trees in *Q. salicina* and 41 candidate trees in *Q. glauca* were finally selected as plus trees. In order to compare the selection effect of plus tree, the mean values of six characteristics, height and diameter growth between accepted group (plus trees) and rejected group (Table 7). On average, the accepted group was excellent in all characteristics in both species, i.e., 103.2~112.9% in *Q. salicina* and 105.0~112.8% in *Q. glauca*. This tendency was the same in HT and DBH, even though they were not considered directly at the time of selection. It was interpreted as the result of indirect selection of HT and DBH in relation to selection of SG, AD, or STF as shown in the correlation analysis (Table 5). It was meaningful that similar selection effects were obtained without direct measuring or selecting growth-related traits like other general selection methods.

Table 7. Comparison of means values of six characteristics, height and diameter growth between accepted trees (plus trees) and rejected trees in *Quercus salicina* and *Q. glauca*.

Species	Class	Mean Values of Each Characteristic						Height (HT)	Diameter at Breast Height (DBH)
		Superiority of Growth (SG)	Superiority of Tree Form (STF)	Adaptability To Disturbance (AD)	Adaptability to Environment (AE)	Superiority of Seed Production (SSP)	Potential of Seed Production (PSP)		
Q.salicina	Accepted (A)	4.72	4.48	4.20	4.57	3.89	4.20	14.1	30.9
	Rejected (R)	4.33	4.12	4.07	4.26	3.63	3.72	14.0	28.3
	A/R (%)	109.0	108.7	103.2	107.3	107.2	112.9	100.7	109.2
Q.glauca	Accepted (A)	4.49	4.20	4.05	4.34	4.15	4.10	12.9	27.2
	Rejected (R)	4.00	4.00	3.83	3.93	3.68	3.88	12.1	24.5
	A/R (%)	112.3	105.0	105.7	110.4	112.8	105.7	106.6	111.0

3.4. Implications is Evergreen Oak Breeding

Evergreen oaks are only distributed in the southernmost part of Korean peninsula and the natural stands are discontinuously scattered with small size [3,4]. The genetic conservation for evergreen oaks is also highly recommended [11,33]. Accordingly, it is difficult to secure large stand for seed production area or seed stand. Thus, establishment of multi-purpose seedling seed orchard (MPSSO) for seed production and genetic conservation by mass selection was suggested as an alternative.

There are several selection methods, but the most important thing is to find the best method for the target species [16]. Comparison tree selection had been widely used in most conifer species as well as deciduous oak species in Korea [23].

In this study, some modified selection criteria and method were applied for plus tree selection. It was primarily due to the characteristics of the target species as mentioned above. Another reason was the results of Stringer et al. [17]. They suggested that selecting several phenotypically above-average candidate trees may be more effective than rigorously selecting a smaller number of phenotypically superior trees in *Quercus rubra*. Reflecting these aspects, the truncation of candidate trees for plus tree selection was performed with baseline value of GVI_w in this study.

Through this process, 44 candidate trees in *Q. salicina* and 41 candidate trees in *Q. glauca* were finally selected as plus trees. Since it was found that this method could provide a selection effect similar to direct selection on growth-related traits, it was expected that application of this methodology could be expanded to other tree species if the conditions were met.

Most tree improvement programs are designed to provide continued gain through each of many cycles of improvement. A selection procedure that involves many cycles of selection and breeding is known as recurrent selection (RS). A number of recurrent selection schemes have been devised to utilize general combining ability (GCA) and, in some cases, specific combining ability (SCA) [12,16].

In case of *Q. salincina* and *Q. glauca*, simple recurrent selection (SRS), was suggested as a strategy for advanced generation. Although SRS is less efficient at achieving genetic gains than other forms of recurrent selection, i.e., RS for GCA and RS for SCA, it is considered to be a more reasonable approach matched the breeding goal and current situation of *Q. salincina* and *Q. glauca*. It could serve the purpose of increasing the average of genetic gain in each generation and maintaining the genetic variation in the selected population as much as possible.

Another reason for making this suggestion is the aspect of forest policy in Korea. The reforestation tree species are classified into two categories, i.e., major vs minor tree species. The major tree species is a species with high demand and large planting area, i.e., deciduous oaks, Pines, Larches, and so on. The minor tree species is the opposite. Evergreen oaks and most of other broad-leaved trees are belonged to this category. Accordingly, RS-GCA and RS-SCA strategy are mainly applied in major tree species. In practical aspects, it is difficult to invest a lot of efforts and expenses into minor tree species [29,34].

Nevertheless, to compensate the disadvantage of SRS and use of GCA information for advanced generation breeding, operation of several MPSSO was suggested as an alternative. At present, a discussion is underway to determine the location and number of sites for establishment of MPSSO in Korea.

4. Conclusions

Although evergreen oaks, including *Q. salicina* and *Q. glauca*, are restricted to subtropical regions on the southernmost part of the Korean peninsula, it is considered as an alternative species for adaptation to climate change in the future. So, Korea Forest Service is being pursed to expand the selection of plus trees and establishment of seed orchard to simultaneously consider conservation and utilization of evergreen oaks. However, a selection methodology for evergreen oaks has not been established to support this. Through this study, we tried to find a suitable methodology for the plus tree selection of evergreen oaks. The usefulness of our method was confirmed through application of plus tree selection in *Q. salicina* and *Q. glauca*. Thus, it is expected that all processes and results thus far will be used as a guideline for plus tree selection as well as conservation of evergreen oaks.

Author Contributions: Conceptualization and methodology, I.S.K. and K.M.L.; investigation, K.M.L., J.J.K. and I.S.K.; data curation, J.J.K. and K.M.L.; formal analysis, D.S. and H.-I.K.; writing—original draft preparation, I.S.K. and K.M.L.; writing—review and editing, D.S., H.-I.K. and I.S.K. All authors have read and agreed to the published version of the manuscript.

Funding: This research received no external funding.

Acknowledgments: We are greatly indebted to W.Y. Choi, J.C. Lee, and S.S. Chang for their valuable suggestions and helping in field investigation. This study is supported by National Institute of Forest Science, Republic of Korea (FG0400-2019-01).

Conflicts of Interest: The authors declare no conflict of interest.

References

1. Nixon, K.C. Global and neotropical distribution and diversity of oak (genus *Quercus*) and oak forests. In *Ecology and Conservation of Neotropical Montane Oak Forests*; Kappelle, M., Ed.; Springer Publication: Berlin, Germany, 2006; pp. 3–13.
2. Tantray, Y.R.; Wani, M.S.; Hussain, A. Genus *Quercus*: An overview. *Int. J. Adv. Res. Sci. Eng.* **2017**, *6*, 1880–1886.
3. Korea National Arboretum. *Distribution Maps of Vascular Plants in Korea*; Korean National Arboretum: Pocheon, Korea, 2016.
4. National Institute of Forest Science. *Economic Tree Species ② Oaks*; Research New Book; National Institute of Forest Science: Seoul, Korea, 2012; Volume 60.
5. National Institute of Forest Science. *Predicting the Changes of Productive Areas for Major Tree Species under Climate Change in Korea*; Research Report No. 14-21; National Institute of Forest Science: Seoul, Korea, 2014.
6. Park, S.U.; Koo, K.A.; Kong, W.S. Potential impact of climate change on distribution of warm temperate evergreen broad-leaved trees in the Korean peninsula. *J. Korean Geogr. Soc.* **2016**, *51*, 201–217.
7. Williams, M.I.; Dumroese, R.K. Growing assisted migration: Synthesis of a climate change adaptation strategy. In *National Proceedings: Forest and Conservation Nursery Association-2012*; Haase, D.L., Pinto, J.R., Wilkinson, K.M., Eds.; Proceedings RMRS-P-69; USDA, Forest Service, Rocky Mountain Research Station: Fort Collins, CO, USA, 2013; pp. 90–96.
8. Benito-Garzon, M.; Fernandez-Manjarres, J.F. Testing scenarios for assisted migration of forest trees in Europe. *New For.* **2015**, *46*, 979–994. [CrossRef]
9. Lee, H.J.; Park, S.N. Antioxidative effect and active component analysis of *Quercus salicina* Blume extracts. *J. Soc. Cosmet. Sci. Korea* **2011**, *37*, 143–152.
10. Moon, S.H.; Song, C.K.; Kim, T.K.; Oh, D.E.; Kim, H.C. Antifungal activity on the water extracts of five Fagaceae plants. *Korean J. Org. Agric.* **2017**, *25*, 295–310.
11. Son, S.G.; Kim, H.J.; Kang, Y.J.; Oh, C.J.; Kim, C.S.; Byun, K.O. Establishment of breeding population for *Quercus glauca* and climate factors. *Korean J. Agric. For. Meteorol.* **2011**, *13*, 109–114. [CrossRef]
12. White, T.L.; Adams, W.T.; Neal, D.B. *Forest Genetics*; CABI International Publication: London, UK, 2007.
13. Falconer, D.S. *Introduction to Quantitative Genetics*, 2nd ed.; Longman Group Ltd.: London, UK, 1981.

14. Morgenstern, E.K. Tree selection techniques in the Northeast: Some problems and questions. In Proceedings of the Northeastern Forest Tree Improvement Conference; Eckert, R.T., Ed.; Institute of Natural and Environmental Resources, Univ. of New Hampshire Durham: Durham, NH, USA, 1982; pp. 145–157.
15. Van Damme, L. *Plus-tree Selection of Black Spruce in Northwestern Ontario for Superior Growing Space Efficiency*; School of Forestry, Lakehead University: Thunder Bay, ON, Canada, 1985.
16. Zobel, B.; Talbert, J. *Applied Forest Tree Improvement*; John Wiley & Sons: New York, NY, USA, 1984.
17. Stringer, J.W.; Wagner, D.B.; Schlarbaum, S.E.; Houston, D.B. An analysis of phenotypic selection in natural stands of northern red oak (*Quercus rubra* L.). In Proceedings of the 10th Central Hardwood Forest Conference, Morgantown, WV, USA, 5–8 March 1995; pp. 226–237.
18. Lee, T.B. *Illustrated Flora of Korea*; Hyangmunsa: Seoul, Korea, 1989.
19. Han, D.H.; Kim, J.Y.; Choi, I.T.; Lee, K.J. Vegetation structure of evergreen broad-leaved forest in Dongbaekdongsan(Mt.), Jeju-Do, Korea. *Korean J. Environ. Ecol.* **2007**, *21*, 336–346.
20. Yun, J.H.; Hukusima, T.; Kim, M.H.; Yoshikawa, M. The comparative study on the distribution and species composition of forest community in Korea and Japan around the east sea. *Korean J. Ecol.* **2011**, *25*, 327–357.
21. Jeong, H.M.; Kim, H.R.; Cho, K.T.; Lee, S.H.; Han, Y.S.; You, Y.H. Aboveground biomass estimation of *Quercus glauca* in evergreen forest, Kotzawal wetland, Cheju Island. *Korea J. Wetl.* **2014**, *16*, 245–250.
22. Jeon, C.H.; Won, H.K.; Kim, H.S.; Cho, Y.J. The vegetation structure of evergreen broad-leaved forest of Jeju experimental forest, Korea. *J. Korean Isl.* **2016**, *28*, 221–238.
23. National Institute of Forest Science. *Genetic Tests and Improvement of Genetic Gain in Major Tree Species*; Research Report No. 19-11; National Institute of Forest Science: Seoul, Korea, 2019.
24. Morgenstern, E.K. Review of the principles of plus-tree selection. In *Plus-tree Selection: Review and Outlook*; Morgenstern, E.K., Holst, M.J., Teich, A.H., Yeatman, C.W., Eds.; Department of the Environment, Canadian Forestry Service Publication: Ottawa, ON, Canada, 1975; pp. 1–27.
25. Han, J.W.; Kamber, M.; Pei, J. *Data Mining: Concepts and Techniques*, 3rd ed.; Morgan Kaufmann Pub.: New York, NY, USA, 2012.
26. Suarez-Alvarez, M.M.; Phan, D.T.; Prostov, M.Y.; Prostov, Y.I. Statistical approach to normalization of feature vectors and clustering of mixed datasets. *Proc. R. Soc. A* **2012**, *468*, 2630–2651. [CrossRef]
27. Muralidharan, K. A note on transformation, standardization and normalization. *Int. J. Oper. Quant. Manag.* **2014**, *9*, 116–122.
28. Khan, A.; Rayner, G.D. Robustness to non-normality of common tests for the many-sample location problem. *J. Appl. Math. Decis. Sci.* **2003**, *7*, 187–206. [CrossRef]
29. Wright, J.W. *Introduction to Forest Genetics*; Academic Press Inc.: New York, NY, USA, 1976.
30. Dupuy, J.M.; Chazdon, R.L. Interacting effects of canopy gap, understory vegetation and leaf litter on tree seedling recruitment and composition in tropical secondary forests. *For. Ecol. Manag.* **2008**, *255*, 3716–3725. [CrossRef]
31. Tang, C.Q.; Ohsawa, M. Ecology of subtropical evergreen broad-leaved forests of Yunnan, southwestern China as compared to those of southwestern Japan. *J. Plant Res.* **2009**, *122*, 335–350. [CrossRef] [PubMed]
32. Ito, S.; Ohtsuka, K.; Yamashita, T. Ecological distribution of seven evergreen *Quercus* species in southern and eastern Kyushu, Japan. *Veg. Sci.* **2007**, *24*, 53–63.
33. Kim, G.U.; Jang, K.S.; Lim, H.W.; Kim, E.H.; Lee, K.H. Genetic diversity of *Quercus gilva* in Je-ju Island. *J. Korean Soc. For. Sci.* **2018**, *107*, 151–157.
34. Rosvall, O.; Mullin, T. Introduction to breeding strategies and evaluation alternatives. In *Best Practice for Tree Breeding in Europe*; Mullin, T.J., Lee, S.J., Eds.; Skogforsk Publication: Uppsala, Sweden, 2013; pp. 7–27.

Article

Genetic Diversity of *Paeonia rockii* (Flare Tree Peony) Germplasm Accessions Revealed by Phenotypic Traits, EST-SSR Markers and Chloroplast DNA Sequences

Xin Guo, Fangyun Cheng * and Yuan Zhong

Beijing Advanced Innovation Center for Tree Breeding by Molecular Design, Peony International Institute, Beijing Key Laboratory of Ornamental Plants Germplasm Innovation & Molecular Breeding, National Engineering Research Center for Floriculture, Key Laboratory of Genetics and Breeding in Forest Trees and Ornamental Plants of Ministry of Education, School of Landscape Architecture, Beijing Forestry University, Beijing 100083, China; freshxiaoxinxin@163.com (X.G.); zhongyuanbjfu@126.com (Y.Z.)

* Correspondence: chengfy8@263.net; Tel.: +86-010-62338027

Received: 16 May 2020; Accepted: 10 June 2020; Published: 12 June 2020

Abstract: Research Highlights: This study, based on the first collection of cultivated *Paeonia rockii* (flare tree peony, FTP) germplasm across the main distribution area by our breeding desires, comprehensively evaluates these accessions by using phenotypic traits, expressed sequence tag (EST)-simple sequence repeat (SSR) markers and chloroplast DNA sequences (cpDNA). The results show that these accessions collected selectively by us can represent the genetic background information of FTP as a germplasm of tree crops. Background and Objectives: FTP has high cultural, ornamental and medicinal value traditionally, as well as recently presenting a significance as an emerging edible oil with high α-linolenic acid contents in the seeds. The objectives of this study are to reveal the characteristics of the genetic diversity of FTP, as well as to provide scientific suggestions for the utilization of tree peony breeding and the conservation of germplasm resource. Materials and Methods: Based on the phenotypic traits, EST-SSR markers and chloroplast DNA sequence variation, we studied the diversity of a newly established population of 282 FTP accessions that were collected and propagated by ourselves in our breeding project in recent years. Results: (1) There was an abundant variation in phenotype of the accessions, and the phenotypic variation was evenly distributed within the population, without significant hierarchical structure, (2) the EST-SSR data showed that these 282 accessions had relatively high genetic diversity, in which a total of 185 alleles were detected in 34 pairs of primers. The 282 accessions were divided into three distinct groups, and (3) the chloroplast DNA sequences (cpDNA) data indicated that these accessions had a higher genetic diversity than the population level and a lower genetic diversity than the species level of wild *P. rockii*, and the existing spatial genetic structure of these accessions can be divided into two branches. Conclusions: From the results of the three analyses, we found that these accessions can fully reflect the genetic background information of FTP germplasm resources, so their protection and utilization will be of great significance for genetic improvement of woody peonies.

Keywords: *Paeonia rockii* (flare tree peony) germplasm accessions; phenotypic traits; EST-SSR markers; chloroplast DNA sequences; genetic diversity

---

1. Introduction

Tree peony or Mudan, known as "the king of flowers" in China, has high cultural, ornamental and medicinal value traditionally, as well as recently presenting a significance as an emerging edible oil with high unsaturated fatty acid (>90%) and α-linolenic acid (>40%) contents in the seeds [1,2].

Tree peony belonging to the section *Moutan* in the genus *Paeonia* and all wild species are endemic to China. It has been cultivated since the Tang Dynasty (618–906 AD) and is now widespread in temperate regions of China and other countries. At present, there are more than 1500 germplasm accessions of tree peony in the world, included in about 17 cultivar groups. In China, long-term and repeated domestication and selection breeding have formed ten groups, about 1000 germplasm accessions, with diverse genetic background [3].

As the representative of the most-widely cultivated tree peonies, *Paeonia suffruticosa* Andrews or the common tree peony (CTP), includes the germplasm accessions traditionally cultivated in China and Japan, while the germplasm accessions originated from *P. rockii* or the flare tree peony (FTP) have established a distinct group from CTP and are known well, with a colored flare at the base of petals [3]. Compared to CTP, FTP is more fertile in most cases as well as more resistant to stress conditions like coldness, drought and poor soil. FTP is distinct from CTP in many other features, like more fragrant blossoms, and taller and stronger plants with more longevity (Figure 1) [1,4]. Since many studies [1,5–11] have confirmed that FTP is a valuable genetic resource of tree peonies, greatly promising in the breeding and industry of peonies, the further in-depth study of FTP has become obviously required for the utilization and protection of FTP as a germplasm resource of tree crops.

Figure 1. The morphological characteristics of *P. rockii* plant: (**a**) *P. rockii* has more fragrant blossoms, and taller and stronger plants with more longevity, (**b**) there is a colored flare at the base of petals, and (**c**) with strong fertility, the ripe follicles contain a large number of seeds.

The analysis of genetic diversity can provide beneficial data for collection of germplasm resources, breeding application and researches on origin and evolution [12]. Phenotypic variation analysis is one of the most important, basic, intuitive but indispensable methods for the development and application of plant germplasm resources. In tree peonies (*Paeonia* Sect. *Moutan*), researchers have focused on the genetic diversity of phenotypic traits in different species and cultivated germplasm, including *P. delavayi* [13], *P. ostii* [14,15], *P. delavayi* var *lutea* [16], *P. suffruticosa* [17–19] and so on. In recent years, the emergence of molecular markers has greatly accelerated the development of genetics and genomics and is expected to be a powerful tool to accelerate the research of tree peony breeding. Researchers had used random amplification polymorphic DNA (RAPD), inter-simple sequence repeat (ISSR), sequence-related amplified polymorphism (SRAP), conserved DNA-derived polymorphism (CDDP), inter-primer binding site (iPBS), target region amplified polymorphism (TRAP), expressed sequence tags microsatellite markers (EST-SSR), chloroplast DNA sequences (cpDNA) and EST-SSR with cpDNA to analyze the genetic diversity of *P. suffruticosa* [20–26], *P.* x *yananensis* [27], *P. rockii* [28,29], *P. delavayi* [30], *P. ludlowii* [31], *Paeonia* subsect. *Delavayanae* [32,33], *P. decomposita* [34], *P. jishanensis* [35], *P. ostii* [36] and *P. quii* [37]. In *P. rockii*, Yuan et al. [28,29] used cpDNA and SSR to analyze the genetic diversity in 20 wild populations, showing that wild *P. rockii* had high genetic diversity at the level of species and can be divided into three genetically distinct clusters corresponding to three geographically distributed regions. Xu et al. [37] used cpDNA and EST-SSR to analyze the genetic diversity of *P. rockii*, *P. jishanensis* and *P. qiui* to discover that the genetic diversity of *P. rockii* among the three tree peony species was relatively low. In the population of 462 cultivated *P. rockii* seedlings, Wu et al. [38] used 40 EST-SSR markers to reveal a medium genetic diversity and the genetic structure composed of three subgroups. These studies provided us with very useful references to carry out further research on the genetic diversity of tree peony germplasm.

With high polymorphism, codominance and ease of operation, EST-SSR molecular markers have been applied more and more in the study of genetic diversity [39]. Chloroplast genome is maternal lineage and retains ancestral patterns of genetic diversity longer than nuclear DNA [40]. Moreover, in the cross-breeding system, the effective population size of cpDNA is only half of that of nuclear DNA, which may lead to higher genetic differentiation and more significant genetic structure [41]. Chloroplast non-coding region sequences have higher mutation rates than coding region sequences, which can be used for population genetics research [42,43].

Based on the phenotypic traits, EST-SSR markers and chloroplast DNA sequence variation, this paper reports the research of genetic diversity in the newly established population of 282 FTP accessions that were collected, propagated and mostly named as cultivars by us in our breeding project in recent years. Our objectives are to reveal the characteristics of genetic diversity of these accessions, as well as to provide scientific suggestions for the utilization in peony breeding and the conservation of FTP germplasm resource.

2. Materials and Methods

2.1. Plant Materials and DNA Extraction

Germplasm resources of FTP had been selectively collected by our breeding desires across the main cultivation area of the FTP, Lanzhou city, Zhang county, Longxi county, Lintao county and Linxia county of Gansu province in Northwest China since 1990, propagated by grafting in Guose Peony Garden of Yan Qing district, Beijing, China (40°46′ N, 116°07′ E) since 2011 (Figure 2). The plants of the 282 accessions used in this study are randomly planted in the same nursery field and maintained under the same condition of water and fertilizer throughout the year.

In March 2018, the young leaves of all accessions were individually collected and dried in silica gel, from which total genomic DNA was extracted by using a DNAsecure plant kit (Tiangen Biotech, Beijing, China). The extracted DNA was detected by electrophoresis using 2% agarose gels and a UnicoUV–visible Spectrophotometer (Agilent, Palo Alto, CA, USA), respectively. The samples were diluted to 20–30 ng/μL with deionized water and then stored in the refrigerator at −20 °C.

A total of 282 accessions, of those collected above, were used for EST-SSR analysis, and the phenotypic traits were investigated only in the completely matured 15-year-old plants of 180 accessions. The color of flowers is genetically stable for every specific accession and in the studied accessions can be classified into five groups: 89 white, 33 pink, 32 red, 116 purplish red and 12 purplish black. By this color group, we selected the accessions for cpDNA sequencing, including 24 white, 10 pink, 8 red, 35 purplish red and 3 purplish black accessions, with the proportions of 26.97%, 30.30%, 25.00%, 30.17% and 25.00%, respectively (Supplementary Table S1).

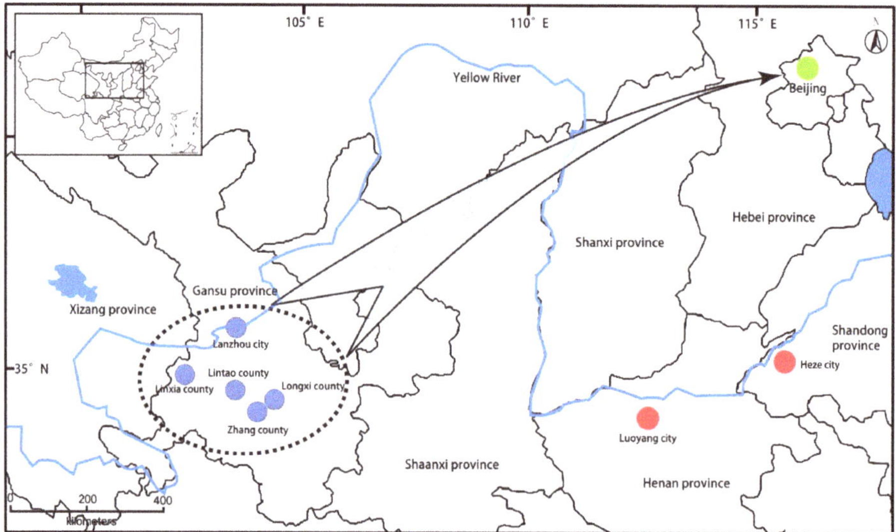

Figure 2. Geographic origins of the flare tree peony (FTP) accessions used in this study. These accessions were selectively collected in the cultivation center (dotted area) of FTP in the middle of Gansu province where the blue dots represent the concrete places, and then propagated clonally and transplanted into the conservation site of Beijing (the green dot). The red dots represent Luoyang and Heze, the two traditional cultivation centers of common tree peony (CTP) in China.

2.2. Measurement of Phenotypic Traits

We recorded 28 phenotypic traits according to the measurement standard (Table 1) during the flowering time from April to May and the seed ripening time from August to September in 2019. In April to May, three plants were randomly selected for each accession as three replicates and three branches with flowers were randomly selected for each plant. The number of flowers, plant height, crown breadth and tiller number of three plants were measured first, then nine flower diameters were measured, and nine carpel numbers were counted. A random outer petal was selected from each flower to measure the length and width of each petal and flare. Then, the number of petals of each flower was counted. Finally, the second or third compound leaf from bottom to top of each branch was selected for leaf trait investigation. From August to September, we chose the three plants and three branches of each plant that we had measured in the blooming period. First, we counted the fruit number of three plants and fruit weight per plant. Then, we measured individual fruit weight, number of seeds per fruit, seed weight per fruit and effective carpel number. Lastly, we randomly selected a single carpel of each follicle to measure fruit length, fruit width, fruit height and pericarp thickness.

Table 1. Traits investigated of 180 accessions and the measurement standard.

Type	Code	Trait	Measurement Standard
Floral trait	1	Petal length	The length of the outer petal (cm)
	2	Petal width	The width of the outer petal (cm)
	3	Flare length	The length of flare (cm)
	4	Flare width	The width of flare (cm)
	5	Flower diameter	The maximum width of flower at full bloom (cm)
	6	Petal number	The number of petals in the whole flower (score)
	7	Carpel number	The number of carpels in the whole flower (score)
Branch and leaf trait	8	Plant height	The aboveground height of the whole plant (cm)
	9	East-west crown breadth	The width of a plant in the east-west direction (cm)
	10	North-south crown breadth	The width of a plant in the north-south direction (cm)
	11	Fruit number	The number of follicles per plant (score)
	12	Flower number	The number of flowers per plant (score)
	13	Tiller number	The number of branches at the base of a plant (score)
	14	Compound leaf length	The length from tip of compound leaf to petiole (cm)
	15	Compound leaf width	The length at widest part of compound leaf (cm)
	16	Petiole length	The length of petiole (cm)
	17	Leaflet number	The number of leaflets in a compound leaf (score)
	18	Terminal leaflet length	The length of terminal leaflet (cm)
	19	Terminal leaflet width	The width of terminal leaflet (cm)
Fruit trait	20	Fruit weight per plant	The weight of fruits per plant (g)
	21	Individual fruit weight	The weight of one fruit (g)
	22	Number of seeds per fruit	The number of seeds per fruit (score)
	23	Seed weight per fruit	The weight of seeds per fruit (g)
	24	Effective carpel number	The number of carpels that produce seeds (score)
	25	Fruit length	The length of a single carpel of mature follicles (mm)
	26	Fruit width	The width of a single carpel of mature follicles (mm)
	27	Fruit height	The height of a single carpel of mature follicles (mm)
	28	Pericarp thickness	The thickness of pericarp of mature follicles (mm)

2.3. Microsatellites Markers

From 72 pairs of EST-SSR primers distributed in 5 linkage groups of the first high-density genetic map of *P. suffruticosa* [44,45], we screened out 34 pairs of primers with high amplification efficiency and rich polymorphism. The sequences of the primers used in this study are listed in Table 2, and all the forward primers were respectively labeled with 5-Carboxyfluorescein (5-FAM), 5-Hexachlorofluorescein (5-HEX), 5-Carboxytetramethylrhodamine (5-TAMRA), 5-Carboxy-X-rhodamine (5-ROX). The polymerase chain reaction (PCR) was conducted in a 10 μL solution, including: 5 μL 2× Power Taq PCR Master MIX (Aidlab Biotechnologies, Beijing, China), 0.5 μL and 10 μmol/L each of forward and reverse primer, 1 μL and 20–25 ng/μL genomic DNA and 3 μL ddH_2O. The SSR-PCR amplification procedure was run as described by Wu et al. [45]. The amplified fragment results were detected by capillary electrophoresis using an ABI3730xl DNA Analyzer with a GeneScan-500LIZ size standard (Applied Biosystems, Carlsbad, CA, USA).

Table 2. Characterization of 34 simple sequence repeat loci based on 282 accessions.

Locus	Primer Sequence	Tm (°C)	Expected Size (bp)	Modified Type	*Na*	*Ne*	*I*	*Ho*	*He*	F_{IS}	PIC	*p*
PS004	F: GTGCTTAGCCTCTAATCTG R: CTTTGCTCCAAGTCTGTC	50.5	274	5-FAM	7	2.859	1.330	0.688	0.650	−0.058	0.611	0.000 ***
PS026	F: TTCCCTCCATTCTAACAC R: ACCCTAGCCTCTGACATT	54	187	5-FAM	5	1.761	0.715	0.475	0.432	−0.100	0.358	0.000 ***
PS030	F: ACCCTCCACCACCATCTT R: TACTCCATCTCGTGACCC	57	237	5-HEX	3	1.290	0.414	0.216	0.225	0.038	0.203	0.000 ***
PS047	F: AGACGACGAGCAAAGATAT R: AAAGGGCAAGATTGGAAAT	54	126	5-FAM	4	1.643	0.735	0.397	0.391	−0.015	0.356	0.343 ns
PS061	F: CTCCTCCAACATTGACCC R: CACCCTCCCAAACATCTC	57	154	5-FAM	7	2.178	0.901	0.532	0.541	0.017	0.436	0.000 ***
PS068	F: CTTTGGCATTCTCATTCA R: GGTGGTATTGGGCTTCTT	52.5	174	5-FAM	6	3.231	1.335	0.738	0.690	−0.068	0.641	0.000 ***
PS073	F: GTCGGTGAATGAAGGGTT R: ATTTCTGGTCAATGTGGC	53.5	269	5-HEX	5	2.066	0.780	0.574	0.516	−0.113	0.400	0.000 ***
PS074	F: TGCCTTGCTCCTCCTTGT R: CGGTTAGCCATGAATCCC	57	236	5-HEX	13	2.947	1.457	0.702	0.661	−0.063	0.621	0.000 ***
PS095	F: TCCCAAGACCTCAAACAAC R: CCATCAATACGAGCCAAC	55	394	5-TAMRA	9	5.394	1.857	0.819	0.815	−0.006	0.792	0.000 ***
PS119	F: GCAAAGACAACAGCCTCG R: CTCACCATCCAATCCCAC	57	289	5-HEX	4	3.627	1.331	0.745	0.724	−0.028	0.673	0.000 ***
PS139	F: CAACAATTTAACACGCAGAG R: GCCTTAGACGGAGACCAG	56.5	482	5-FAM	7	1.765	0.853	0.447	0.433	−0.031	0.391	0.000 ***
PS144	F: CAACCTACAATCCGACAATG R: CGACTTCCCTTCAATACA	54.5	317	5-TAMRA	6	1.378	0.553	0.273	0.274	0.005	0.252	0.000 ***
PS149	F: AGTCGCCTCCTACACCTC R: TCCGTAAAGCCCACAATAC	55.5	173	5-FAM	5	1.777	0.798	0.426	0.437	0.027	0.391	0.741 ns
PS157	F: CTCCCTGAACTCCCTACC R: CTTTCTAAACAGCCAACG	56	322	5-TAMRA	4	2.565	1.091	0.677	0.610	−0.110	0.551	0.000 ***
PS158	F: TTTCCCTGCTTCTTCTGAC R: CACCTCCTTCCTTTCTTACT	55	423	5-ROX	7	2.976	1.298	0.638	0.664	0.039	0.608	0.000 ***
PS159	F: CCTCCATTCATTCCTGTC R: GCAATAAATAGCCGTCCT	52	124	5-FAM	11	2.384	1.134	0.567	0.581	0.023	0.506	0.000 ***
PS166	F: TTCAGTGGGCAAGACCTAC R: TAGCCAATACAGAACAAACC	55	337	5-TAMRA	10	3.698	1.455	0.805	0.730	−0.103	0.683	0.000 ***
PS180	F: CCCCGAAATGGAGGAGTC R: AGGGCAGTAGCAGAAGAAAGTC	60	188	5-FAM	6	1.331	0.485	0.238	0.248	0.044	0.226	0.992 ns
PS187	F: AAGCGGCGTCCATCATAC R: TCACAAGCCCAACCCAGA	57	233	5-HEX	4	2.399	1.000	0.784	0.583	−0.344	0.498	0.000 ***

Table 2. *Cont.*

Locus	Primer Sequence	Tm (°C)	Expected Size (bp)	Modified Type	*Na*	*Ne*	*I*	*Ho*	*He*	F_{IS}	PIC	*p*
PS221	F: GATACAAGGCGGAAAGTG R: AGAGTTGGGAACCAGACC	56	301	5-TAMRA	5	2.488	1.007	0.823	0.598	−0.375	0.514	0.000 ***
PS260	F: ATTCACGCCAGTATCAAAG R: TGTAAATGCCCATGTCTAG	53	349	5-TAMRA	2	1.595	0.560	0.376	0.373	−0.007	0.304	0.905 ns
PS265	F: TTTTATGGGTCCTGTTGC R: GAAGAGTAAGCCTTTGTCG	54	290	5-TAMRA	4	1.897	0.759	0.564	0.473	−0.192	0.386	0.008 **
PS271	F: AGAATCCACCTCCTGTCAC R: AACCCTGCCCTAAACTAAAC	56.5	406	5-ROX	4	3.516	1.317	0.716	0.716	−0.001	0.664	0.903 ns
PS276	F: CTGTATCCTATCGGTTCTT R: CCTCATCTGCCTTTATCT	52.5	447	5-ROX	4	2.029	0.749	0.436	0.507	0.140	0.389	0.000 ns
PS296	F: CTCTTTCGCTGCCACAAC R: CTCTGCTCTTCCCGTCTT	57.5	419	5-ROX	4	1.985	0.937	0.521	0.496	−0.051	0.456	0.002 **
PS309	F: AAGCAAAGCCGTGGAGAT R: GTGCGTGAAAAGGAGACAGAAC	55	257	5-HEX	7	1.321	0.536	0.255	0.243	−0.051	0.230	0.000 ***
PS311	F: AACGCCACCATCACCTTT R: CACCTGAACTCACCCTCC	60	277	5-HEX	2	1.988	0.690	0.915	0.497	−0.841	0.374	0.000 ***
PS323	F: CTCACCCGTTCTAAAGTCA R: CCTCCTCCCTGTTCTTCT	53	466	5-FAM	2	1.004	0.013	0.004	0.004	−0.002	0.004	0.976 ns
PS335	F: TAATCACCCAATGAGCCA R: CGTCGTCGCCGAATACTT	50	395	5-ROX	5	1.172	0.340	0.096	0.147	0.347	0.141	0.000 ***
PS337	F: ATCCTCTTCACGGCAATC R: CGTCCACTCTTCCTCCTC	53.5	448	5-ROX	5	1.638	0.695	0.461	0.390	−0.183	0.338	0.013 *
PS339	F: TGAGGCAGCCAAAGAATT R: GGCAGGTGTAGGGTATGTT	50	175	5-FAM	3	1.025	0.074	0.021	0.025	0.134	0.024	0.000 ***
PS345	F: TGAAGTGAATCGAAGCAT R: CAACAGGCAGAAGAAAGG	48	366	5-ROX	4	2.078	0.944	0.589	0.519	−0.135	0.469	0.000 ***
PS356	F: TCAAGCCCAAGGTCATTC R: ACTTGCTCACCTCGCTCT	53	354	5-TAMRA	7	3.647	1.476	0.730	0.726	−0.006	0.683	0.000 ***
PS367	F: AGACGGACGGAAATAGGG R: ACGAGCGATCTCAACCAT	53.5	265	5-HEX	4	3.238	1.267	0.993	0.691	−0.437	0.636	0.000 ***
Mean					5.441	2.291	0.908	0.537	0.489	−0.074	0.611	
SE					0.429	0.166	0.072	0.043	0.035	0.034	0.358	

Note: F = forward primer, R = reverse primer, Tm = annealing temperature; ns = not significant, * $p < 0.05$, ** $p < 0.01$, *** $p < 0.001$.

2.4. Chloroplast DNA Sequences

Since the same chloroplast gene fragment has different evolutionary rates in different plants, the optimal chloroplast fragment should be selected for different species [46–48]. From the reported chloroplast fragments, three pairs of chloroplast gene spacer fragments, *pet*B-*pet*D, *acc*D-*psa*I and *psb*E-*pet*L, which can be amplified stably and have high polymorphism, were screened out on the basis of existing common primers [26,49]. Primer information is shown in Table 3. Three chloroplast DNA primers were selected to access the genetic diversity of 80 accessions that can represent the typical flower colors from 282 accessions. The PCR amplification reaction was conducted in a 50 μL solution, including: 25 μL 2× Power Taq PCR Master MIX (Aidlab Biotechnologies, Beijing, China), 2.5 μL and 10 μmol/L each of forward and reverse primer, 2 μL and 20–25 ng/μL genomic DNA and 18 μL ddH_2O. The PCR procedure was as follows: 5 min at 94 °C, followed by 35 cycles of 30 s at 94 °C, 30 s at 52–54 °C and 50 s at 72 °C, and lastly, 7 min at 72 °C. The amplified fragment results were detected by capillary electrophoresis using an ABI3730xl DNA Analyzer (Applied Biosystems, Carlsbad, CA USA).

Table 3. Information of 3 chloroplast DNA sequence (cpDNA) primers used in 80 accessions.

Number	Loci	Primer Pairs	Primer Sequences (5′-3′)	Tm (°C)
1	*pet*B-*pet*D	*petB* *petD*	CTATCGTCCRACCGTTACWGAGGCT CAAAYGGATAYGCAGGTTCACC	54
2	*acc*D-*psa*I	*accD-psaI*-21F *accD-psaI*-747R	AACATTGAATAAGACAGTACCTGAG GTAAGTTAAGAGTTGTCATAGGATGG	52
3	*psb*E-*pet*L	*psbE-petL*-356F *psbE-petL*-1219R	CCTTCTTCTGACACAGCAATG TTACCATTATAGACAGCACTAACAA	52

2.5. Statistical Analyses

2.5.1. Analysis of Phenotypic Data

The average values per plant were used in the statistical analysis and the data were analyzed using the statistics software SPSS version 18.0 (IBM Inc., Chicago, IL, USA) [50]. One-way analysis of variance (ANOVA; with post hoc Duncan' s multiple range test with a probability $p < 0.05$ and $p < 0.01$) was used to analyze the 28 traits' variation and Pearson's correlations between traits were calculated. Prior to the ANOVA and Pearson's correlation analysis, all data were tested for normality with the Shapiro–Wilk W test and for homogeneity of variance with the Levene's test, and the non-normal data were logarithmic transformed. We adjusted the *p*-value for Pearson's correlation using the Benjamini–Hochberg (BH) false discovery rate (FDR) correction. Coefficient of variation (%) = (standard deviation/mean) × 100. Principal component analysis (PCA) was performed for 28 traits. Maximum deviation method was used for factor rotation, and the principal components were extracted according to Kaiser criterion (characteristic root > 1). Then, with the first principal component (PC-1) and the second principal component (PC-2) as the main coordinates, the distribution diagrams of 28 quantitative traits and 180 accessions were plotted, respectively.

2.5.2. Analysis of EST-SSR Data

GeneMarker V2.2.0 was used to analyze the capillary electrophoresis data. GenAIEx version 6.5 [51] was used to calculate the following indicators: Number of Different Alleles (Na), Number of Effective Alleles (Ne), Shannon's Information Index (I), Observed Heterozygosity (Ho), Expected Heterozygosity (He), Inbreeding coefficients (FIS) and Nei's genetic diversity (GD). Polymorphism information content (PIC) was calculated for each locus by using a Microsatellite Toolkit. GENEPOP version 4.2 [52] was used to detect whether microsatellite loci deviated from the Hardy–Weinberg equilibrium (HWE). Additionally, a Neighbor-Joining phylogenetic tree based on Nei's unbiased genetic distance was constructed with MEGA-X [53]. Finally, Principal Coordinates Analysis (PCoA) was carried out using GenAIEx version 6.5 based on Nei's genetic distance of all accessions.

2.5.3. Analysis of Chloroplast DNA Sequences

MAFFT version 7 [54] was used to conduct multi-sequence alignment and adjustment. SequenceMatrix 1.7.8 [55] was used to splice three chloroplast non-coding regions, *accD-psaI*, *psbE-petL* and *petB-petD*, into a sequence. Values of haplotype diversity (Hd), haplotype number (Hap) and nucleotide diversity (Pi) were calculated by DnaSP 5.0 [56]. PopART [57] and the Median-Joining network (MJ) algorithm were used to construct the association map of the chloroplast DNA haplotypes. MEGA-X [53] was used to construct the phylogenetic tree of 3 haplotypes based on the maximum likelihood method (ML). Meanwhile, phylogenetic trees were constructed for the sequences of 80 accessions based on the Maximum Parsimony method (MP) with PAUP 4.0 [58]. The consistency index was 1.000000, the retention index was 1.000000, and the composite index was 1.000000 for all sites and parsimony-informative sites (in parentheses).

3. Results

3.1. Genetic Diversity Based on Phenotypic Traits

3.1.1. Phenotypic Traits Variation

ANOVA analysis on phenotypic variation showed that in addition to the number of flowers per plant, the other 27 quantitative traits showed very significant differences in 180 accessions ($p < 0.01$) (Table 4). Since all investigated plants were adult with stable characteristics growing in a consistent cultivation condition, it could be considered that the phenotypic variation was mainly due to genotypic differences. Descriptive statistical analysis was conducted at the whole population level, and it was found that the coefficient of variation of flower traits, stem and leaf traits and fruit traits ranged from 12.33% to 108.73%, with a very high degree of variation. Among the flower traits, the coefficient of variation ranged from 12.33–108.73%. The variation of petal number (108.73%) was very obvious, but the other flower traits varied weakly in various accessions, among which the variation coefficients of flower diameter (12.33%) and petal length (12.33%) were the smallest. As for the branch and leaf traits, the variation coefficients of tillers' number (90.53%), fruits (66.86%) and flowers (57.43%) per plant were relatively higher. The variation coefficient of the fruit traits ranged from 17.84% to 99.93% (average 48.50%). Among them, the variation coefficient of seed weight per fruit was the largest (99.93%), followed by seed number per fruit (94.30%). So, significant phenotypic differences of phenotypic variation in the FTP accessions provide valuable germplasm resources for selective breeding of tree peonies.

3.1.2. Distribution of Phenotypic Traits Variation in the Population

PCA on 28 quantitative traits showed that the cumulative contribution rate of 8 principal components reached 70.516% (Table 5). The explanatory contribution rate of the first principal component was 17.722%. The traits that determined the first principal component were, in order, seed weight per fruit, individual fruit weight, number of seeds per fruit, fruit width, fruit height, fruit length and fruit weight per plant, all of which were fruit traits. Therefore, the first principal component represented the variation of fruit traits, which can be defined as the fruit character factor. The traits that determined the second principal component were fruit number, flower number, east-west crown breadth, north-south crown breadth and fruit weight per plant. The third principal component mainly consisted of petal length, petal width, flower diameter, flare length and flare width, all of which were floral traits, which can be defined as the flower character factor. The fourth principal component consisted of compound leaf length, petiole length, compound leaf width, terminal leaflet length and terminal leaflet width, all of which were stem and leaf traits. Length and width of the terminal leaflet majorly affected the character of the fifth principal component. The fourth and fifth principal components can be defined as the branch and leaf character factor. The PCA of quantitative traits explained the variation distribution of phenotypic traits in the whole population.

Table 4. Descriptive statistics for quantitative traits measured in 180 accessions.

Trait	Mean	Standard Deviation	Minimum	Maximum	F-Value	p-Value	Variable Coefficient
1	7.54	0.93	4.80	10.07	15.96	0.00	12.33
2	7.41	1.19	2.73	9.80	18.11	0.00	16.06
3	2.71	0.63	1.57	4.37	16.09	0.00	23.25
4	1.97	0.58	0.67	3.77	18.43	0.00	29.44
5	15.41	1.90	9.67	19.63	6.97	0.00	12.33
6	32.55	35.39	9.00	238.33	22.72	0.00	108.73
7	5.28	1.05	0.00	10.33	8.08	0.00	19.89
8	125.92	22.56	71.33	206.67	7.42	0.00	17.92
9	149.02	32.42	78.67	276.00	4.15	0.00	21.76
10	163.84	34.41	75.33	277.00	5.37	0.00	21.00
11	28.24	18.88	0.00	97.33	2.99	0.00	66.86
12	32.04	18.40	4.33	86.00	1.56	0.08	57.43
13	21.02	19.03	1.00	169.33	7.71	0.00	90.53
14	36.98	6.24	20.67	55.84	9.46	0.00	16.87
15	23.13	5.57	11.67	45.07	7.26	0.00	24.08
16	12.00	2.52	6.97	21.57	9.96	0.00	21.00
17	15.15	3.41	9.00	31.00	14.87	0.00	22.51
18	6.95	1.46	3.53	11.13	10.99	0.00	21.01
19	7.06	2.07	3.13	15.30	5.21	0.00	29.32
20	478.26	408.66	25.33	2661.64	4.95	0.00	85.45
21	62.26	31.57	16.89	237.01	3.75	0.00	50.71
22	33.66	31.74	0.00	151.33	4.15	0.00	94.30
23	15.01	15.00	0.00	79.67	3.32	0.00	99.93
24	5.49	1.20	2.33	12.33	19.08	0.00	21.86
25	40.75	7.27	23.74	75.64	11.29	0.00	17.84
26	15.43	3.27	7.60	25.24	10.12	0.00	21.19
27	14.79	2.91	6.97	29.07	10.73	0.00	19.68
28	2.51	0.64	0.79	4.52	12.65	0.00	25.50

Note: $p < 0.05$: Significant difference; $p < 0.01$: Very significant difference. The traits corresponding to the serial number are shown in Table 1.

In addition, among the principal coordinate distributions of all quantitative traits, we found that fruit traits were distributed along the horizontal axis, stem and leaf traits along the vertical axis, and flower traits were distributed near the origin (Figure 3a). The 180 accessions were evenly distributed on the coordinate axes composed of the first principal component and the second principal component (Figure 3b), indicating that the variation of these phenotypic traits was evenly distributed in the population and no obvious hierarchical structure was formed.

Table 5. The principal component analysis (PCA) of 28 quantitative traits.

Trait	Principal Component							
	1	2	3	4	5	6	7	8
1	−0.019	−0.068	0.831	0.150	0.076	−0.065	−0.064	−0.017
2	0.100	0.019	0.776	0.043	0.134	0.153	−0.277	0.148
3	−0.025	−0.044	0.669	−0.164	−0.204	0.238	0.165	−0.273
4	0.019	−0.010	0.543	−0.182	−0.030	0.525	0.125	−0.138
5	0.211	0.063	0.688	0.146	0.040	−0.211	−0.133	0.163
6	−0.293	0.053	−0.196	0.065	0.027	−0.077	0.633	0.286
7	−0.047	−0.007	0.167	−0.069	−0.053	0.001	−0.629	0.413
8	0.180	0.453	0.129	0.377	0.068	0.379	0.203	0.222
9	0.203	0.669	0.184	0.083	0.084	0.006	0.398	0.176
10	0.163	0.641	0.342	0.144	0.103	0.030	0.308	0.092
11	−0.053	0.839	−0.153	−0.001	−0.057	−0.160	−0.074	−0.136
12	−0.076	0.796	−0.144	−0.073	0.054	−0.017	−0.126	0.042
13	−0.005	0.352	0.077	0.026	−0.144	−0.684	0.140	−0.099
14	0.237	0.053	0.059	0.866	0.120	0.006	−0.012	−0.036
15	0.240	0.098	0.160	0.569	−0.029	0.466	−0.042	−0.060
16	0.110	−0.029	−0.018	0.842	0.043	−0.175	0.129	−0.038
17	0.181	0.153	0.071	0.206	−0.675	0.193	−0.137	0.073
18	0.116	0.156	0.079	0.438	0.716	0.179	0.083	−0.049
19	0.169	0.279	−0.015	0.315	0.494	0.251	−0.070	−0.165
20	0.562	0.633	−0.031	0.133	−0.018	−0.193	−0.102	−0.144
21	0.904	0.063	0.074	0.249	0.115	0.032	−0.050	0.017
22	0.846	−0.024	−0.032	0.047	−0.247	−0.033	−0.196	−0.066
23	0.909	0.033	−0.003	0.105	−0.130	−0.031	−0.132	−0.022
24	−0.070	0.011	0.001	−0.077	−0.078	0.030	0.009	0.795
25	0.648	0.060	0.160	0.287	0.311	0.107	0.210	−0.128
26	0.842	0.076	0.077	0.015	0.117	0.086	0.072	0.034
27	0.730	0.059	0.088	0.136	0.297	0.085	0.080	−0.032
28	0.408	0.036	0.144	0.062	0.566	0.105	−0.106	0.091
Total variance/%	17.722	11.066	10.356	9.328	7.026	5.387	5.207	4.423
Cumulative total variance/%	17.722	28.789	39.144	48.472	55.499	60.886	66.093	70.516

Note: The traits corresponding to the serial number are shown in Table 1.

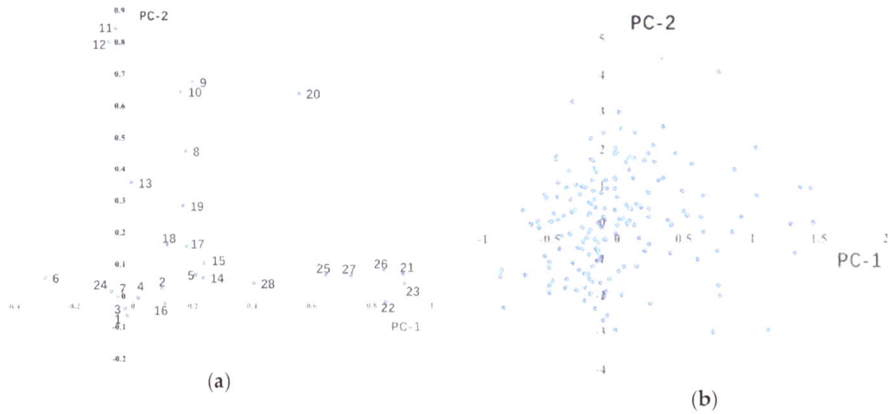

Figure 3. Principal component (PC) coordinate distribution of traits and accessions: (**a**) Distribution of 28 quantitative traits of the population on the PC-1 and PC-2 axis. The traits corresponding to the serial number are shown in Table 1. (**b**) Distribution of 180 accessions on the PC-1 and PC-2 axis.

3.1.3. Correlation Analysis of Quantitative Traits

Correlation analysis on 28 quantitative traits showed that there were different degrees of correlation among all traits (Table 6). Among the 378 pairs of combinations, 205 pairs of all traits showed significant correlation ($p < 0.05$), among which 191 pairs showed very significant correlation ($p < 0.01$). In tree peony breeding for ornamental uses, we mainly focused on the improvement of the flower diameter and the petal number. There were 17 traits that were very significantly positively correlated with flower diameter, including all floral traits except petal numbers and all fruit traits except effective carpel number. Petal number was negatively correlated with flower diameter, and only positively correlated with effective carpel number. This indicated that the traits of flower and fruit were closely related and the two ornamental traits we want to improve are opposite. For the breeding of oil tree peony, we need accessions that have more fruit number per plant and more seed weight per fruit. In Table 6, we can find that the trait that was very significantly positively correlated with these two traits was fruit weight per plant. For conventional breeders, it is not necessary to peel off the pericarp to weigh the seeds, but simply to weigh the fruit weight per plant to make a preliminary selection of high seed yield accessions.

3.2. Genetic Diversity Based on the EST-SSR

3.2.1. EST-SSR Polymorphism

Amplification of 282 accessions using 34 SSR pairs generated 185 alleles and the number of alleles detected per locus was in a range of two to thirteen, with an average of 5.441 alleles (Table 2). Observed heterozygosity (Ho) and expected heterozygosity (He) ranged from 0.004 to 0.993 and 0.004 to 0.815, with means of 0.537 and 0.489, respectively. Among all 34 SSR loci, PS004, PS026, PS047, PS068, PS073, PS074, PS095, PS119, PS139, PS157, PS166, PS187, PS221, PS260, PS265, PS296, PS309, PS311, PS337, PS345, PS356 and PS367 had lower He than Ho. Overall, the mean of Ho was higher than that of He, indicating that there was a heterozygote excess in these accessions. The effective number of alleles (Ne) was 2.291 ± 0.166. Shannon diversity index (I) was 0.908 ± 0.072, and the range of PIC at 34 SSR pairs of 282 accessions was 0.004~0.792. According to the Shannon's Information Index, the genetic diversity of PS095 (I = 1.857) was the highest, followed by primers PS074 (I = 1.457), PS166 (I = 1.455) and PS356 (I = 1.476). Primer PS323 has the lowest genetic diversity. The mean value of SSR marker polymorphism information content (PIC) in this study was 0.611. In addition, 24 SSR sites deviated significantly from the HWE ($p < 0.001$), and 2 SSR sites deviated significantly from the HWE ($p < 0.01$).

3.2.2. Genetic Relationships among 282 Accessions

It was found that 282 accessions were firstly divided into 3 major branches (Figure 4a), comprising 175, 87 and 20 accessions. Branch I; included 61 white, 25 pink, 21 red, 59 purplish red and 9 purplish black accessions, branch II included 23 white, 8 pink, 9 red, 44 purplish red and 3 purplish black accessions, and branch III included 5 white, 2 red and 13 purplish red accessions. The accessions of each color were relatively evenly distributed on each cluster, which showed the same results as the PCoA analysis (Figure 4b) and showed no regular distribution. Some accessions with similar traits have a relatively long genetic distance, while some with great phenotypic differences have relatively short genetic distances. For example, accessions '73' and '74', both of which had single white flowers, were very distantly related, while accessions '40' and '75', one purplish red double and one white single, were particularly closely related. This also indicated that there was no direct relationship between the relative distance and the color of the accessions.

Table 6. Pearson correlation coefficients of phenotypic traits of 180 accessions.

Trait	1	2	3	4	5	6	7	8	9	10	11	12	13	14	15	16	17	18	19	20	21	22	23	24	25	26	27	28
1																												
2	0.657 **																											
3	0.505 **	0.294 **																										
4	0.354 **	0.469 **	0.655 **																									
5	0.519 **	0.539 **	0.309 **	0.272 **																								
6	−0.167 *	−0.287 **	−0.081	−0.122	0.170 **																							
7	0.102	0.220 **	0.163 *	0.115	0.148 **	−0.111 *																						
8	0.135 *	0.232 **	0.034	0.180 **	0.146 **	0.016	0.048																					
9	0.160 *	0.126	0.062	0.052	0.117 *	0.049	0.013	0.577 **																				
10	0.192 **	0.146 *	0.096	0.095	0.165 **	0.031	0.050	0.652 **	0.708 **																			
11	−0.119	−0.041	−0.123	−0.058	0.010	0.018	−0.046	0.176 **	0.291 **	0.281 **																		
12	−0.134	−0.021	−0.171	−0.087	0.008	0.056	−0.013	0.284 **	0.317 **	0.357 **	0.581 **																	
13	−0.037	−0.118	−0.108	−0.191 **	−0.044	0.001	0.105	0.106	0.203 **	0.124 *	0.167 **	0.199																
14	0.167 **	0.115 **	−0.008	0.039	0.160 **	0.027	0.014	0.211 **	0.150 **	0.212 **	−0.025	−0.067	−0.005															
15	0.164 **	0.191 **	0.097 *	0.160 **	0.143 **	0.030	0.015	0.345 **	0.241 **	0.265 **	−0.022	−0.034	−0.065	0.570 **														
16	0.102 **	−0.032	0.046	−0.084*	0.044	0.027	−0.091	0.056	0.049	0.081	−0.021	0.046	0.089	0.723 **	0.310 **													
17	0.048	0.050	−0.017	−0.024	0.081	0.162 **	0.066	0.165 **	0.154 **	0.144 **	0.106	0.057	0.108	0.043	0.054 *	−0.056 *												
18	0.114 **	0.110 **	0.040	0.131 **	0.084	0.145 *	−0.036	0.218 * *	0.143 **	0.181 **	−0.001	0.078	−0.086	0.532 **	0.389 **	0.310 **	−0.327 **											
19	0.054	0.121 **	0.028	0.11 2**	0.133 *	0.026	−0.032	0.222 **	0.197 **	0.143 **	0.130 *	0.162	−0.038	0.357 **	0.298 **	0.185 **	−0.164 **	0.538 **										
20	0.029	0.109	0.067	0.007	0.164 **	−0.181 **	−0.076	0.229 **	0.345 **	0.310 **	0.570 **	0.381 **	0.059	0.237 **	0.163 **	0.144 **	0.099	0.107 *	0.162 **									
21	0.147 *	0.207 **	0.021	0.111	0.266 **	−0.206 **	−0.085	0.242 **	0.181 **	0.198 **	0.057	0.017	−0.082	0.402 **	0.333 **	0.224 **	0.044	0.239 **	0.199 **	0.615 **								
22	0.023	0.040	−0.034	0.008	0.155 **	−0.211 **	−0.096	0.005	0.006	−0.038	0.081	0.000	0.038	0.174 **	0.191 **	0.106 *	0.133 *	0.048	0.019	0.467 **	0.696 **							
23	0.061	0.094	−0.031	0.027	0.185 **	−0.199 **	−0.101	0.070	0.064	0.031	0.074	0.036	0.019	0.262 **	0.231 **	0.148 **	0.093	0.059	0.085	0.541 **	0.817 **	0.952 **						
24	0.085 *	0.115 **	0.114 *	0.082	0.029	0.167 **	0.397**	0.053	0.025	−0.014	−0.042	0.088	−0.069	0.003	0.014	−0.027	−0.017	0.029	0.018	−0.007	−0.003	−0.094 *	−0.079					
25	0.197 **	0.182 **	0.059	0.154 **	0.186 **	0.073	−0.163 **	0.254 **	0.217 **	0.232 **	0.011	−0.020	−0.066	0.316 **	0.279 **	0.152 **	−0.009	0.258 **	0.171 **	0.359 **	0.675 **	0.367 **	0.467 **	0.160 **				
26	0.161 **	0.204 **	−0.012	0.115 **	0.165 **	−0.185 * *	−0.086	0.182 **	0.137 **	0.176 **	0.064	−0.016	−0.127*	0.248 **	0.207 **	0.115 **	0.082**	0.122 **	0.135 **	0.453 **	0.684 **	0.463 **	0.534 **	−0.108 **	0.652 **			
27	0.178 **	0.212 **	0.007	0.124 **	0.184 **	−0.171 **	−0.115	0.184 **	0.119 *	0.173 **	−0.010	0.006	0.189 **	0.251 **	0.201 **	0.121 **	0.052	0.153 **	0.147 **	0.394 **	0.674 **	0.358 **	0.463 **	−0.151 **	0.625 **	0.718 **		
28	0.220 **	0.286 **	0.042	0.197 **	0.193 **	0.149 **	0.009	0.136 **	0.130 **	0.183 **	−0.063	−0.025	0.159 **	0.263 **	0.189 **	0.081 **	−0.031	0.231**	0.178 **	0.285 **	0.502 **	0.087 *	0.199 **	0.001	0.450 **	0.518 **	0.515 **	1

Note: ns = not significant, * $p < 0.05$, ** $p < 0.01$. The traits corresponding to the serial number are shown in Table 1.

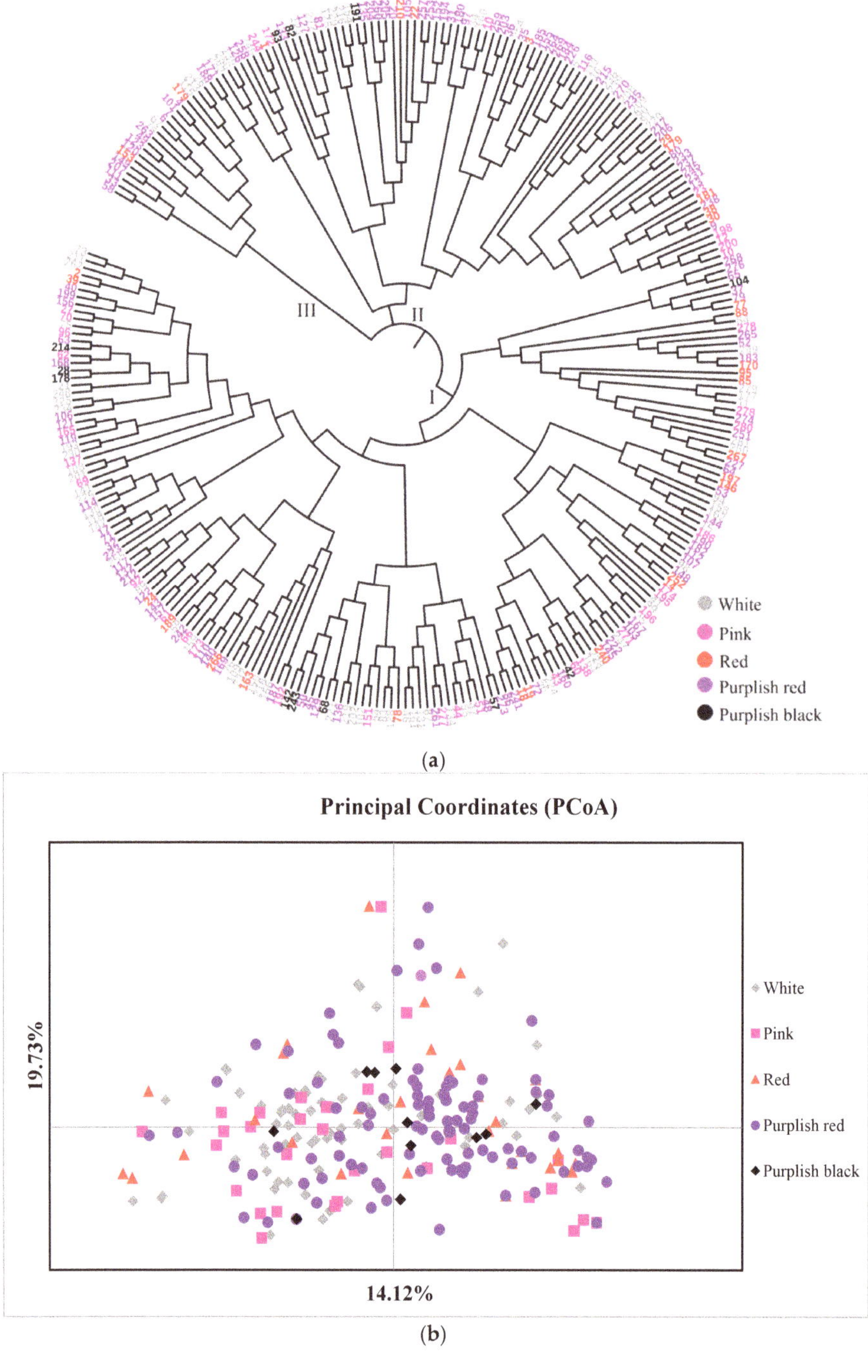

Figure 4. Relationships among 282 accessions: (**a**) Neighbor-Joining (NJ) phylogenetic tree based on the data of 34 EST-SSR markers, (**b**) Principal coordinates analysis (PCoA) based on Nei's unbiased genetic distance.

3.3. Genetic Diversity Based on the cpDNA

The *acc*D-*psa*I, *psb*E-*pet*L and *pet*B-*pet*D fragments of three chloroplast non-coding regions were used to detect 80 accessions. After combination, the total length of the three fragments was 1580 bp. A total of 4 mutation sites were detected in the combined fragment, all of which were single base mutations and corresponding to three haplotypes. The *pet*B-*pet*D fragment had 2 variation loci, the other two fragments were each one, respectively (Table 7). According to the calculation results of the three combined fragments by DnaSP 5.0, the total Hd was 0.164 and Pi was 0.28×10^{-3}.

Table 7. The variations in *acc*D-*psa*I, *psb*E-*pet*L and *pet*B-*pet*D regions among 80 accessions.

Domain	Variation Site	Hap 1	Hap 2	Hap 3
*acc*D-*psa*I	1	A	C	C
*psb*E-*pet*L	2	G	A	A
*pet*B-*pet*D	3	T	C	C
	4	G	G	A

Note: A, T, C and G are bases, respectively.

In the haplotype spectrum diagram (Figure 5a), Haplotype 1 was located in the center of the network, which was an obvious widespread haplotype. It appeared in 73 accessions, which may be the original haplotype. Haplotype 2 and 3 contained 6 and 1 accessions, which were '98', '120', '140', '146', '150', '187' and '128', respectively. According to Figure 5a, Haplotype 1 was closer to Haplotype 3 than Haplotype 2, showing the same result with phylogenetic relationship of 3 haplotypes (Figure 5b). Three haplotypes were clustered into two branches in the tree, and Haplotype 2 formed a single branch.

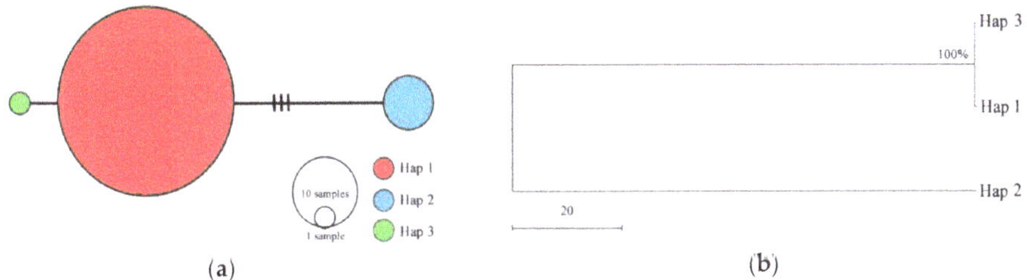

Figure 5. Analysis of cpDNA haplotypes: (**a**) Network of 3 cpDNA haplotypes (Hap 1–Hap 3), (**b**) Phylogenetic relationship of 3 haplotypes based on the Maximum Parsimony (MP) method analysis.

Meanwhile, phylogenetic trees were constructed for the sequences of 80 accessions. The accessions were divided into two distinct branches (Figure 6). The first branch consisted of 6 germplasms of Haplotype 2, including '98', '120', '140', '146', '150' and '187', and the second branch consisted of 74 accessions of Haplotype 1 and Haplotype 3, which also indicated that the phylogenetic relationship of Haplotype 1 and 3 were closer than that of Haplotype 2.

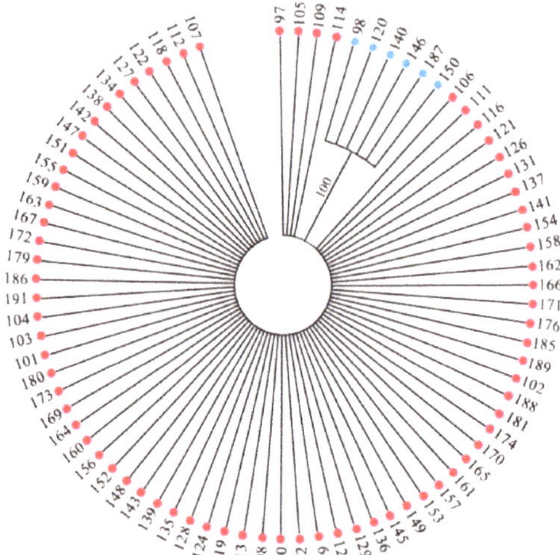

Figure 6. Strict consensus tree of 80 accessions using the Maximum Parsimony (MP) method (the numbers in the figure represent the number of the accessions).

4. Discussion

4.1. Diversity of Phenotypic Traits

Because the phenotype is the result of the interaction of genotype and external environment and the diversity of phenotypic traits is the manifestation of genotype difference in morphology, the phenotypic variation in the population mainly depends on the genotype difference when all accessions used in this study were grown in the same condition. In FTP, Pang et al. [59] reported phenotypic variation coefficients ranging from 10% to 30% based on 32 traits of 150 accessions, but Wu et al. [60], on the diversity of 29 quantitative traits of 462 accessions, found that the coefficient of genetic variation was 9.52–112.1%. By the genetic diversity of 28 quantitative traits of 180 accessions, this study revealed a genetic variation coefficient of 12.33–108.73%, which was similar to Wu et al. [60]. As the plants used both by Wu et al. and by this study are widely selected and collected from the main cultivation area of the FTP, Gansu province (Figure 2), we felt that the accession population of this study should be large enough to represent the overall situation of FTP as a germplasm resource. Moreover, each accession collected in this study has been clonally propagated and most of them were named as the cultivars (Supplementary Table S1) and can be released into the industry as crop germplasm resources.

Studies on phenotypic variation can not only improve our understanding of the biological basis of plants, but also have very important breeding value [61]. In FTP used in this study, the fact that the number of petals varied from 9 petals to 238.33 petals (Table 4) indicates that these accessions have diverse flower types, from single through to semi-double to double [1]. Such various flower types are really valuable for breeding new cultivars for ornamental plants like tree peonies. Moreover, the tree peonies have been recently developed rapidly as an emerging oil crop and the selection of high-yield accessions has become very crucial for increasing cultivation. Among the accessions used in this study, some of them with single flower, in most cases, are fertile for the formation of fruits (seeds) and can be screened as the high-yield oil germplasms, in which the maximum fruit weight per plant was up to 2661.64 g (Table 4).

4.2. Genetic Diversity Based on SSR Markers

The primary task of evaluating all kinds of samples in germplasm resources is to determine their genetic constitution in the repository. We can also compare newly collected cultivated or wild plants from various locations with existing germplasm accessions to facilitate identification. With the advent of molecular technologies and methods, it has become more efficient and scientific to use genetic information at the genome level to analyze accession differences in genetic resources and compare these samples with germplasm collections elsewhere [62].

This study detected a total of 185 alleles in 34 pairs of EST-SSR primers in 282 FTP accessions, and the average number of different alleles (Na) was 5.441 (Table 2), which was smaller than in 20 populations (Na = 9.15) of 335 *P. rockii* wild individuals [29], but larger than in 40 pairs of primers in 462 *P. rockii* seedlings (Na = 4.5) [38]. Such differences might be caused by the different primers used in the different studies, but it was very clear that the wild germplasm resources of *P. rockii* do have more abundant allele variation than that of cultivated FTP. The average inbreeding coefficients (FIS) was negative, and the expected heterozygosity (He) of 22 pairs of primers was lower than the observed heterozygosity (Ho) (Table 2), indicating that there was a surplus of heterozygotes in this FTP population [29]. Yuan et al. [63] believed that the higher observed heterozygote meant higher diversity of the population, as their results showed that the wild *P. rockii* (Ho = 0.475) has significant diversity compared to CTP (Ho = 0.61) and FTP (Ho = 0.58). In FTP used in this study, the observed heterozygote was 0.537, which was larger in wild *P. rockii* (Ho = 0.475) but smaller in CTP (Ho = 0.61), and closer to FTP of Yuan et al. (Ho = 0.58). This study by Neighbor-joining tree based on microsatellite analysis identified the 282 FTP accessions into three distinct groups and there was no direct relationship between the genetic distance and the flower color. The grouping was consistent with the studies on 335 wild *P. rockii* individuals of 20 populations and 462 *P. rockii* seedlings, which were also divided into three groups [29,38]. Guo et al. [64] used SRAP to classify 16 accessions of *P. suffruticosa*, which also indicated that the genetic distance had no correlation with the color: white, green, yellow and black accessions formed a group, while blue, pink and/or multi-colored flowers formed another group.

To sum up, SSR data showed that the number of alleles for FTP used in this study was less than that of wild *P. rockii*, but more than that of cultivated *P. rockii* seedlings. The heterozygosity of FTP is less than that of CTP, larger than that of wild *P. rockii*, and close to that of cultivated *P. rockii* seedlings. This also showed that wild plants have abundant allele variation but low population heterozygosity, and cultivated germplasm have less allele variation but high population heterozygosity, which just indicates that gene exchange is more frequent and more heterozygosity is retained artificially under cultivation conditions [65]. Most woody perennial species are obligately outcrossing, resulting in high heterozygosity [66]. The longer the culture history of tree peony, the higher the heterozygous, and these individuals maintained high heterozygosity by asexual reproduction or clonal propagation under cultivation conditions. The FTP accessions were divided into three groups, which were consistent with the wild *P. rockii* and cultivated *P. rockii* seedlings. So, we speculate that the germplasm resources of FTP in this study, to a large extent, can represent the genetic background information of species tree peony *P. rockii* germplasm.

4.3. Genetic Diversity Based on the cpDNA

According to Yuan et al. [28], the mean nucleotide haplotype diversity (Hd) and the mean nucleotide diversity (Pi) of the chloroplast (cpDNA) in 335 *P. rockii* wild individuals at the population level was 0.0686 and 0.765×10^{-4} respectively, while Hd and Pi at the species level was 0.887 and 0.185×10^{-2}, which was significantly higher than population level. Based on the analysis of three cpDNA sequences, Xu et al. [37] measured the diversity of 214 individuals in 24 populations of *P. rockii* and obtained the same results as Hd and Pi at the species level, 0.874 and 0.294×10^{-2}, which were significantly higher than Hd (0.1097) and Pi (0.383×10^{-4}) at the population level. This shows that wild *P. rockii* has a low genetic diversity at the population level and a higher genetic diversity at the species level. This situation was consistent with other tree peony species, *P. delavayi* [33], *P. ludlowii* [33], *P. qiui* [37] and *P. jishanensis* [37]. However, Hd and Pi in FTP accessions in this study were 0.164 and 0.28×10^{-3}, which were higher than the population level and lower than the species level in populations of wild *P. rockii*. Considering these accessions as a population to compare with wild

P. rockii, we found that the genetic diversity of the FTP under cultivation conditions is really different from its ancestor species in the habitats in forming genetic diversity or evolution. This is probably because wild plants of *P. rockii* reproduce by seeds in the same population, while gene exchange does not occur due to geographical isolation between populations. The chloroplast diversity of all individuals in the same population was lower, while it was higher at the species level because each population has distinct haplotypes. In this study, the accessions of FTP were selected and collected from various places of Gansu province without geographic isolation, and gene exchange may have occurred during the formation of germplasms through repeated hybridization, which finally resulted in a higher chloroplast diversity at the population level after long cultivation.

Meanwhile, based on the results of phylogenetic tree of chloroplast gene fragments (Figure 6) and haplotype spectrum (Figure 5b) analysis, the existing spatial genetic structure of the FTP can be divided into two branches, which was consistent with Yuan et al. [27] using chloroplast fragment to divide the genetic structure of wild *P. rockii* into two branches.

4.4. Conversation and Utilization of FTP Germplasm Accessions

Research on genetic diversity and structure is the theoretical basis for the protection and utilization of species. One of the goals of species protection is to maximize the conservation of genetic diversity of the species [67,68]. The accessions we investigated have abundant phenotypic variations and represent FTP germplasm, from which we have selected to name some cultivars for ornamental oil uses. The next step is to establish multi-site repositories and a multi-year, comprehensive phenotypic database of FTP germplasm resources. Germplasm accessions are crucial resources for studying crucial characteristics like plant resistance and flowering time through multi-year and multi-site observations. Similarly, using these germplasms to study phenotypic variation may be significant to researching phenotypic limberness as they react to changing climatic and different living environments.

With the results from SSR and cpDNA analysis, the FTP accessions can represent the genetic background information of *P. rockii* germplasm resources and have relatively high genetic diversity. In order to protect this diversity, a germplasm resource bank should be established to conserve as many accessions as possible. Analysis of chloroplast sequence variation showed that three haplotypes were found in 80 accessions. One haplotype consists of 73 accessions, but the other two haplotypes only consist of six and one, respectively. This suggests that the presence of seven accessions is closely related to the genetic diversity of *P. rockii* germplasms, and their disappearance will bring about an irreversible loss of diversity. Therefore, special protection should be given to these unique accessions.

Due to the rapid change of global climate caused by human activities, the selection pressure of woody ornamental plant commercialization is increasing, and the diversity of existing germplasm resources has more important breeding value [69]. As a kind of high-quality germplasm resource with strong resistance, high ornamental value and great potential for oil use, *P. rockii* accessions play an important role in the improvement of tree peony. At the level of DNA, this cultivated *P. rockii* population we have established can represent the basic level of *P. rockii* germplasm resources. In terms of phenotypic diversity, the variation of each trait is also extremely rich. Therefore, the value of these accessions is especially obvious for breed improvement and development of tree peonies. Meanwhile, the germplasm resources contain various plant genetic resources, which is of great significance to the basic research of plant biology. In order to explore the potential excellent characters within these accessions, we are required to continue to explore the phenotypes, genotypes and preservation of these resources. In this study, we planted these accessions from the traditional cultivation area of Gansu province to Beijing. It is necessary to continuously observe and measure these characters to explore the adaptability of the FTP to the different environment. Ultimately, germplasm collections should not simply be viewed as tools for conservation, but as resources with unique benefits that hold immense value for sustaining the future of woody perennials and their wild relatives. The comprehensive evaluation and utilization of these cultivated accessions is to protect the germplasm of *P. rockii* tree peony and its derivatives.

5. Conclusions

This study is the first to analyze the genetic diversity of FTP accessions newly selected by us using phenotypic traits, EST-SSR markers and chloroplast DNA sequences. Phenotypic data indicated that these accessions have abundant phenotypic variations and are representative of woody peony *P. rockii* germplasm. SSR data showed that the FTP accessions used in this study had relatively high genetic diversity and were composed of three distinct groups, like their wild relatives. The cpDNA data indicated that the genetic diversity of chloroplast genome in the FTP accessions had a higher genetic diversity than the population level and a lower genetic diversity than the species level of wild *P. rockii*, and they can be divided into two branches based on this. The FTP accessions fully reflected the genetic diversity and current situation of species tree peony *P. rockii* germplasm to a large extent, so they can be regarded as the core germplasm resources and their protection and utilization will be of great significance both for genetic improvement and biological studies of tree peonies and for sustaining development of the peony industry.

Supplementary Materials: The following are available online at http://www.mdpi.com/1999-4907/11/6/672/s1: Table S1: The numbers, names, flower colors and grouping results of 282 accessions.

Author Contributions: Conceptualization, X.G. and F.C.; validation, X.G. and F.C.; formal analysis, X.G.; investigation, X.G.; data curation, X.G.; writing—original draft preparation, X.G.; writing—review and editing, X.G., F.C. and Y.Z.; visualization, X.G.; supervision, F.C.; project administration, F.C. All authors have read and agreed to the published version of the manuscript.

Funding: This research was funded by the National High Technology Research and Development Program of China, grant number 2011AA100207 and the Science and Technology Project of Beijing, grant number Z181100002518001.

Acknowledgments: We would like to thank Xinyun Cheng and Xiwen Tao for their efforts in maintaining living plant materials for this study.

Conflicts of Interest: The authors declare no conflict of interest. In this study, Beijing Guose Peony Technologies Co., Ltd. carried out the collection and management of plant materials.

References

1. Cheng, F.Y.; Li, J.J.; Chen, D.Z.; Zhang, Z.S. *Chinese Paeonia rockii*; Chinese Forestry Publishing House: Beijing, China, 2005; pp. 20–30.
2. Han, P.; Ruan, C.J.; Ding, J.; Wu, B.; Zhang, W.C.; Ruan, D.; Xiong, C.W.; Liu, W.H.; Wang, G.H. Polygene regulation of high accumulation of carbon 18 unsaturated fatty acids in seeds of *Paeonia suffruticosa*. *Mol. Plant Breed.* **2019**, *17*, 2101–2108. (In Chinese)
3. Cheng, F.Y. Advances in the breeding of tree peonies and a cultivar system for the cultivar group. *Int. J. Plant Breed.* **2007**, *1*, 90–104.
4. Cheng, F.Y.; Yu, X.N. Flare tree peonies (*Paeonia rockii* hybrids) and the origin of the cultivar group. *Acta Horticulturae* **2008**, *766*, 375–382. [CrossRef]
5. Li, J.J. *Chinese Tree Peony (Xibei, Xinan, Jiangnan Volume)*; Chinese Forestry Publishing House: Beijing, China, 2005; pp. 73–82.
6. Zhang, Q.Y.; Niu, L.X.; Yu, R.; Zhang, X.X.; Bai, Z.Z.; Duan, K.; Gao, Q.H.; Zhang, Y.L. Cloning, characterization, and expression analysis of a gene encoding a putative lysophosphatidic acid acyltransferase from seeds of *Paeonia rockii*. *Appl. Biochem. Biotech.* **2017**, *182*, 721–741. [CrossRef]
7. Shi, Q.Q.; Li, L.; Zhang, X.X.; Luo, J.R.; Li, X.; Zhai, L.J.; He, L.X.; Zhang, Y.L. Biochemical and Comparative Transcriptomic Analyses Identify Candidate Genes Related to Variegation Formation in *Paeonia rockii*. *Molecules* **2017**, *22*, 1364. [CrossRef]
8. Zhang, Q.Y.; Yu, R.; Sun, D.Y.; Bai, Z.Z.; Li, H.; Xue, L.; Zhang, Y.L.; Niu, L.X. PrLPAAT4, a putative lysophosphatidic acid acyltransferase from *Paeonia rockii*, plays an important role in seed fatty acid biosynthesis. *Molecules* **2017**, *22*, 1694. [CrossRef]
9. Cui, H.L.; Chen, C.R.; Huang, N.Z.; Cheng, F.Y. Association analysis of yield, oil and fatty acid content, and main phenotypic traits in *Paeonia rockii* as an oil crop. *J. Hortic. Sci. Biotechnol.* **2017**, *93*, 425–432. [CrossRef]

10. Shi, Q.Q.; Zhang, X.X.; Li, X.; Zhai, L.J.; Luo, X.N.; Luo, J.R.; He, L.X.; Zhang, Y.L.; Li, L. Identification of microRNAs and their targets involved in *Paeonia rockii* petal variegation using high-throughput sequencing. *J. Am. Soc. Hortic. Sci.* **2019**, *144*, 118–129. [CrossRef]
11. Liu, N.; Cheng, F.Y.; Zhong, Y.; Guo, X. Comparative transcriptome and coexpression network analysis of carpel quantitative variation in *Paeonia rockii*. *BMC Genomics* **2019**, *20*, 683. [CrossRef]
12. Soltis, P.S.; Soltis, D.E. Genetic variation in endemic and widespread plant species. *Aliso J. Syst. Evolut. Bot.* **1991**, *13*, 215–223. [CrossRef]
13. Wang, X.Q. Studies on Genetic Diversity of *Paeonia delavayi* in Shangri-la. Ph.D. Thesis, Beijing Forestry University, Beijing, China, 2009. Available online: https://kns.cnki.net/KCMS/detail/detail.aspx?&dbcode=CDFDdbname=CDFD0911&filename=2009134629.nh&v=MTA1MDExMjdGN0s3R3RmT3BwRWJQSVI4ZVgxTHV4WVM3RGgxVDNxVHJXTTFGckNVUjdxZlplUnFGQ25sVzd2S1Y= (accessed on 12 June 2020). (In Chinese).
14. Wang, J. Genetic Diversity of *Paeonia ostii* and Germplasm Resources of Tree Peony Cultivars from Chinese Jiangnan Area. Ph.D. Thesis, Beijing Forestry University, Beijing, China, 2009. (In Chinese).
15. Peng, L.P.; Cheng, F.Y.; Zhong, Y.; Xu, X.X.; Xian, H.L. Phenotypic variation in cultivar populations of *Paeonia ostii*. *Plant Sci. J.* **2018**, *36*, 170–180. (In Chinese)
16. Li, Z.Y.; Zhang, H.Y. Morphological variation and diversity in populations of *Paeonia lutea*. *J. Northwest. For. Univ.* **2011**, *26*, 117–122. (In Chinese)
17. Li, B.Y. Studies on Genetic Diversity and Construction of Core Collection of the Tree Peony Cultivars from Chinese Central Plains. Ph.D. Thesis, Beijing Forestry University, Beijing, China, 2007. (In Chinese).
18. Yuan, T.; Wang, L.Y. Discussion on the origination of Chinese tree-peony cultivars according to pollen grain morphology. *J. Beijing Fore. Univ.* **2002**, *24*, 5–11. (In Chinese)
19. Zhou, B.; Jiang, H.D.; Zhang, X.X.; Xue, J.Q.; Shi, Y.T. Morphological diversity of some introduced tree peony cultivars. *Biodiv. Sci.* **2011**, *19*, 543–550. (In Chinese) [CrossRef]
20. Hosoki, T.; Kimura, D.; Hasegawa, R.; Nagasako, T.; Nishimoto, K.; Ohta, K.; Sugiyama, M.; Haruki, K. Comparative study of Chinese tree peony cultivars by random amplified polymorphic DNA (RAPD) analysis. *Sci. Hortic.* **1997**, *70*, 67–72. [CrossRef]
21. Yu, H.P.; Cheng, F.Y.; Zhong, Y.; Cai, C.F.; Wu, J.; Cui, H.L. Development of simple sequence repeat (SSR) markers from *Paeonia ostii* to study the genetic relationships among tree peonies (Paeoniaceae). *Sci. Hortic.* **2013**, *164*, 58–64. [CrossRef]
22. Wang, X.W.; Fan, H.M.; Li, Y.Y.; Sun, X.; Sun, X.Z.; Wang, W.L.; Zheng, C.S. Analysis of genetic relationships in tree peony of different colors using conserved DNA-derived polymorphism markers. *Sci. Hortic.* **2014**, *175*, 68–73. [CrossRef]
23. Suo, Z.L.; Li, W.Y.; Yao, J.; Zhang, H.J.; Zhang, Z.M.; Zhao, D.X. Applicability of leaf morphology and intersimple sequence repeat markers in classification of tree peony (Paeoniaceae) cultivars. *Hortscience* **2005**, *40*, 329–334. [CrossRef]
24. Zhang, J.J.; Shu, Q.Y.; Liu, Z.A.; Ren, H.X.; Wang, L.S.; De Keyser, E. Two EST-derived marker systems for cultivar identification in tree peony. *Plant Cell Rep.* **2012**, *31*, 299–310. [CrossRef]
25. Duan, Y.B.; Guo, D.L.; Guo, L.L.; Wei, D.F.; Hou, X.G. Genetic diversity analysis of tree peony germplasm using iPBS markers. *Genet. Mol. Res.* **2015**, *14*, 7556–7566. [CrossRef]
26. Suo, Z.L.; Zhang, C.H.; Zheng, Y.Q.; He, L.X.; Jin, X.B.; Hou, B.X.; Li, J.J. Revealing genetic diversity of tree peonies at micro-evolution level with hyper-variable chloroplast markers and floral traits. *Plant Cell Rep.* **2012**, *31*, 2199–2213. [CrossRef] [PubMed]
27. Yuan, J.H.; Cheng, F.Y.; Zhou, S.L. Hybrid origin of *Paeonia* × *yananensis* revealed by microsatellite markers, chloroplast gene sequences, and morphological characteristics. *Int. J. Plant Sci.* **2010**, *171*, 409–420. [CrossRef]
28. Yuan, J.H.; Cheng, F.Y.; Zhou, S.L. The phylogeographic structure and conservation genetics of the endangered tree peony, *Paeonia rockii* (Paeoniaceae), inferred from chloroplast gene sequences. *Conserv. Genet.* **2011**, *12*, 1539–1549. [CrossRef]
29. Yuan, J.H.; Cheng, F.Y.; Zhou, S.L. Genetic structure of the tree peony (*Paeonia rockii*) and the Qinling Mountains as a geographic barrier driving the fragmentation of a large population. *PLoS ONE* **2012**, *7*, e34955. [CrossRef] [PubMed]
30. Ren, X.X.; Zhang, Y.; Xue, J.Q.; Zhu, F.Y.; Shi, F.R.; Wang, S.L.; Zhang, X.X. Genetic diversity analysis of natural populations in *Paeonia delavayi*. *J. Plant Gen. Res.* **2015**, *16*, 772–780. (In Chinese)

31. Tang, Q.; Zeng, X.L.; Liao, M.A.; Pan, G.T.; Zha, X.; Gong, J.H.; Ci Ren, Z.G. SRAP analysis of genetic diversity of *Paeonia ludlowii* in Tibet. *Sci. Silva. Sin.* **2012**, *48*, 70–76. (In Chinese)
32. Zhang, J.M.; Liu, J.; Sun, H.L.; Yu, J.; Wang, J.X.; Zhou, S.L. Nuclear and chloroplast SSR markers in *Paeonia delavayi* (Paeoniaceae) and cross-species amplification in *P. ludlowii*. *Am. J. Bot.* **2011**, *98*, e346–e348. [CrossRef]
33. Zhang, J.M.; López-Pujol, J.; Gong, X.; Wang, H.F.; Vilatersana, R.; Zhou, S.L. Population genetic dynamics of Himalayan-Hengduan tree peonies, *Paeonia* subsect. *Delavayanae*. *Mol. Phylogenet. Evol.* **2018**, *125*, 62–77. [CrossRef]
34. Tong, F.; Xie, D.F.; Zeng, X.M.; He, X.J. Genetic diversity of *Paeonia decomposita* and *Paeonia decomposita* subsp. *rotundiloba* detected by ISSR markers. *Acta Bot. Boreal.-Occident. Sin.* **2016**, *36*, 1968–1976. (In Chinese)
35. Xu, X.X.; Cheng, F.Y.; Xian, H.L.; Peng, L.P. Genetic diversity and population structure of endangered endemic *Paeonia jishanensis* in China and conservation implications. *Biochem. Syst. Ecol.* **2016**, *66*, 319–325. [CrossRef]
36. Peng, L.P.; Cai, C.F.; Zhong, Y.; Xu, X.X.; Xian, H.L.; Cheng, F.Y.; Mao, J.F. Genetic analyses reveal independent domestication origins of the emerging oil crop *Paeonia ostii*, a tree peony with a long-term cultivation history. *Sci. Rep.* **2017**, *7*, 5340–5352. [CrossRef] [PubMed]
37. Xu, X.X.; Cheng, F.Y.; Peng, L.P.; Sun, Y.Q.; Hu, X.G.; Li, S.Y.; Xian, H.L.; Jia, K.H.; Abbott, R.J.; Mao, J.F. Late Pleistocene speciation of three closely related tree peonies endemic to the Qinling-Daba Mountains, a major glacial refugium in Central China. *Ecol. Evol.* **2019**, *9*, 7528–7548. [CrossRef] [PubMed]
38. Wu, J.; Cheng, F.Y.; Cai, C.F.; Zhong, Y.; Jie, X. Association mapping for floral traits in cultivated *Paeonia rockii* based on SSR markers. *Mol. Genet. Genom.* **2016**, *292*, 187–200. [CrossRef] [PubMed]
39. Weber, J.L.; May, P.E. Abundant class of human DNA polymorphisms which can be typed using the polymerase chain reaction. *Am. J. Hum. Genet.* **1989**, *44*, 388–396. [PubMed]
40. Dumolin, S.; Demesure, B.; Petit, R.J. Inheritance of chloroplast and mitochondrial genomes in pedunculate oak investigated with an efficient PCR method. *Theor. Appl. Genet.* **1995**, *91*, 1253–1256. [CrossRef]
41. Comes, H.P.; Kadereit, J.W. The effect of Quaternary climatic changes on plant distribution and evolution. *Trends Plant Sci.* **1998**, *3*, 1360–1385. [CrossRef]
42. Qiu, Y.X.; Fu, C.X.; Comes, H.P. Plant molecular phylogeography in China and adjacent regions: Tracing the genetic imprints of Quaternary climate and environmental change in the world's most diverse temperate flora. *Mol. Phylogenet. Evol.* **2011**, *59*, 225–244. [CrossRef]
43. Cheng, Y.P.; Hwang, S.Y.; Lin, T.P. Potential refugia in Taiwan revealed by the phylogeographical study of *Castanopsis carlesii* Hayata (Fagaceae). *Mol. Ecol.* **2005**, *14*, 2075–2085. [CrossRef]
44. Cai, C.F. High Density Genetic Linkage Map Construction and QTLs Analyses for Phenotypic Traits in Tree Peony. Ph.D. Thesis, Beijing Forestry University, Beijing, China, 2015. (In Chinese).
45. Wu, J.; Cai, C.F.; Cheng, F.Y.; Cui, H.L.; Zhou, H. Characterisation and development of EST-SSR markers in tree peony using transcriptome sequences. *Mol. Breed.* **2014**, *34*, 1853–1866. [CrossRef]
46. Bousquet, J.; Strauss, S.H.; Doerksen, A.H.; Price, R.A. Extensive variation in evolutionary rate of rbcL gene sequences among seed plants. *Proc. Natl. Acad. Sci. USA* **1992**, *89*, 7844–7848. [CrossRef]
47. Gaut, B.S.; Muse, S.V.; Clark, Z.W.D.; Clegg, M.T. Relative rates of nucleotide substitution at the rbcl locus of monocotyledonous plants. *J. Mol. Evol.* **1992**, *35*, 292–303. [CrossRef]
48. Baker, W.J.; Hedderson, T.A.; Dransfield, J. Molecular phylogenetics of subfamily *Calamoideae* (Palmae) based on nrDNA ITS and cpDNA rps16 intron sequence data. *Mol. Phylogenet. Evol.* **2000**, *14*, 195–217. [CrossRef] [PubMed]
49. Grivet, D.; Heinze, B.; Vendramin, G.G.; Petit, R.J. Genome walking with consensus primers application to the large single copy region of chloroplast DNA. *Mol. Ecol. Notes* **2001**, *1*, 345–349. [CrossRef]
50. Davis, J.W. *Handbook of univariate and multivariate data analysis and interpretation with SPSS*; CRC Press: Boca Raton, FL, USA, 2008. [CrossRef]
51. Peakall, R.; Smouse, P.E. GenAlEx 6.5: Genetic analysis in Excel. Population genetic software for teaching and research—An update. *Bioinformatics* **2012**, *28*, 2537–2539. [CrossRef] [PubMed]
52. Rousset, F. Genepop'007: A complete re-implementation of the genepop software for Windows and Linux. *Mol. Ecol. Resour.* **2008**, *8*, 103–106. [CrossRef]
53. Kumar, S.; Stecher, G.; Li, M.; Knyaz, C.; Tamura, K.; Battistuzzi, F.U. MEGA X: Molecular evolutionary genetics analysis across computing platforms. *Mol. Biol. Evol.* **2018**, *35*, 1547–1549. [CrossRef]

54. Katoh, K.; Misawa, K.; Kuma, K.-i.; Miyata, T. MAFFT a novel method for rapid multiple sequence alignment based on fast Fourier transform. *Nucleic Acids Res.* **2002**, *30*, 3059–3066. [CrossRef]
55. Vaidya, G.; Lohman, D.J.; Meier, R. SequenceMatrix: Concatenation software for the fast assembly of multi-gene datasets with character set and codon information. *Cladistics* **2011**, *27*, 171–180. [CrossRef]
56. Librado, P.; Rozas, J. DnaSP v5: A software for comprehensive analysis of DNA polymorphism data. *Bioinformatics* **2009**, *25*, 1451–1452. [CrossRef]
57. Leigh, J.W.; Bryant, D.; Nakagawa, S. Popart: Full-feature software for haplotype network construction. *Methods Ecol. Evol.* **2015**, *6*, 1110–1116. [CrossRef]
58. Swofford, D.L. *Paup**: Phylogenetic Analysis Using Parsimony (*and other Methods)*; Version 4; Sinauer: Sunderland, MA, USA, 2003; pp. 233–234.
59. Pang, L.Z.; Cheng, F.Y.; Zhong, Y.; Cai, C.F.; Cui, H.L. Phenotypic analysis of association population for flare tree peony. *J. Beijing Fore. Univ.* **2012**, *34*, 115–120. (In Chinese)
60. Wu, J. Dissection of Allelic Variation Underling Imoortant Traits in *Paeonia rockii* by Using Association Mapping. Ph.D. Thesis, Beijing Forestry University, Beijing, China, 2016. (In Chinese).
61. Warschefsky, E.; Penmetsa, R.V.; Cook, D.R.; Von Wettberg, E.J.B. Back to the wilds: Tapping evolutionary adaptations for resilient crops through systematic hybridization with crop wild relatives. *Am. J. Bot.* **2014**, *101*, 1791–1800. [CrossRef]
62. Bernard, A.; Lheureux, F.; Dirlewanger, E. Walnut: Past and future of genetic improvement. *Tree Genet. Genomes* **2017**, *14*, 1–28. [CrossRef]
63. Yuan, J.H.; Cornille, A.; Giraud, T.; Cheng, F.Y.; Hu, Y.H. Independent domestications of cultivated tree peonies from different wild peony species. *Mol. Ecol.* **2014**, *23*, 82–95. [CrossRef] [PubMed]
64. Guo, D.L.; Hou, X.G.; Zhang, J. Sequence-related amplified polymorphism analysis of tree peony (*Paeonia suffruticosa* Andrews) cultivars with different flower colours. *J. Hortic. Sci. Biotechnol.* **2015**, *84*, 131–136. [CrossRef]
65. Nielsen, R. Molecular signatures of natural selection. *Annu. Rev. Genet.* **2005**, *39*, 197–218. [CrossRef] [PubMed]
66. Miller, A.J.; Gross, B.L. From forest to field: Perennial fruit crop domestication. *Am. J. Bot.* **2011**, *98*, 1389–1414. [CrossRef] [PubMed]
67. Margules, C.R.; Pressey, R.L. Systematic conservation planning. *Nature* **2000**, *405*, 243–253. [CrossRef]
68. Rodrigues, L.; van den Berg, C.; Póvoa, O.; Monteiro, A. Low genetic diversity and significant structuring in the endangered *Mentha cervina* populations and its implications for conservation. *Biochem. Syst. Ecol.* **2013**, *50*, 51–61. [CrossRef]
69. Migicovsky, Z.; Warschefsky, E.; Klein, L.L.; Miller, A.J. Using living germplasm collections to characterize, improve, and conserve woody perennials. *Crop Sci.* **2019**, *59*, 2365–2380. [CrossRef]

Article

Genetic Structure and Pod Morphology of *Inga edulis* Cultivated vs. Wild Populations from the Peruvian Amazon

Alexandr Rollo [1,2,†], Maria M. Ribeiro [3,4,5,†], Rita L. Costa [6], Carmen Santos [7], Zoyla M. Clavo P. [8], Bohumil Mandák [9], Marie Kalousová [1,10], Hana Vebrová [1,10], Edilberto Chuqulin [11], Sergio G. Torres [11,12], Roel M. V. Aguilar [13], Tomáš Hlavsa [14] and Bohdan Lojka [1,*]

1 Department of Crop Sciences and Agroforestry, Faculty of Tropical Forest AgriSciences, Czech University of Life Sciences, Kamýcká 129, 165 00 Prague, Czech Republic; rollo@ftz.czu.cz (A.R.); marie.kalousova@gmail.com (M.K.); Hani.V@seznam.cz (H.V.)
2 Crop Research Institute, Drnovská 507/73, 161 06 Prague, Czech Republic
3 Department of Natural Resources and Sustainable Development, Instituto Politécnico de Castelo Branco, Escola Superior Agrária, 6001-909 Castelo Branco, Portugal; mataide@ipcb.pt
4 Forest Research Center, School of Agriculture, University of Lisbon, 1349-017 Lisbon, Portugal
5 Research Center for Natural Resources, Environment and Society (CERNAS), Polytechnic Institute of Castelo Branco, 6000-084 Castelo Branco, Portugal
6 INIAV, Instituto Nacional de Investigação Agrária e Veterinária, I.P., Av. República, Quinta do Marquês, 2780-157 Oeiras, Portugal; rita.lcosta@iniav.pt
7 Instituto de Tecnologia Química e Biológica António Xavier, Universidade Nova de Lisboa, Av. da República, 2780-157 Oeiras, Portugal; css@itqb.unl.pt
8 IVITA, Facultad de Medicina Veterinaria, Universidad Nacional Mayor de San Marcos, Jr. Daniel Alcides Carrión 319, Pucallpa, Ucayali 25001, Peru; zclavop@unmsm.edu.pe
9 Department of Ecology, Faculty of Environmental Sciences, Czech University of Life Sciences, Kamýcká 129, 165 00 Prague, Czech Republic; mandak@fzp.czu.cz
10 Students for the Living Amazon o.p.s., Národní obrany 984/18, 160 00 Prague, Czech Republic
11 Facultad de Agronomia, Universidad National Agraria de la Selva, Carratera Central km 1.21, Tingo María 10131, Peru; ecbangel@gmail.com (E.C.); garciats@outlook.com (S.G.T.)
12 ONG Students for the Living Amazon Perú, Jr Uruguay, m/d, lote 18, Yarinacocha, Ucayali 25000, Peru
13 Facultad de Ciencias Forestales, Universidad Nacional de Ucayali, Car. Federico Basadre Km 6.2, Pucallpa, Ucayali 25004, Peru; rmva25@hotmail.com
14 Department of Statistics, Faculty of Economics and Management, Czech University of Life Sciences, Kamýcká 129, 165 00 Prague, Czech Republic; hlavsa@pef.czu.cz
* Correspondence: lojka@ftz.czu.cz
† These authors contributed equally to this work.

Received: 5 May 2020; Accepted: 4 June 2020; Published: 8 June 2020

Abstract: *Research Highlights*: This study assesses the genetic diversity and structure of the ice-cream-bean (*Inga edulis* Mart.; Fabaceae) in wild and cultivated populations from the Peruvian Amazon. This research also highlights the importance of protecting the biodiversity of the forest in the Peruvian Amazon, to preserve the genetic resources of species and allow further genetic improvement. *Background and Objectives*: Ice-cream-bean is one of the most commonly used species in the Amazon region for its fruits and for shading protection of other species (e.g., cocoa and coffee plantations). Comprehensive studies about the impact of domestication on this species' genetic diversity are needed, to find the best conservation and improvement strategies. *Materials and Methods*: In the current study, the genetic structure and diversity were assessed by genotyping 259 trees, sampled in five wild and 22 cultivated *I. edulis* populations in the Peruvian Amazon, with microsatellite markers. Pod length was measured in wild and cultivated trees. *Results*: The average pod length in cultivated trees was significantly higher than that in wild trees. The expected genetic diversity and the average number of alleles was higher in the wild compared to the cultivated populations; thus, a loss of genetic diversity was confirmed in the cultivated populations. The cultivated trees in the Loreto region had the highest pod length and lowest allelic richness; nevertheless, the wild populations' genetic structure was not clearly differentiated (significantly different) from that of the cultivated populations. *Conclusions*: A loss of genetic diversity was confirmed in the cultivated populations. The species could have been simultaneously domesticated in multiple locations, usually from local origin. The original *I. edulis* Amazonian germplasm should be maintained. Cultivated populations' new germplasm influx from wild populations should be undertaken to increase genetic diversity.

Keywords: agroforestry; domestication; *Inga edulis*; amazon forest; microsatellite markers; genetic diversity

1. Introduction

The Peruvian rainforest, due to its large and relatively continuous area of primary forest, is a worldwide biodiversity hotspot, which is suffering intense disturbance and deforestation from human exploitation and global change [1]. Amazonian inhabitants have used these natural resources through millennia and modified the natural environment, but how human management practices resulted in Amazonian forests' domestication is not known, in particular the germplasm source [2]. Moreover, the species' gene pool could have been reduced due to farmers' selection, thus strategies for genetic resource conservation and management are needed [3]. The Peruvian rainforest that remains a large and relatively continuous area of primary forest has major conservation value and is considered a priority in nearly all global biodiversity inventories, due to its biodiversity and the disturbance and deforestation rates [4,5]. Despite the international recognition of its major conservation value resulting from its uniqueness and global importance, the impacts of human activities throughout the region remain poorly understood [5]. Indeed, information about the species' genetic structure will assist in tree breeding programmes and conservation strategies, in particular in tropical trees, and, also, to study the implications of human impact on genetic resources [6,7].

The genus *Inga* (Fabaceae) comprises ca. 300 species of neotropical rainforest trees [8], but earlier studies suggest that the diversification of *Inga* species in Amazonia is recent, during the past 2–10 million years [9,10]. Ice-cream-bean (*Inga edulis* Mart.; Fabaceae) is a lowland rainforest light-demanding species, distributed in Colombia and tropical South America to the east of the Andes, extending from south to north-western Argentina (Figure 1). The species' natural altitudinal range is mostly below 750 m a.s.l., though it has been occasionally identified at 1200 m in Roraima, Brazil [8]. The flowers are hermaphrodite and the pollination is provided by hawkmoths, bats and hummingbirds that may carry the pollen grains across large areas [11]. The species is diploid, 2n = 26 [12], and it is believed to be self-incompatible [13]. Fruiting occurs, in three year-old trees, as a long pod containing recalcitrant seeds, covered by a white, fleshy and slightly sweet edible sarcotesta [14]. Seed dispersal is performed by mammals and birds after eating the sarcotesta [11,15]. The species is widely cultivated for its edible fruit throughout South and Central America [16]. It is one of the most widely distributed and economically useful species in the whole Amazon region [14,17]. In Amazonian Peru, the fruits from cultivated trees may exceed 2 m in length and 5–6 cm in diameter. The wild trees have smaller pods than cultivated trees, rarely exceeding 50 cm in length [8]. This fast growing, symbiotic nitrogen-fixing tree, with umbrella-like canopy, is commonly used as a shade tree for cocoa, coffee, coca and tea plantations, in agroforestry systems and in "home grown" multi-purpose cultivation uses [8].

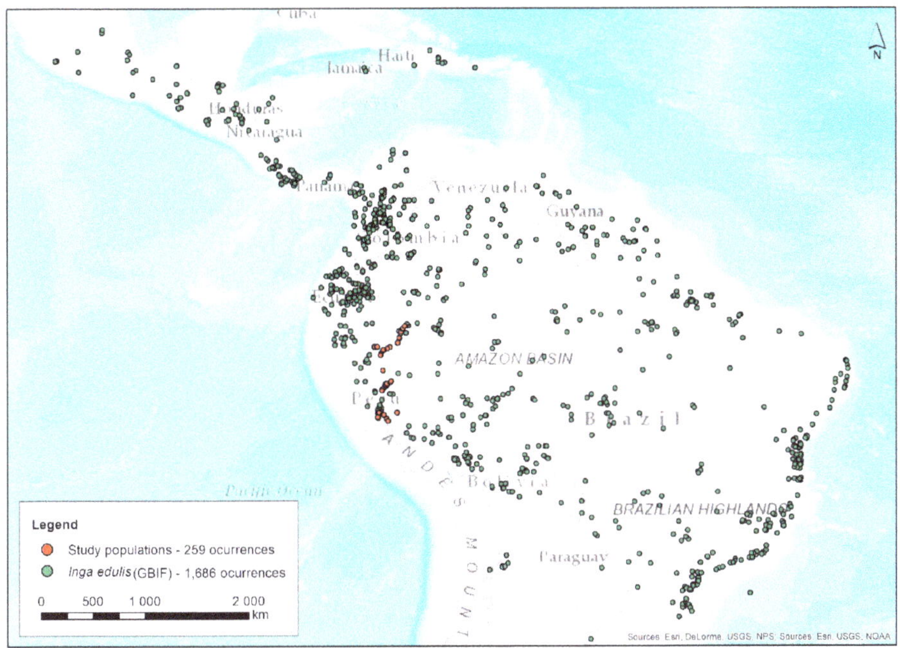

Figure 1. Distribution map of *Inga edulis*. The green dots represent the 1686 occurrences, and the red dots the trees sampled in the current study (259 occurrences). GBIF.org (10th October 2018) GIBF Occurrence. Download: https:doi.org/10.15468/dl.ik3uki.

Amazonia was a major centre of crop domestication, with at least 83 native species containing populations domesticated to some degree, which expanded rapidly in the Mid-Holocene [18], including the *Inga* genus. The historical records for *Inga edulis* show that this species has been cultivated in Peru for its edible fruit since the pre-Colombian time and has become a commonly used tree species in the Amazon region [19]. The origin of the cultivated populations of *I. edulis* is uncertain [8], however, León [17] and Clement [20] claimed West Amazonia as a probable origin. The species' genetic structure was not studied in detail, yet a reduction of allelic richness in cultivated relative to natural populations was found in *I. edulis* from the Peruvian Amazon [13,21]. *Inga edulis* has become a model species to evaluate the maintenance of genetic resources in agroforestry systems, and the putative genetic diversity reduction associated with domestication [22]. More recently, Cruz-Neto et al. [23], using microsatellite markers, observed high levels of genetic diversity within *I. vera* populations from the Atlantic forest of north-eastern Brazil. They concluded that cultivated populations compared to natural populations displayed reduced genetic diversity. Nevertheless, maintaining high levels of genetic variation within agroforestry trees is important for two main reasons: genetic variation in agricultural landscapes helps farmers to manage their inputs in more efficient ways and because they provide the ability for tree species to adjust to new environments, such as the shifting climate and weather conditions, allowing local adaptation and the migration of better-suited provenances along ecological gradients [3]. In addition, a stronger emphasis on the genetic quality of the trees planted by smallholders is needed, which means paying attention both to domestication and to the systems by which improved germplasm is delivered to farmers for the management of tree genetic resources and the livelihoods of rural communities in the tropics [24].

In the present study, the objectives were to (i) explore differences in pod length between wild trees and cultivated *I. edulis* trees from different geographical regions in the Peruvian Amazon; (ii) compare the wild and cultivated *I. edulis* populations' genetic structure using microsatellite markers; and (iii) determine if the cultivated populations' genetic structure reflects the different uses and cultivation practices throughout the species' use history, to help design practical measures to preserve *I. edulis* genetic resources.

2. Materials and Methods

2.1. Plant Material Sampling

The leaves and mature fruits from 259 individuals and 27 populations of *I. edulis* were sampled in the Peruvian Amazon, between 2009 and 2012. Each tree was identified according to the morphological aspects detailed by Rollo et al. [25] and, additionally, with the help of locals. Each sampled tree's young leaves were preserved in micro test-tubes with silica gel for further DNA extraction. One to ten mature pods were sampled per tree, from opposite sides and different heights of the crown, according to the availability of mature fruits on the tree. The pod length was measured from the base to the top of the pod apex. The mature fruits had the following phenological characteristics: seeds from creamy white to purple black up to vivipary; and the sarcotesta membranous creamy to generally flashy white, watery, soft and slightly sweet [8]. A total of 448 mature pods were measured in the 259 trees from the 27 *I. edulis* populations. We chose to study the pod length, as we can draw a null hypothesis based on the domestication process, since the trees were selected for their pod length (H0: Is the pod length of the domesticated trees higher than that of the wild trees?). Indeed, this trait has economic importance in the species domestication and agroforestry value.

Each sampled tree's geographical coordinates were recorded, and the minimum distance between any two trees was 200 m. Voucher specimens were kept in the Regional Herbarium of Ucayali IVITA-Pucallpa, Peru, with the code AR1-384. Each population was numbered from 1 to 27 and coded, e.g., 1 SRc, 23 RPw (the two capital letters are taken from the initial letters of the geographic origin of the population, e.g., San Ramón or River Pacaya, and the third letter meaning either c = cultivated—managed by humans or w = wild—growing spontaneously) (Table 1).

Table 1. The sampling region (Site), population code (Pop.), sample size (N), geographic location (GPS coordinates in WGS84; latitude S and longitude W) and altitude (in metres above sea level) of the 27 sampled *Inga edulis* populations (cultivated and wild).

I. edulis	Site	Pop.	N	Latitude S	Longitude W	Altitude (m)
Cultivated Selva Central	San Ramon	1 SRc	10	11°08′	75°21′	828–1200
	Villa Rica	2 VRc	5	10°44′	75°16′	1467–1494
	Pichanaqui	3 PIc	10	10°55′	74°52′	497–631
	Satipo	4 SAc	10	11°16′	74°38′	550–677
	San Martín de Pangoa	5 SMc	10	11°26′	74°30′	788–949
Cultivated Ucayali	Atalaya	6 ATc	10	10°43′	73°45′	223–244
	Von Humboldt	7 VHc	8	8°51′	75°00′	210–243
	Campo Verde-Tournavista	8 CTc	18	8°35′	74°46′	180–207
	Campo Verde	9 CVc	12	8°31′	74°47′	198–210
	Antonio Raimondi	10 ARc	11	8°29′	74°49′	147–158
	Yarinacocha	11 YAc	5	8°20′	74°36′	144–154
	Santa Sofia	12 SSc	8	8°09′	74°15′	152–159
Cultivated Loreto	Bretaña	13 BRc	16	5°15′	74°20′	103–108
	Jenaro Herrera	14 JHc	5	4°54′	73°40′	100–127
	Lagunas	15 LAc	10	5°14′	75°37′	108–135
	Nauta	16 NAc	5	4°30′	73°34′	106–139
	Ex Petroleros	17 EPc	5	4° 5′	73° 27′	97–108
	El Dorado	18 EDc	12	3° 57′	73° 25′	109–151
	Manacamiri	19 MAc	5	3° 43′	73° 17′	95–97
	Santa Clotylda	20 SCc	5	3° 40′	73° 15′	93–128
	Indiana	21 INc	7	3°29′	73°02′	92–108
	Mazán	22 MZc	10	3°30′	73°04′	93–122
Wild	Pacaya River	23 RPw	12	5° 41′	74° 57′	110–131
	Samiria River	24 RSw	6	5° 14′	75° 28′	105–123
	Utiquinia River	25 RUw	12	8° 10′	74° 17′	150–160
	Macuya	26 MAw	27	8° 53′	75° 0′	216–233
	Sierra del Divisor	27 SDw	5	7° 13′	74° 57′	196–231

Cultivated trees were sampled in 22 geographically different populations, in home gardens and agricultural landscapes surrounding the urban areas in the Selva Central, Ucayali and Loreto regions. Wild trees were sampled in five geographically different populations, in lowland forests; four populations (23 RPw, 24 RSw, 26 MAw, 27 SDw) in protected natural areas in original forest vegetation and one (25 RUw) in secondary forest as described in Rollo et al. [25]. Details on the sampled populations are displayed in Table S1. Finally, no *I. edulis* wild trees were found in the Selva Central region's original vegetation, since the species is a lowland rainforest species and has been only occasionally recorded above 750 m [8].

2.2. DNA Extraction and Amplification

Total DNA was extracted from dried young leaves, using the Invitek, Invisorb® Spin Plant Mini Kit following the manufacturer's instructions. Four microsatellite markers were used to genotype all the individuals, Pel5 [26] and Inga03, Inga08 and Inga33 [21]. For microsatellite detection, each forward primer was fluorescently labelled at the 5′ end (6-FAM, NED or VIC). Amplification conditions were performed according to the conditions described by Rollo et al. [25]. The amplified products were separated on an ABI PRISM 310 Genetic Analyzer (Applied Biosystems, Foster City, CA, USA) and ran according to the manufacturer's protocol. Fragment sizes were determined using the ROX500 internal size standard and the global southern algorithm implemented by ABI PRISM GeneMapper® software version 4.0 (Applied Biosystems).

2.3. Data Analysis

2.3.1. Morphological Data

The pod length was measured in the 448 pods sampled from the 259 trees of the *I. edulis* populations from the Peruvian Amazon, and all the individual values were used in the following analysis (not averaged per tree). The pod lengths' normality (in both cultivated populations originated in Selva Central, Ucayali and Loreto regions and in wild populations) was tested using the Kolmogorov–Smirnov test, and all the groups displayed normal distribution except for the wild trees group. A non-parametric Kruskal–Wallis test (k independent samples) was performed to check for significant differences in the groups' average followed by the non-parametric Mann–Whitney U post-hoc test [27]. The statistical analyses were performed using the IBM SPSS v.22 statistics software.

2.3.2. Molecular Data

The estimated genetic diversity parameters included the average number of alleles per locus (A), the effective number of alleles (Ne), the number of private alleles (Pa), the mean allelic richness (R_S) that uses a rarefaction index to consider differences in sample size [28], the observed (H_O) and expected (H_E) heterozygosities [29] and the fixation index F_{IS}. Comparisons of the genetic diversity parameters between groups (i.e., cultivated and wild populations) were performed with 10,000 permutations. The estimates were made using FSTAT 2.9.3.2 [30] and GenAlEx v. 6.501 [31]. Genepop 4.3 software [32] was used to test the heterozygote deficiency for each population and to compute the average frequency of null alleles.

Genetic variation at the level of populations and groups (i.e., cultivated and wild populations) was investigated with a hierarchical analysis of molecular variance (AMOVA), which partitions the total variance into covariance components due to inter-group differences, inter-populations within groups differences and inter-population differences, in the Arlequin software [33]. Levels of significance were determined by computing 1000 random permutation replicates.

A Bayesian clustering method was performed in the STRUCTURE v.2.3.4. software [34] to infer population genetic structure. The number of genetic clusters (K) was estimated and the individuals sampled from cultivated and wild populations were fractionally assigned to the inferred groups. Afterwards, the allele frequencies were estimated in each of the K groups as well as the proportion of the genome derived from each group for each tree. We applied the model allowing population admixture and correlated allele frequency [34]. However, due to the weak population structure found in the *I. edulis* populations, we used a model that incorporated a priori sampling location information [35], i.e., a "locprior" model. This improved model has the advantage of allowing cryptic structures to be detected at a lower level of divergence and does not bias towards detecting structures spuriously when none is present, helpful in situations when the standard structure models do not provide a clear signal of structure [35]. Two groups of populations were used as priors, i.e., cultivated and wild populations (see Table 1). The alternative ancestry prior $1/K$ was used due to unbalanced population sampling [36]. The number of clusters (K) was set from one through twenty-seven and the simulation was run ten times at each K value to confirm the repeatability of the results. Each run comprised a burn-in period of 25,000, followed by 100,000 Markov chain Monte Carlo (MCMC) steps. We used the ΔK distribution statistic of Evanno et al. [37] to determine the most appropriate number of genetic clusters. Hence, the STRUCTURE output data were parsed using the STRUCTURE HARVESTER [38] to determine the optimal K value following the method referred to above. Alignment of cluster assignments across replicate analyses was then conducted in CLUMPP 1.1.2 [39] and subsequently visualized using STRUCTURE PLOT [40]. The results of Bayesian clustering were further mapped in the ArcGIS® Desktop version 10.2 software [41].

3. Results

3.1. Pod Length

From a total of 259 individual trees, 448 fruits were collected and measured: 329 and 119 fruits from 197 cultivated trees and 62 wild trees, respectively. The longest pod (148 cm) was found in 18 EDc, a cultivated population in the Loreto region around El Dorado village (Table S2). The Kruskal–Wallis non-parametric test, using the four groups of populations (regions), showed significant group differences for pod length (Figure 2). The Mann–Whitney U post-hoc test produced three homogeneous groups, which indicated that the pod length average (78 cm) in the Selva Central region was not significantly different ($p < 0.05$) from the Ucayali region's 80 cm long average, and both values were significantly different from the Loreto value (90 cm) (Figure 2).

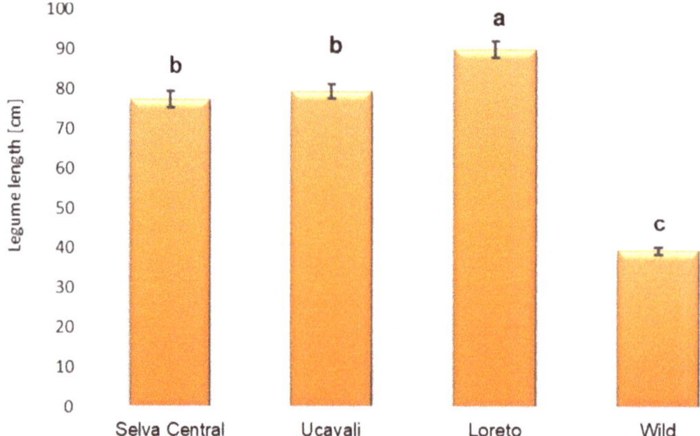

Figure 2. Pod length comparison among the cultivated and wild populations. From Selva Central, 94 mature pods were collected from 45 trees from 5 populations; from Ucayali, 120 mature pods were collected from 72 trees from 7 populations; and from Loreto, 115 mature pods were collected from 80 trees from 10 populations. In the wild populations, 119 mature pods were collected from 62 trees from 5 populations. Significantly different means are followed by different letters ($p < 0.05$).

The Selva Central and Ucayali regions could be one group, considering the cultivated trees' average pod length. The Loreto region's cultivated trees produced the highest pod length average. The average pod length of 83 ± 1.17 cm (mean ± standard error) in the cultivated trees was significantly higher than the 39 ± 0.95 cm pod length average in the wild trees.

3.2. Genetic Diversity

We identified a total of 71 alleles using the four microsatellite markers after genotyping all the individuals from the 27 populations. The average A was 5.7, the R_S was 4.4, the H_O was 0.59, and H_E was 0.69. The overall inbreeding coefficient F_{IS} was 0.11 (Table 2).

Table 2. Summary of the genetic diversity of the 27 *I. edulis* populations. Sample size (N), average number of alleles per locus (A), allelic richness (Rs), effective number of alleles (Ne), expected heterozygosity (H_E), observed heterozygosity (H_O) and fixation index (F_{IS}) averaged over loci. Sig. refers to the significance resulting from the heterozygote deficiency test (a conservative α value for the test of at least $p < 0.01$ was used, due to the low number of individuals per population: NS, not significant, ** $p < 0.01$ and *** $p < 0.001$, significant). *F-null* refers to the average estimate of the null frequency over the loci. Standard errors in brackets.

I. edulis	Population	N	A	R_S	N_e	H_O	H_E	F_{IS}	Sig	f-Null
Cultivated Selva Central	1 SRc	10	5.5	3.9	2.91 (0.42)	0.53 (0.13)	0.67 (0.06)	0.18 (0.16)	NS	0.06
	2 VRc	5	5.5	5.5	4.09 (0.52)	0.70 (0.17)	0.83 (0.03)	0.06 (0.22)	NS	0.08
	3 PIc	10	5.0	4.0	3.32 (0.91)	0.55 (0.13)	0.67 (0.08)	0.15 (0.16)	NS	0.06
	4 SAc	10	5.8	4.1	2.96 (0.72)	0.58 (0.15)	0.63 (0.09)	0.06 (0.19)	NS	0.05
	5 SMc	10	6.3	4.4	3.39 (0.84)	0.60 (0.12)	0.67 (0.10)	0.07 (0.08)	NS	0.04
Cultivated Ucayali	6 ATc	10	4.8	3.7	2.72 (0.51)	0.53 (0.18)	0.61 (0.09)	0.16 (0.19)	NS	0.09
	7 VHc	8	5.3	4.5	3.86 (0.55)	0.63 (0.09)	0.77 (0.04)	0.15 (0.09)	**	0.06
	8 CTc	18	7.3	4.6	4.17 (0.78)	0.65 (0.08)	0.76 (0.05)	0.12 (0.07)	**	0.06
	9 CVc	12	5.8	4.5	4.15 (0.78)	0.67 (0.12)	0.76 (0.05)	0.11 (0.12)	NS	0.05
	10 ARc	11	5.5	4.3	3.48 (0.21)	0.61 (0.15)	0.74 (0.02)	0.14 (0.21)	***	0.10
	11 YAc	5	5.0	5.0	3.22 (0.77)	0.65 (0.15)	0.66 (0.16)	−0.10 (0.03)	NS	0.00
	12 SSc	8	5.5	4.6	4.03 (0.99)	0.50 (0.21)	0.75 (0.07)	0.36 (0.25)	***	0.15
Cultivated Loreto	13 BRc	16	8.0	4.8	4.52 (1.45)	0.64 (0.12)	0.71 (0.10)	0.08 (0.09)	NS	0.04
	14 JHc	5	3.8	3.8	2.55 (0.26)	0.50 (0.19)	0.66 (0.05)	0.20 (0.31)	NS	0.11
	15 LAc	10	6.8	4.8	4.42 (1.26)	0.65 (0.12)	0.73 (0.11)	0.03 (0.14)	NS	0.04
	16 NAc	5	4.3	4.3	2.93 (0.64)	0.60 (0.18)	0.64 (0.15)	−0.04 (0.16)	NS	0.03
	17 EPc	5	3.8	3.8	2.79 (0.31)	0.55 (0.19)	0.69 (0.06)	0.18 (0.29)	NS	0.09
	18 EDc	12	5.0	3.8	3.03 (0.68)	0.46 (0.17)	0.62 (0.12)	0.33 (0.26)	**	0.11
	19 MAc	5	3.8	3.8	2.59 (0.61)	0.50 (0.13)	0.61 (0.11)	0.10 (0.15)	NS	0.04
	20 SCc	5	3.8	3.8	2.58 (0.33)	0.45 (0.13)	0.66 (0.07)	0.25 (0.20)	NS	0.12
	21 INc	7	3.8	3.3	2.60 (0.84)	0.43 (0.21)	0.50 (0.17)	0.20 (0.27)	NS	0.06
	22 MZc	10	5.5	4.3	3.76 (0.72)	0.60 (0.16)	0.73 (0.07)	0.10 (0.10)	**	0.11
	Mean		5.3	4.2	3.37 (0.16)	0.57 (0.03)	0.69 (0.02)	0.14 (0.04)		0.07
Wild	23 RPw	12	8.3	5.2	5.06 (1.17)	0.63 (0.17)	0.72 (0.13)	0.09 (0.18)	NS	0.07
	24 RSw	6	6.5	5.8	5.32 (1.37)	0.75 (0.08)	0.79 (0.13)	−0.08 (0.09)	NS	0.02
	25 RUw	12	7.3	5.2	4.58 (1.15)	0.67 (0.14)	0.76 (0.07)	0.11 (0.17)	NS	0.06
	26 MAw	27	11.0	5.4	5.98 (1.99)	0.66 (0.16)	0.75 (0.12)	0.12 (0.10)	NS	0.06
	27 SDw	5	4.0	4.0	2.77 (0.94)	0.70 (0.13)	0.60 (0.11)	−0.30 (0.07)	NS	0.06
	Mean		7.4	5.1	4.74 (0.64)	0.68 (0.06)	0.72 (0.05)	−0.01 (0.06)		0.04
Cultivated and Wild	Mean		5.7	4.4	3.62 (0.18)	0.59 (0.03)	0.69 (0.02)	0.11 (0.03)		0.07

From the results of the current study, the population with the highest and lowest expected heterozygosity possessed the highest and the lowest allelic richness values, in both cultivated and wild populations. The allelic richness parameter correlated well with the populations' genetic diversity parameters, which is not surprising since the number of individuals sampled per population was unevenly distributed. The population with the highest H_E was 2 VRc, 0.83, a cultivated population from the Selva Central region, and the lowest was 21 INc (0.50), a cultivated population from the Loreto region, and they both possessed the highest and the lowest R_S, 5.5 and 3.3, respectively. The wild populations 24 RSw and 27 SDw displayed the highest and lowest H_E values, 0.79 and 0.60, similarly with the highest and lowest R_S values, 5.8 and 4.0. Interestingly, the population with the highest A and Ne was 26 MAw, which could be partially explained by the highest number of sampled individuals (27). The cultivated populations 8 CTc and 13 BRc also had a high number of sampled individuals, which was also reflected in the A and N_e parameters. The average expected genetic diversity is slightly higher in the wild compared to the cultivated populations, 0.72 and 0.69, respectively. The average number of alleles is much higher in the wild (7.4) than in the group of cultivated populations (5.3), but when we consider the allelic richness and effective number of alleles, these differences are reduced (Table 2).

Six cultivated populations out of 22 had significant heterozygote deficiency, but this parameter was not significant in the wild populations (Table 2). The Selva Central group of *cultivated* populations lacked populations with significant inbreeding coefficient, and the Loreto and the Ucayali regions had only two populations and more than half of the populations with heterozygote deficiency, respectively. Positive and significant F_{IS} values mirror differences between observed and expected heterozygosity, due to putative heterozygosity loss because of non-random mating of the parents. We should also emphasize that the cultivated populations are an assembly of individuals and we should not expect them to be in Hardy–Weinberg Equilibrium. Additionally, the presence of null alleles is an unlikely explanation since the estimated frequency is very low across populations (Table 2). Nevertheless, when we compared the overall inbreeding coefficient from the wild with the cultivated populations, no significant differences were found between them. Conversely, the allelic richness and the observed heterozygosity were significantly lower in the cultivated populations (Table S3).

Seven private alleles (Pa) were identified in three wild populations, the highest Pa per population was found in the 26 MAw population (3) and two in both 23 RPw and 25 RUw. Only one Pa was identified in four different *I. edulis cultivated* populations (2 VRc, 9 CVc, 14 JHc and 22 MZc). The locus Inga08 had the highest Pa (7 across all populations) and Inga33 and Pel5 only one (data not shown).

The cultivated populations possessed 13 exclusive alleles compared to the wild ones, and only two had a frequency lower than 5%. The regions with the highest number of cultivated populations with exclusive alleles was Selva Central, 80%, followed by Ucayali, 60%, and the Loreto region had the lowest number of populations with those alleles (40%) (data not shown).

3.3. Population Structure

The population genetic structure was investigated by a hierarchical analysis of molecular variance (AMOVA), which revealed that most of the genetic diversity existed within populations (92%). The differentiation between the cultivated and wild group populations ($\Phi_{CT} = 0.010$) was low (~1%), and not significant ($p < 0.0958$), and the variation among populations within groups was appreciable, ca. 7% ($\Phi_{SC} = 0.073$), and significant ($p < 0.0001$) (Table 3).

Table 3. Hierarchical AMOVA between the cultivated and wild population groups, among populations within the cultivated and wild population groups and within *I. edulis.* populations. df = degrees of freedom; SS = sum of squared deviation; Φ statistics = fixation indexes; P = level of probability of obtaining a more extreme component estimate by chance alone. The significance of the variance components were tested by a permutation test.

Source of Variation	d.f.	SS	Variance Components	% of Variation	Φ Statistics	*p*
Between groups (*cultivated* vs. *wild*)	1	7.697	0.01486	0.97	Φct = 0.010	<0.0958
Among populations within groups	25	86.53	0.11103	7.24	Φsc = 0.073	<0.0001
Within populations	491	690.914	1.40716	91.79	Φst = 0.082	<0.0001
Total	517	785.141	1.53304			

The *I. edulis* genetic structure was further estimated using a Bayesian approach. Using the method of Evanno et al. [37], the most appropriate number of genetic clusters (*K*) is 2, referred to as red and green (Figure 3; Figure 4 and Figure S1). The red cluster was predominant in the wild populations and in the cultivated populations in the northernmost region (Loreto). Conversely, the green cluster was predominant in the southernmost region (Selva Central). The Ucayali region displayed a mixture of both types of cultivated populations, probably a mixture from the southern and the northern regions (Figure 3; Figure 4).

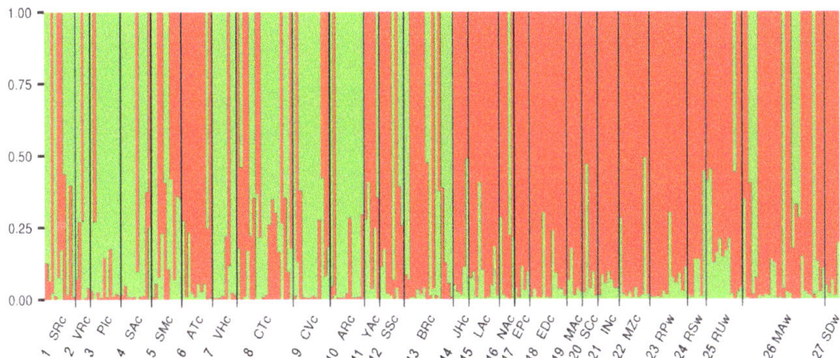

Figure 3. Proportion of genotype membership q (y-axis) based on STRUCTURE cluster analysis. Plots of proportional group membership for the 259 trees for $K = 2$. Each tree is represented by a single vertical line, which is divided into different colours based on the genotype affinities to each K cluster (red and green). Divisions between populations are made with black lines.

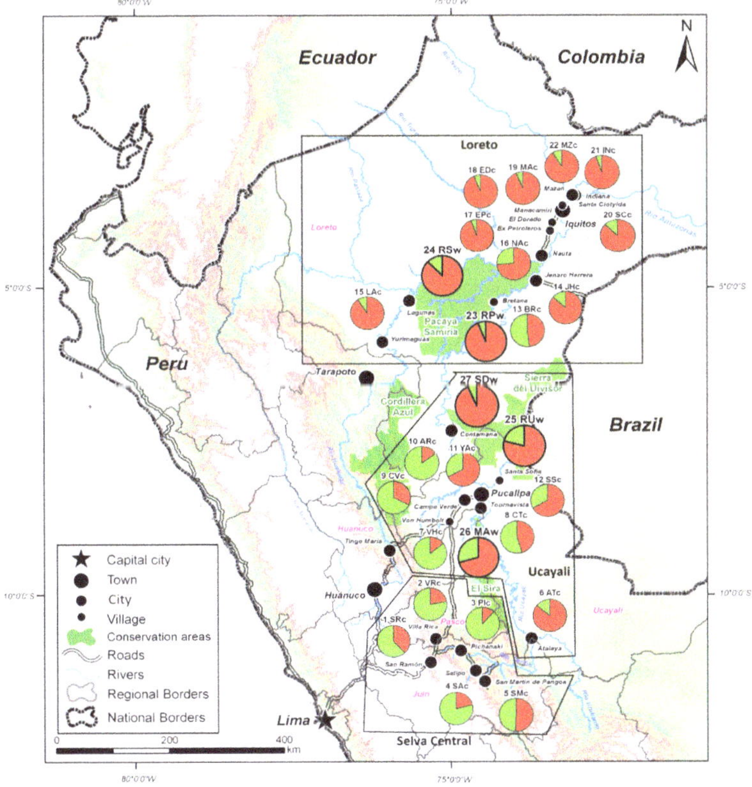

Figure 4. *Inga edulis* populations investigated in this study plotted on the map of Peru. Bayesian clustering for $K = 2$. Populations assigned to two clusters (red and green) corresponding to the *I. edulis* wild (bigger pie charts outlined in black) and cultivated populations (smaller pie charts).

For $K = 2$, the highest proportion of red cluster was observed in cultivated populations along the navigable river watersheds in the Loreto and Ucayali regions (e.g., 6 ATc, 11 YAc, 14 JHc, 15 LAc, 16 NAc, 19 MAc, 20 SCc, 21 INc and 22 MZc). Moreover, the green cluster was found to be prevalent in populations cultivated on the Andean foothills and "terra firme" in the Selva Central and Ucayali regions (e.g., 2 VRc, 3 PIc, 4 SAc, 7 VHc and 10 ARc) (Figure 4).

4. Discussion

4.1. Influence of Domestication on Fruit Length

Although the history of cultivation of *I. edulis* is not well documented, a crop domestication study suggested that humans have domesticated this species over a considerable period of time [20]. Indeed, Amazonia is a major world centre of plant domestication, where selection began in the Late Pleistocene to Early Holocene in peripheral parts of the basin [18]. The origin of cultivated *I. edulis* trees is uncertain, though probably Amazonian [8]; nevertheless, some authors have suggested it was started by European settlers in west Amazonia [17,20]. Since this tree was cultivated mainly for fruit production, domestication is expected to increase pod length [8,16,20]. To our knowledge, no study has been made comparing both types of populations, cultivated vs. wild, considering this morphologic characteristic (pod length). Certainly, the higher values found in cultivated trees compared to the wild trees clearly support the domestication of *I. edulis* for food supply. Plant domestication is a long-term process in which natural selection interacts with human selection, driving changes that improve usefulness to humans and adaptations to domesticated landscapes [18].

In the current study, maximum pod length in the wild and the cultivated populations was 73 and 148 cm, respectively, in agreement with Pennington [8]. This author reported that wild trees' pods rarely exceed 50 cm and cultivated trees could, exceptionally, produce pods exceeding 2 m. The average pod length was higher in the Loreto region's cultivated trees, compared to Ucayali and Selva Central regions; the smallest fruits were observed in Selva Central. The species' different cultivation and uses, and differences in ecological conditions, could explain these results. Indeed, in Selva Central, the species was mainly used to shade coffee or cocoa rather than to produce large fruits [42,43]. Farmers were focused mainly on the cash crop yield, rather than the fruit yield provided by shade trees. Additionally, large fruits could be more attractive to uninvited guests, which could then cause damage to the cash crop due to *Inga* fruit collection. Another supporting argument is the wild *I. edulis* local name among the Selva Central region inhabitants. The local name for the cultivated *I. edulis* in Selva Central is "pacay soga", whereas in the Ucayali and Loreto regions, the name "Guaba" is used for the cultivated type and "guabilla" or "guabilla del monte" for the wild tree (A. Rollo, pers. communication). The difference in local names in these regions might be related to the species abundance, both in the wild and cultivated form. Locals in Selva Central informed us that *I. edulis* was hard to find in the surrounding wild vegetation; indeed, the species is rarely seen above 750 m [8]. We were also unable to find and sample wild trees in the Selva Central region.

4.2. Genetic Diversity of Wild and Cultivated Populations of *I. edulis* in the Peruvian Amazon

The overall H_E (0.69) was slightly higher than the H_O (0.59), inducing an overall inbreeding coefficient index of 11%. In a meta-study for microsatellites and outcrossing species, the author showed a similar value of H_E (0.65), but slightly higher H_O (0.63) [44].

The results from our study further indicate that all the genetic diversity estimates were lower in the case of the cultivated populations compared to the wild ones, as well as the average inbreeding coefficient. These results confirm a loss of genetic diversity in the cultivated populations, in agreement with the studies by Hollingsworth et al. [21] and Dawson et al. [13] on the same species. These authors concluded that cultivated stands possessed lower total allelic richness than neighbouring wild populations, but the expected genetic diversity remained unchanged, indicating that the process of domestication reduced the number of alleles. Both authors stated that the wild plant material they studied was collected from nearby cultivated populations, in old-growth, primary forest, but due to (i) the long history of the species' use, (ii) the habits of slash-and-burn in primary forest and (iii) gene flow among nearby stands, the wildness of the trees could be questioned [2,18]. Nevertheless, Dawson et al. [13] observed marked differences in the haplotype composition between natural and cultivated stands. In our case, the wild material was sampled in natural vegetation in protected areas and secondary forest, and unless extensive long-distance gene flow existed, no ambiguities in distinguishing both types existed. In addition, the results regarding pod length clearly distinguish the wild from cultivated material. We found an important effect of the domestication on the natural resources of a species, which is an expected phenomenon when a species is used by humans [23,45]. In some cases, the expected heterozygosity might be higher or similar in the cultivated population than that displayed in the wild population, due to a putative "melting pot" phenomenon in the former populations (introduced alleles from different origins). Nevertheless, the allelic richness and observed heterozygosity found, in our study, in the cultivated populations was lower than in the wild ones, indicating the loss of rare alleles during selection as observed by other authors [23,46]. Some cultivated populations from the current study had a significant heterozygote deficit, particularly in the Ucayali region. The consequences of the inbreeding effect in fruit trees, such as *I. edulis,* might impact fruit production due to inbreeding depression, which would directly impact farmers' yield [15,23]. The fact that the species is self-incompatible [13] excludes the possibility of heterozygote deficiency due to self-pollination; probably, related trees were introduced in these populations and the value reflect biparental inbreeding.

4.3. Population Structure

The genetic variance partition in our study (92% of the variance was observed within populations and a low genetic structure, 7%, was detected among populations) is usual in outcrossing tropical forest tree species with high levels of gene flow [6]. The hierarchical AMOVA showed that the Φct between wild and cultivated populations was 1%, yet not significant.

Dawson et al. [13] found low genetic structure, similarly to our results, in *I. edulis* natural and cultivated stands, with nuclear but not chloroplast microsatellite data. Nevertheless, the authors used only two chloroplast loci, which might have biased the results, since the smaller effective population size of the chloroplast genome makes it more susceptible to genetic drift and species differentiation [47]. Conversely, a high genetic structure was found between natural and cultivated stands of *I. vera*, and the authors reasoned that the cultivated populations were derived from seeds coming from different mother trees, but a different geographic origin was also possible [23].

In the current study, for $K = 2$ (Figures 3 and 4), the wild populations displayed identical composition, with the predominant red cluster. The uniform composition of the studied wild material could be due to the relatively recent speciation of the genus [9] and, also, to regional wild populations sampling [8]. The red cluster prevailed in the northern cultivated populations: the Loreto region and along the Ucayali river in the Ucayali region, which could express large population centres occupying the margins of main rivers with extensive trade networks [43]. A tiny green genetic cluster is present in the *wild* populations and in the cultivated populations of the Loreto region. Conversely, the green cluster is relevant in Selva Central and Ucayali cultivated populations. The green cluster increases in the sub-Andean Selva Central region and in the higher elevated sites from the Ucayali region, where the *I. edulis* trees were traditionally used on coca fields before the Conquest, and for shading protection of cocoa, coffee and tea plantations after the Spanish settlement [43]. In the Loreto region, the cultivated population 13 BRc possessed a higher proportion of green cluster than others from this region. This population is near Bretaña village, which was named after the Europeans, who arrived from the Andes and the coastal regions of Peru, during the rubber boom at the end of the 19th century [48].

Iquitos, in the Loreto region, is referred to as a crop domestication centre in Amazonia, created as populations expanded, and providing strong evidence that pre-conquest human populations had intensively transformed their plant resources [18,20]. Indeed, the *I. edulis* domestication was probably achieved by selecting from the local wild population and possibly started in the Loreto region, since the genetic structure of the cultivated populations from this region do not differ much from the wild ones. Moreover, they have bigger pods and lower allelic richness than the other cultivated populations, which could indicate that the selection intensity was higher here. Indeed, some authors claim that the possible origin of *I. edulis* domestication was in this region, which was also the location for the domestication of other species [18,22]. Additionally, the crop is probably recently domesticated, since when a crop is in an initial process of domestication no clear genetic structuring occurs, as in Brazil nut [22]. The genetic differentiation between the wild and cultivated populations is low and with admixture; the cultivated populations seem to originate from the wild ones. Conversely, the results of Dawson et al. [13] on chloroplast haplotype composition displayed a completely different pattern between natural and cultivated populations. The authors explained these results by a non-local origin of the *I. edulis* cultivated material. Our results do not support this theory. Instead, we inferred that the cultivated populations had a local germplasm origin, yet without representative sampling, which is expected, since a few trees were probably selected in nearby wild populations. Indeed, a possible genetic drift effect (change in the frequency of the allele in a population due to random sampling of organisms) in *the cultivated* populations is expected.

4.4. Practical Measures to Maintain I. edulis Genetic Resources

The *I. edulis* germplasm management should focus on both the wild and the cultivated stands. In the case of wild material, the protection of the original Amazonian vegetation remnants is key to maintaining the species' genetic resources in the region. In modern-day Amazonia, increasing deforestation for the establishment of pastures has become a global concern due to its impacts on biodiversity [43]. Considering the cultivated stands, the villages and indigenous settlements are the units of interest because they are the domesticated plant population keepers. Consequently, the fate of the village will determine the maintenance of the crop genetic resources. For example, the post-Colombian population collapse that resulted in a loss of village units and corresponded to the loss in human numbers (ca. 90–95% population decline), was quickly reflected in the loss of crop diversity [20,43]. The cultivated populations with low genetic diversity and/or high inbreeding estimates (e.g., 7 VHc, 10 ARc, 12 SSc, 14 JHc, 17 EPc, 18 EDc, 19 MAc and 21 INc) should be supported with new germplasm sources (from wild populations) to eliminate the risk of biparental inbreeding and diversity loss, which might be reflected in the future crop value (inbreeding depression, flower abortion, and crop yield failure).

5. Conclusions

The results of the current study on *I. edulis* show a significantly higher value for average pod length in cultivated trees than in wild trees. The wide-scale infusion from wild stands into farms could negatively affect fruit size and weaken domestication efforts over time. Additionally, the Loreto region displayed the highest average pod length, as well as having populations with a lower allelic richness when compared to the cultivated populations of other regions.

The cultivated stands in the Selva Central and Ucayali region could, additionally, be a germplasm material source, and thus could provide a long-term safeguard for on-farm conservation since the Loreto region possesses the populations with the lowest values of allelic richness. Hybridization programmes using such a germplasm source and local wild material with backward selection could help increase the crop yield and genetic diversity in the cultivated populations. Additionally, new selection should consider the needs of modern agriculture and forest management practices, as well as global warming.

Supplementary Materials: The following are available online at http://www.mdpi.com/1999-4907/11/6/655/s1, Figure S1. Representation of the ΔK distribution statistic. The method of Evanno et al. [37] was used to determine the most appropriate number of genetic clusters (K); Table S1. *Inga edulis* 27 populations (cultivated and wild), region (Geographic region), sampling region (Site), population code (Pop.) and sampling details (Sampling description); Table S2. Average fruit length per population. The minimal and maximal pod length values (in brackets) and the number of pods (Nl) per population, in a total of 448 pods. Table S3. Diversity parameters comparison between cultivated and wild populations. Allelic richness (R_S), observed heterozygosity (H_O) and inbreeding coefficient (F_{IS}). P = probability values for differences between groups for two-sided t-test after 1000 permutations. * = significant test ($p < 0.05$).

Author Contributions: Conceptualization, A.R., B.L. and M.M.R.; field data collection, A.R., Z.M.C.P., H.V., E.C., S.G.T. and R.M.V.A.; methodology, A.R., M.M.R., R.L.C., C.S.; formal analysis, writing—original draft preparation, M.M.R., A.R., B.M., M.K. and T.H.; writing—review and editing, M.M.R., A.R., B.L., B.M., M.K. and T.H.; funding acquisition, A.R., B.L. and M.M.R. All authors have read and agreed to the published version of the manuscript.

Funding: This study was funded by the Bilateral Project ´Morphological and genetic diversity of indigenous tropical trees in the Amazon—model study of Inga edulis Mart. in Peruvian Amazon´; Czech Academy of Sciences and CONCYTEC, Peru 2011–2012; The Internal Grant Agency of CULS Prague (No. 20185004, No. 20205003); The Scholarship National University of Ucayali, Peru; European Union Lifelong Learning Programme Erasmus Consortium—Practical Placement Scholarship (Certificate No. CZ-01-2009). The Foundation Nadace Nadání Josefa, Marie a Zdeňky Hlávkových, Czech Republic; Supported by the Ministry of Agriculture of the Czech Republic, institutional support MZE-RO0418; Foundation for Science and Technology, Portugal, UIDB/00239/2020 and UIDB/00681/2020 supported MMR.

Acknowledgments: We thank the Servicio Nacional de Áreas Naturales Protegidas and to José Grocio Gil Navarro director of Pacaya Samiria National Reservation for the investigation authorization (N° 004-2012-SERNANP-RNPS-J). Thanks are extended to N. Roque for help with Figure 1, to D. Petrus for help with Figure 4, and to I. Salavessa for editing the English of the manuscript. We want to thank two anonymous reviewers for the very helpful comments and suggestions, which considerably improved the manuscript.

Conflicts of Interest: The authors declare no conflict of interest.

References

1. Bodmer, R.; Fang, T.; Antunez, M.; Puertas, P.; Chota, K.; Pittet, M.; Kirkland, M.; Walkey, M.; Rios, C.; Perez-Peña, P.; et al. Impact of Climate Change on Wildlife and Indigenous Communities in Flooded Forests of the Peruvian Amazon. In *The Lima Declaration on Biodiversity and Climate Change: Contributions from Science to Policy for Sustainable Development*; Rodríguez, L., Anderson, I., Eds.; Secretariat of the Convention on Biological Diversity: Montreal, QC, Canada, 2017; Volume 89, pp. 81–90.
2. Levis, C.; Flores, B.M.; Moreira, P.A.; Luize, B.G.; Alves, R.P.; Franco-Moraes, J.; Lins, J.; Konings, E.; Peña-Claros, M.; Bongers, F.; et al. How People Domesticated Amazonian Forests. *Front. Ecol. Evol.* **2018**, *5*. [CrossRef]
3. Dawson, I.K.; Lengkeek, A.; Weber, J.C.; Jamnadass, R. Managing genetic variation in tropical trees: Linking knowledge with action in agroforestry ecosystems for improved conservation and enhanced livelihoods. *Biodivers. Conserv.* **2009**, *18*, 969–986. [CrossRef]
4. Brooks, T.M.; Mittermeier, R.A.; da Fonseca, G.A.B.; Gerlach, J.; Hoffmann, M.; Lamoreux, J.F.; Mittermeier, C.G.; Pilgrim, J.D.; Rodrigues, A.S.L. Global biodiversity conservation priorities. *Science* **2006**, *313*, 58–61. [CrossRef] [PubMed]
5. Oliveira, P.J.C.; Asner, G.P.; Knapp, D.E.; Almeyda, A.; Galván-Gildemeister, R.; Keene, S.; Raybin, R.F.; Smith, R.C. Land-use allocation protects the Peruvian Amazon. *Science* **2007**, *317*, 1233–1236. [CrossRef] [PubMed]
6. Finkeldey, R.; Hattemer, H.H. *Tropical Forest Genetics*; Springer-Verlag: Berlin/Heidelberg, Germany, 2007; p. 315. [CrossRef]
7. Wee, A.K.S.; Li, C.H.; Dvorak, W.S.; Hong, Y. Genetic diversity in natural populations of *Gmelina arborea*: Implications for breeding and conservation. *New Forests* **2012**, *43*, 411–428. [CrossRef]
8. Pennington, T.D. *The Genus Inga: Botany*; Royal Botanic Gardens, Kew: London, UK, 1997; p. 854.
9. Richardson, J.E.; Pennington, R.T.; Pennington, T.D.; Hollingsworth, P.M. Rapid diversification of a species-rich genus of neotropical rain forest trees. *Science* **2001**, *293*, 2242–2245. [CrossRef]
10. Lavin, M. Floristic and geographical stability of discontinuous seasonally dry tropical forests explains patterns of plant phylogeny and endemism. In *Neotropical Savannas and Seasonally Dry Forests: Plant Diversity,*

Biogeography, and Conservation; Pennington, R.T., Lewis, G.P., Ratter, J.A., Eds.; Crc Press-Taylor & Francis Group: Boca Raton, FL, USA, 2006; pp. 433–447.

11. Cruz-Neto, O.; Machado, I.; Duarte, J., Jr.; Lopes, A. Synchronous phenology of hawkmoths (Sphingidae) and *Inga* species (Fabaceae-Mimosoideae): Implications for the restoration of the Atlantic forest of northeastern Brazil. *Biodivers. Conserv.* **2011**, *20*, 751–765. [CrossRef]
12. Figueiredo, M.; Bruno, R.; Barros e Silva, A.; Nascimento, S.; Oliveira, I.; Felix, L. Intraspecific and interspecific polyploidy of Brazilian species of the genus *Inga* (Leguminosae: Mimosoideae). *Genet. Mol. Res.* **2014**, *13*, 3395–3403. [CrossRef]
13. Dawson, I.; Hollingsworth, P.; Doyle, J.; Kresovich, S.; Weber, J.; Sotelo Montes, C.; Pennington, T.; Pennington, R. Origins and genetic conservation of tropical trees in agroforestry systems: A case study from the Peruvian Amazon. *Conserv. Genet.* **2008**, *9*, 361–372. [CrossRef]
14. Pennington, T.D. Inga management. In *The Genus Inga: Utilization*; Pennington, T., Fernandes, E., Eds.; Royal Botanic Gardens, Kew: London, UK, 1998; pp. 159–167.
15. Koptur, S. Outcrossing and pollinator limitation of fruit-set: Breeding systems of Neotropical *Inga* trees (Fabaceae: Mimosoideae). *Evolution* **1984**, *38*, 1130–1143. [CrossRef]
16. Reynel, C.; Pennington, T.D. *El Género Inga en el Perú: Morfología, Distribución y Usos*; Royal Botanic Gardens, Kew: London, UK, 1997.
17. León, J. *Botánica de los Cultivos Tropicales*; Servicio Editorial IICA: San José, Costa Rica, 1987; p. 445.
18. Clement, C.R.; Denevan, W.M.; Heckenberger, M.J.; Junqueira, A.B.; Neves, E.G.; Teixeira, W.G.; Woods, W.I. The domestication of Amazonia before European conquest. *Proc. R. Soc. Lond. B Biol. Sci.* **2015**, *282*, 20150813. [CrossRef]
19. Nichols, J.D.; Carpenter, F.L. Interplanting *Inga edulis* yields nitrogen benefits to *Terminalia amazonia*. *For. Ecol. Manag.* **2006**, *233*, 344–351. [CrossRef]
20. Clement, C.R. 1492 and the loss of Amazonian crop genetic resources. I. The relation between domestication and human population decline. *Econ. Bot.* **1999**, *53*, 188. [CrossRef]
21. Hollingsworth, P.M.; Dawson, I.K.; Goodall-Copestake, W.P.; Richardson, J.E.; Weber, J.C.; Sotelo Montes, C.; Pennington, R.T. Do farmers reduce genetic diversity when they domesticate tropical trees? A case study from Amazonia. *Mol. Ecol.* **2005**, *14*, 497–501. [CrossRef] [PubMed]
22. Clement, C.; De Cristo-Araújo, M.; Coppens D'Eeckenbrugge, G.; Alves Pereira, A.; Picanço-Rodrigues, D. Origin and domestication of native Amazonian crops. *Diversity* **2010**, *2*, 72. [CrossRef]
23. Cruz-Neto, O.; Aguiar, A.V.; Twyford, A.D.; Neaves, L.E.; Pennington, R.T.; Lopes, A.V. Genetic and ecological outcomes of *Inga vera* subsp *affinis* (Leguminosae) tree plantations in a fragmented tropical landscape. *PLoS ONE* **2014**, *9*. [CrossRef]
24. Dawson, I.K.; Leakey, R.; Clement, C.R.; Weber, J.C.; Cornelius, J.P.; Roshetko, J.M.; Vinceti, B.; Kalinganire, A.; Tchoundjeu, Z.; Masters, E.; et al. The management of tree genetic resources and the livelihoods of rural communities in the tropics: Non-timber forest products, smallholder agroforestry practices and tree commodity crops. *For. Ecol. Manag.* **2014**, *333*, 9–21. [CrossRef]
25. Rollo, A.; Lojka, B.; Honys, D.; Mandák, B.; Wong, J.A.C.; Santos, C.; Costa, R.; Quintela-Sabarís, C.; Ribeiro, M.M. Genetic diversity and hybridization in the two species *Inga ingoides* and *Inga edulis*: Potential applications for agroforestry in the Peruvian Amazon. *Ann. For. Sci.* **2016**, *73*, 425–435. [CrossRef]
26. Dayanandan, S.; Bawa, K.; Kesseli, R. Conservation of microsatellites among tropical trees (Leguminosae). *Am J Botany* **1997**, *84*, 1658–1663. [CrossRef]
27. Sokal, R.R.; Rohlf, F.J. *Biometry*, 2nd ed.; W.H. Freeman and Co.: San Francisco, CA, USA, 1981; p. 859.
28. El Mousadik, A.; Petit, R.J. High level of genetic differentiation for allelic richness among populations of the argan tree [*Argania spinosa* (L.) Skeels] endemic to Morocco. *Theor. Appl. Genet.* **1996**, *92*, 832–839. [CrossRef]
29. Nei, M. *Molecular Evolutionary Genetics*; Columbia University Press: New York, NY, USA, 1987; p. 333.
30. Goudet, J. FSTAT (Version 1.2): A computer program to calculate F-statistics. *J. Hered.* **1995**, *86*, 485–486. [CrossRef]
31. Peakall, R.; Smouse, P.E. GenAlEx 6.5: Genetic analysis in Excel. Population genetic software for teaching and research—an update. *Bioinformatics* **2012**, *28*, 2537–2539. [CrossRef] [PubMed]
32. Rousset, F. Genepop'007: A complete re-implementation of the genepop software for Windows and Linux. *Mol. Ecol. Resour.* **2008**, *8*, 103–106. [CrossRef] [PubMed]

33. Excoffier, L.; Lischer, H.E.L. Arlequin suite ver 3.5: A new series of programs to perform population genetics analyses under Linux and Windows. *Mol. Ecol. Resour.* **2010**, *10*, 564–567. [CrossRef] [PubMed]
34. Pritchard, J.K.; Stephens, M.; Donnelly, P. Inference of population structure using multilocus genotype data. *Genetics* **2000**, *155*, 945–959.
35. Hubisz, M.J.; Falush, D.; Stephens, M.; Pritchard, J.K. Inferring weak population structure with the assistance of sample group information. *Mol. Ecol. Resour.* **2009**, *9*, 1322–1332. [CrossRef]
36. Wang, J. The computer program structure for assigning individuals to populations: Easy to use but easier to misuse. *Mol. Ecol. Resour.* **2017**, *17*, 981–990. [CrossRef]
37. Evanno, G.; Regnaut, S.; Goudet, J. Detecting the number of clusters of individuals using the software structure: A simulation study. *Mol. Ecol.* **2005**, *14*, 2611–2620. [CrossRef]
38. Earl, D.A.; vonHoldt, B.M. STRUCTURE HARVESTER: A website and program for visualizing STRUCTURE output and implementing the Evanno method. *Conserv. Genet. Resour.* **2012**, *4*, 359–361. [CrossRef]
39. Jakobsson, M.; Rosenberg, N.A. CLUMPP: A cluster matching and permutation program for dealing with label switching and multimodality in analysis of population structure. *Bioinformatics* **2007**, *23*, 1801–1806. [CrossRef]
40. Ramasamy, R.K.; Ramasamy, S.; Bindroo, B.B.; Naik, V.G. STRUCTURE PLOT: A program for drawing elegant STRUCTURE bar plots in user friendly interface. *Springerplus* **2014**, *3*, 431. [CrossRef] [PubMed]
41. Ormsby, T.; Napoleon, E.; Burke, R.; Groessl, C.; Bowden, L. *Getting to Know ArcGIS Desktop*; ESRI Press: Redlands, CA, USA, 2010; p. 592.
42. León, J. Inga as shade for coffee, cacao, and tea: Historical aspects and present day utilization. In *The Genus Inga: Utilization*; Pennington, T., Fernandes, E., Eds.; Royal Botanic Gardens, Kew: London, UK, 1998; pp. 101–115.
43. Miller, R.P.; Nair, P.K.R. Indigenous agroforestry systems in Amazonia: From Prehistory to Today. *Agrofor. Syst.* **2006**, *66*, 151–164. [CrossRef]
44. Nybom, H. Comparison of different nuclear DNA markers for estimating intraspecific genetic diversity in plants. *Mol. Ecol.* **2004**, *13*, 1143–1155. [CrossRef] [PubMed]
45. Ribeiro, M.M.; Plomion, C.; Petit, R.; Vendramin, G.G.; Szmidt, A.E. Variation in chloroplast single-sequence repeats in Portuguese maritime pine (*Pinus pinaster* Ait.). *Theor. Appl. Genet.* **2001**, *102*, 97–103. [CrossRef]
46. Jones, T.H.; Steane, D.A.; Jones, R.C.; Pilbeam, D.; Vaillancourt, R.E.; Potts, B.M. Effects of domestication on genetic diversity in *Eucalyptus globulus*. *For. Ecol. Manag.* **2006**, *234*, 78–84. [CrossRef]
47. Petit, R.J.; Duminil, J.; Fineschi, S.; Hampe, A.; Salvini, D.; Vendramin, G.G. Comparative organization of chloroplast, mitochondrial and nuclear diversity in plant populations. *Mol. Ecol.* **2005**, *14*, 689–701. [CrossRef]
48. Eidt, R.C. Pioneer settlement in Eastern Peru. *Ann. Assoc. Am. Geogr.* **1962**, *52*, 255–278. [CrossRef]

Article

High Genetic Diversity and Low Differentiation in *Michelia shiluensis*, an Endangered Magnolia Species in South China

Yanwen Deng [1,2], Tingting Liu [1,2], Yuqing Xie [1,2], Yaqing Wei [3,4], Zicai Xie [2], Youhai Shi [3,4,*] and Xiaomei Deng [1,2,*]

1 Guangdong Key Laboratory for Innovative Development and Utilization of Forest Plant Germplasm, Guangzhou 510642, China; hinmentang@gmail.com (Y.D.); tingoo1225@163.com (T.L.); serumia@163.com (Y.X.)

2 College of Forestry and Landscape Architecture, South China Agricultural University, Guangzhou 510642, China; XieZC2020@163.com

3 Key Laboratory of Germplasm Resources Biology of Tropical Special Ornamental Plants of Hainan Province, Haikou 570228, China; hnweiyaqing@163.com

4 College of Forestry, Hainan University, Haikou 570228, China

* Correspondence: shiyouhai@163.com (Y.S.); xmdeng_scau@163.com (X.D.); Tel.: +86-135-1803-0049 (Y.S.); +86-137-1116-8219 (X.D.)

Received: 26 March 2020; Accepted: 17 April 2020; Published: 21 April 2020

Abstract: *Research Highlights*: This study is the first to examine the genetic diversity of *Michelia shiluensis* (Magnoliaceae). High genetic diversity and low differentiation were detected in this species. Based on these results, we discuss feasible protection measures to provide a basis for the conservation and utilization of *M. shiluensis*. *Background and Objectives*: *Michelia shiluensis* is distributed in Hainan and Guangdong province, China. Due to human disturbance, the population has decreased sharply, and there is thus an urgent need to evaluate genetic variation within this species in order to identify an optimal conservation strategy. *Materials and Methods*: In this study, we used eight nuclear single sequence repeat (nSSR) markers and two chloroplast DNA (cpDNA) markers to assess the genetic diversity, population structure, and dynamics of 78 samples collected from six populations. *Results*: The results showed that the average observed heterozygosity (Ho), expected heterozygosity (He), and percentage of polymorphic loci (PPL) from nSSR markers in each population of *M. shiluensis* were 0.686, 0.718, and 97.92%, respectively. For cpDNA markers, the overall haplotype diversity (Hd) was 0.674, and the nucleotide diversity was 0.220. Analysis of markers showed that the genetic variation between populations was much lower based on nSSR than on cpDNA (10.18% and 77.56%, respectively, based on an analysis of molecular variance (AMOVA)). Analysis of the population structure based on the two markers shows that one of the populations (DL) is very different from the other five. *Conclusions*: High genetic diversity and low population differentiation of *M. shiluensis* might be the result of rich ancestral genetic variation. The current decline in population may therefore be due to human disturbance rather than to inbreeding or genetic drift. Management and conservation strategies should focus on maintaining the genetic diversity in situ, and on the cultivation of seedlings ex-situ for transplanting back to their original habitat.

Keywords: nSSR; cpDNA; Magnoliaceae; conservation genetics; fragmentation

1. Introduction

Habitat destruction and environmental pollution caused by human disturbance [1] have limited the distribution and thus promoted the spatial isolation of many wildlife species, threatening their reproduction and development [2,3]. In addition to human disturbance, genetic factors can also represent threats to wild species [4]; such processes as mating systems, genetic drift, gene flow, evolution, and life history greatly affect the genetic diversity and spatial structure of plant populations [5–9]. Under the combined effects of human and genetic factors, the gene flow in natural populations is reduced in the short term, resulting in increased genetic differentiation among populations and genetic drift effects within populations, which together reduce genetic variation [10,11]. Over extended temporal scales, low genetic variation coupled with high differentiation put species at risk of extinction by reducing their evolutionary potential [12]. In order to maintain the adaptability and long-term survival ability of species [13–15], it is necessary to understand the genetic diversity and population structure across a given species' distribution area [16], and thereby develop scientifically grounded conservation strategies [17].

Michelia shiluensis (Magnoliaceae) is an evergreen tree that has been placed on the Red List of Magnoliaceae, which identifies magnolia species that are at risk of extinction throughout the world [18]. This tree is scattered throughout Hainan [19] and southern Guangdong provinces [20], and is characterized by a straight tree shape, bright leaves, elegant and aromatic flowers, and dense wood structure [21,22]. As a desirable ornamental garden plant and an important timber tree species, the wild resources of *M. shiluensis* have been severely misappropriated and plundered. In addition, the development of scenic tourism in recent years has caused damage to habitats, and thus seriously threatened the natural survival and reproduction of *M. shiluensis* [23]. Therefore, there is an urgent need to understand the genetic diversity and population structure of *M. shiluensis* in order to plan conservation actions on its behalf.

Molecular markers have been widely used to study magnolia species. Simple sequence repeat (SSR) markers, also known as microsatellite DNA, are widely distributed throughout the nuclear genome [24]. These markers can display high polymorphism and reproducibility [25], reflect gene flow from seed and pollen [26], and depict fine genetic structure [27]. Unlike the bi-parentally inherited nuclear genome, the chloroplast genome is inherited exclusively from the maternal line in most angiosperms [28,29], and is therefore widely used as a marker for inferring species phylogeography, origin, and historical dynamics [30–32]. The combined analysis of the two markers can be used to comprehensively assess the importance of each population and provide a theoretical basis for conservation strategies [33].

In this study, we used eight pairs of nuclear single sequence repeat (nSSR) loci and two cpDNA gene spacers to study the genetic diversity and population structure of 78 *M. shiluensis* individuals from six populations. The purpose of this study was to: (1) identify genetic diversity at both population and species levels, (2) explore the correlation between genetic diversity and population size, (3) determine the level of population differentiation and dynamics, and (4) discuss possible effective strategies for protection. This research will help the development of conservation and breeding plans for *M. shiluensis*.

2. Materials and Methods

2.1. Plant Material and DNA Extraction

Based on a previous investigation and information from the National Plant Specimen Resource Center (NPSRC), we identified seven possible wild populations for sampling: one located in Guangdong province (YC(YangChun E'HuangZhang Nature Reserve), and the other six in Hainan province (DL (DiaoLuoShan Nature Reserve), BS (BaiSha County), YG (YingGeLing Nature Reserve), WZ (WuZhiShan Nature Reserve), CJ (ChangJiang County), and WN (WanNing County)). However, we failed to locate any individuals from the WN population. Thus, we sampled 78 individuals out of six populations (DL, BS, YG, WZ, CJ, and YC) from two provinces (Table S1). It should be noted that we were unable to locate seedlings from any of the populations except for DL. Therefore, we sampled all individuals from the BS, YG, WZ, CJ, and YC populations, and randomly collected samples from DL (at least 15 m between each individual). Leaves from each individual were placed into Ziploc bags sealed with silica gel, then brought back to our laboratory and stored at 4 °C until DNA extraction.

Total genomic DNA was extracted from dried leaves following the supplied protocol in the Plant Genomic DNA Kit (DP305, TianGen Biotech Co., LTD, Beijing, China). The NanoDropTM 2000 Spectrophotometers (Thermo Scientific, Waltham, MA, USA) were used to test the concentration of extracted DNA solution, and the DNA work solution was prepared at an approximate concentration of 10 ng/μL.

2.2. Primers and Fragment Amplification

In total, 117 nSSR primer pairs were chosen from previous studies on Magnoliaceae; all were screened using eight samples from six populations. Screening revealed 16 pairs of primers that were able to amplify specific bands from selected samples. Among these, eight pairs showed high peak fluorescence signals and rich polymorphism in genotyping results. A total of 10 pairs of universal cpDNA primers were selected for screening; only two displayed rich polymorphisms in sequencing results. All primers (Table S2) were manufactured by Tsingke (Tsingke Biotech Co., Beijing, China).

Two PCR procedures were conducted for nSSR amplification, following the method of Schuelke (2000) [34]. The final volume used in the first PCR cycle was 10 μL; this contained 1 μL of genomic DNA (about 10 ng), 5 μL of 2X Es Taq Master Mix (Cwbiotech, Beijing, China), 1 μL each of forward and reverse primers, and ddH_2O filled in the remainder. In the second PCR cycle, 5 μL of PCR product from the first cycle was used as a template in the final 30 μL volume, which contained 15 μL of 2X Es Taq Master Mix (Cwbiotech, Beijing, China), 3 μL fluorescent dye primer, and ddH_2O filling in the remainder.

The DNA amplification protocol for nSSR was as follows: 5 min at 94 °C for denaturation, followed by cycles of 30 s at 94 °C, 30 s at specific Ta (Table S2), and 30 s at 72 °C, with a final extension step of 72 °C for 7 min. Twenty cycles were conducted for the first PCR procedure, and 35 for the second. The SSR-PCR products were run on the ABI 3730 DNA Analyzer (Biosystems, Foster City, CA, USA), using GS-500 as an internal size standard. Allele size was assessed using GeneMarker 2.2.0 [35]. For cpDNA amplification, cycling parameters were set as follows: initial denaturation at 94 °C for 3 min, followed by 35 cycles of denaturation at 94 °C for 1 min, annealing for 35 s at 56 °C, extension at 72 °C for 1 min, and a final extension at 72 °C for 5 min. Products of cpDNA amplification were sequenced on the ABI 3730 xl DNA Analyzer (Biosystems, Foster City, CA, USA).

2.3. Data Analysis

2.3.1. nSSR Analysis

Genetic diversity parameters were generated using GeneAlEx 6.5.2 [36]. These were: the number of observed alleles (Na), number of effective alleles (Ne), the observed (Ho) and expected (He) heterozygosity, the Shannon diversity index (I), the percentage of polymorphic loci (PPL), and the private alleles (Np). A Mantel test was also carried out in GenAlEx 6.5.2 [36]. PowerMarker v3.25 [37] was used to calculate polymorphism information content (PIC) and linkage disequilibrium (LD). Population genetic structure analysis using mixed models and allele-dependent frequency models was conducted using STRUCTURE 2.3.4 [38]. The parameters in STRUCTURE were set as follows: K = 1–6, Number of Interaction = 10, Burn-in period = 1,000,000, Markov chain Monte Carlo (MCMC) = 200,000. The results from STRUCTURE were compressed and uploaded to Structure Harvester online service; from this, we obtained the most likely K value (ΔK) and L value (K) using Evanno's method [39]. Optional clustering was summarized using CLUMPP v1.1.2 [40]. The neighbor-joining (NJ) tree clustering algorithm was conducted using Mega 7 [41] and edited on iTOL [42]. Additionally, gene flow between populations (Nm), F-statistics for inbreeding coefficient (F_{IS}), global inbreeding coefficient (F_{IT}), and the coefficient of genetic differentiation (F_{ST}) were generated using Popgene 32 [43]. Arelquin v3.5 [44] was used to calculate the pairwise F_{ST} for each population and for analysis of molecular variance (AMOVA).

2.3.2. cpDNA Analysis

Sequencing results were checked; misidentified results were manually corrected using BioEdit 7.0.5.3 [45]. Calculation of haplotype diversity (Hd) and nucleotide diversity (Pi), derivation of polymorphic loci, and a mismatch analysis were performed using DnaSP 6.12.03 [46]. The haplotype network was visualized using TCS 1.21 [47]. Sequence alignment shearing and Maximum Likelihood (ML) mapping were performed using MEGA 7 [41] with *Michelia odora* as an outgroup. We used PERMUT 1.2.1 [48] to calculate H_S, H_T, G_{ST}, and N_{ST}, along with their P values. Arlequin 3.5.2.2 [44] was used to calculate the pairwise F_{ST}, AMOVA, and neutrality tests. A Mantel test was conducted using GenAlEx 6.502 [36].

3. Results

3.1. Genetic Diversity and Population Structure from nSSR Markers

A total of 115 alleles were detected at the eight loci (Table S3). The Na at each locus ranged from 7 to 21, with an average of 14.375 per loci. The average I was 1.506; the lowest was 0.820 (WS18) and the highest was 1.767 (LT116). The PIC values ranged from 0.531 to 0.909, with an average of 0.809 per locus. The Ho values ranged from 0.482 (WS18) to 0.780 (MA3-7), with an average value of 0.686. Four of the eight loci displayed higher He than Ho, with the highest at LT116 (0.797) and the lowest at WS18 (0.445), and an average of 0.718. There was no evidence of significant LD. The average estimated F_{IS}, F_{IT}, F_{ST}, and Nm were 0.042, 0.176, 0.139, and 1.546, respectively (Table S4).

At the population level, the average Na, Ne, I, Ho, and He were 6.021 (4.625–9.500), 4.236 (3.598–5.174), 1.506 (1.316–1.784), 0.686 (0.563–0.786), and 0.718 (0.650–0.773), respectively (Table 1). The PPL in each population ranged from 87.50% to 100%, with an average of 97.92%. The PPL of the WZ accession was 87.50, whereas those of all other populations were 100. All populations contained private alleles, with 24 found in DL, followed by seven in YC, four in YG, three each in BS and WZ, and two in CJ. The inbreeding coefficient (F_{IS}) for CJ was 0.207; those for all of the other populations were less than 0.1. This result indicates that intense breeding had occurred within the CJ population. An examination of the correlation between population genetic diversity parameters and population size showed that Na, Ne, I, and Np were significantly related to the population size; however, Ho and He were not related (Figure 1). At the species level, Na, Ne, I, Ho, He were 14.375, 7.019, 2.120, 0.689, and 0.825, respectively. These parameters show high genetic diversity in *M. shiluensis* at both the species and population levels.

Analysis of nSSR genetic structure shows a Δ K of 167.38 for K = 2. The results show that all 78 individuals comprised two clusters (Figure 2), which separate individuals from DL from the other populations; there was evidence of minor genetic mixing between DL and other populations.

Table 1. Genetic diversity parameters among six populations of *Michelia shiluensis*.

Pop	N	Na ± SE	Ne ± SE	I ± SE	Ho ± SE	He ± SE	Np	F_{IS} ± SE	PPL (%)
DL	49	9.500 ± 1.195	5.174 ± 0.824	1.784 ± 0.137	0.694 ± 0.033	0.773 ± 0.035	24	0.093 ± 0.050	100.00
BS	7	6.250 ± 0.526	4.380 ± 0.470	1.606 ± 0.095	0.786 ± 0.071	0.753 ± 0.026	3	-0.044 ± 0.092	100.00
YG	5	5.375 ± 0.596	4.275 ± 0.619	1.483 ± 0.151	0.750 ± 0.098	0.715 ± 0.058	4	-0.054 ± 0.105	100.00
WZ	5	4.750 ± 0.648	3.598 ± 0.416	1.316 ± 0.197	0.700 ± 0.120	0.650 ± 0.094	3	−0.080 ± 0.096	87.50
CJ	8	5.625 ± 0.885	4.053 ± 0.614	1.447 ± 0.172	0.563 ± 0.085	0.701 ± 0.054	2	0.207 ± 0.099	100.00
YC	4	4.625 ± 0.324	3.936 ± 0.399	1.402 ± 0.109	0.625 ± 0.082	0.715 ± 0.046	7	0.095 ± 0.127	100.00
Mean	13	6.021 ± 0.376	4.236 ± 0.233	1.506 ± 0.061	0.686 ± 0.035	0.718 ± 0.023	7.17	0.039 ± 0.040	97.92
Overall	78	14.375 ± 1.647	7.019 ± 0.971	2.120 ± 0.153	0.689 ± 0.025	0.825 ± 0.039	-	0.156 ± 0.033	100.00

Notes: DL, DiaoLuoShan Nature Reserve; BS, BaiSha County; YG, YingGeLing Nature Reserve; WZ, WuZhiShan Nature Reserve; CJ, ChangJiang County; YC, YangChun E'HuangZhang Nature Reserve; N, number of samples; Na, number of alleles; Ne, number of effective alleles; I, Shannon index; Ho, observed heterozygosity; He, expected heterozygosity; Np, number of private alleles; F_{IS}, population inbreeding coefficient; PPL, percentage of polymorphic loci.

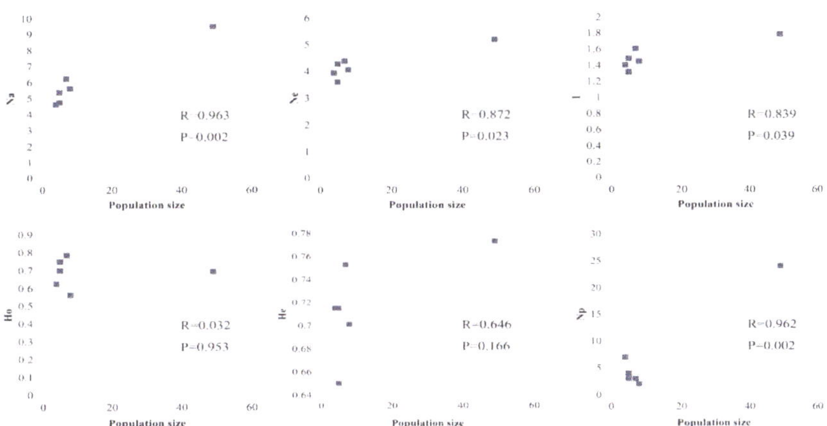

Figure 1. The correlation coefficient between population genetic diversity parameters and population size of *Michelia shiluensis*.

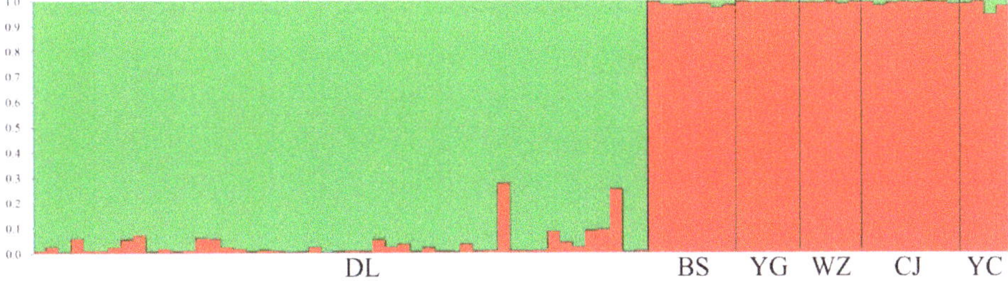

Figure 2. Genetic relation among six populations of *Michelia shiluensis* analyzed by STRUCTURE based on nuclear single sequence repeat (nSSR) markers.

The AMOVA analysis of the *M. shiluensis* populations revealed a high proportion of variance within the population (89.92%, $p < 0.001$), and a lower proportion of variance across populations (10.18%, $p < 0.001$; Table 2).

Table 2. AMOVA result for six populations from nSSR markers and cpDNA haplotypes.

Markers	Source of Variation	Degree of Freedom	Sum of Squares	Variance Component	Percentage of Variation	*p* Value
nSSR	Among populations	5	47.296	0.353	10.18	<0.01
	Within populations	150	467.370	3.116	89.92	<0.01
	Total	155	514.667	3.469	100	
cpDNA	Among populations	5	66.612	1.437	77.56	<0.01
	Within populations	72	29.927	0.416	22.44	<0.01
	Total	77	96.539	1.853	100	

The tree diagram obtained by the NJ clustering method shows that the 78 individuals from six accessions were divided into two clusters (Figure 3). Forty-six individuals from DL and one from BS comprise the green cluster, and three individuals from DL and the remaining individuals comprise the red cluster. This result is very similar to that obtained through STRUCTURE analysis.

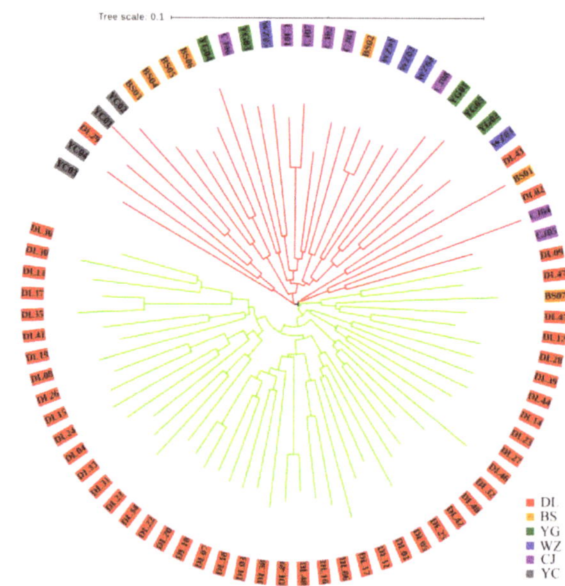

Figure 3. Neighbor joining (NJ) tree among individuals was constructed by MEGA 7.0 using nSSR data.

The Mantel test (Figure 4) confirmed that genetic distance and geographic distance were significantly correlated ($p = 0.04$, R = 0.66), indicating that an increase in geographic distance leads to increased genetic differentiation among populations. However, this correlation becomes insignificant upon removal of the YC population, which is located far from the main distribution area ($p = 0.20$, R = 0.33).

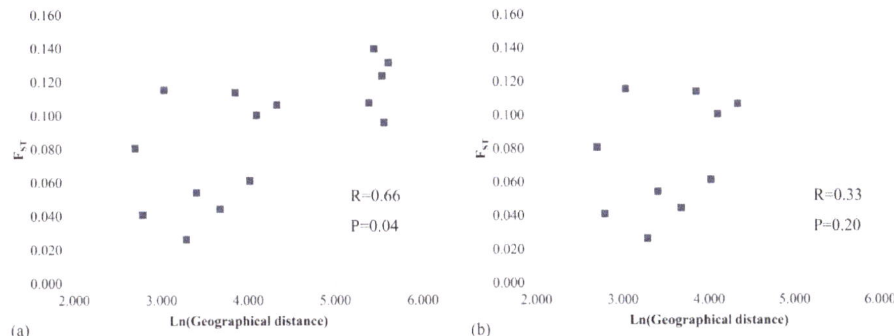

Figure 4. Mantel test among populations of *Michelia shiluensis* from nSSR markers. (**a**) Includes YC. (**b**) Excludes YC.

3.2. Genetic Diversity and Population Structure from cpDNA Marker

The aligned sequences consist of 1142 bp from two chloroplast DNA regions: *trnH-psbA* (400 bp) and *trnK59-matK* (742 bp). A total of 12 nucleic acid substitutions were observed within the region, and a total of six haplotypes were identified (Table S5). Haplotype diversity is highest in YG, followed by YC and DL; the lowest was found in BS, WZ, and CJ. Haplotype diversity ranges from 0 to 0.800, with an average of 0.674, reflecting the degree of difference among the haplotypes in each population. The YG accession shows moderate nucleotide diversity (0.00333), while DL (0.00074) and YC (0.00117) both show lower diversity. Haplotype and nucleotide diversity are both zero in three populations (BS, WZ, and CJ), which reflects the absence of variation between individuals within each of these populations (Table 3).

Table 3. Distribution of cpDNA haplotypes, Hd, and Pi among six populations of *Michelia shiluensis*.

Haplotype		H1	H2	H3	H4	H5	H6	Sum	Hd	Pi
Pops	DL	34	12	3	0	0	0	49	0.464	0.00074
	BS	0	0	7	0	0	0	7	0	0
	YG	0	0	2	2	1	0	5	0.800	0.00333
	WZ	0	0	5	0	0	0	5	0	0
	CJ	0	0	8	0	0	0	8	0	0
	YC	0	0	2	0	0	2	4	0.667	0.00117
	Overall	34	12	27	2	1	2	78	0.674	0.00220

The haplotype H3 was shared among all populations (Figure 5), and the remaining haplotypes were private in populations. The haplotype H3 was central in the haplotype network, and can mutate into any of the other five haplotypes through base substitution. Haplotypes H1 and H2 were exclusive to DL; H4 and H5 were exclusive to YG; and H6 was exclusive to YC. Based on its widespread distribution in all populations, we surmise that H3 may represent the ancestral haplotype of *M. shiluensis*.

The analysis of the ML tree (Figure 6) shows that the haplotypes H3, H4, and H5 comprised one cluster; H1 and H2 comprised another cluster; and H6 comprised a third cluster.

Total genetic diversity (H_T = 0.856) was higher than that within each population (H_S = 0.137). The permutation test revealed a higher value for N_{ST} (0.728) than G_{ST} (0.534), with $P < 0.05$, indicating a clear phylogenetic structure among the populations.

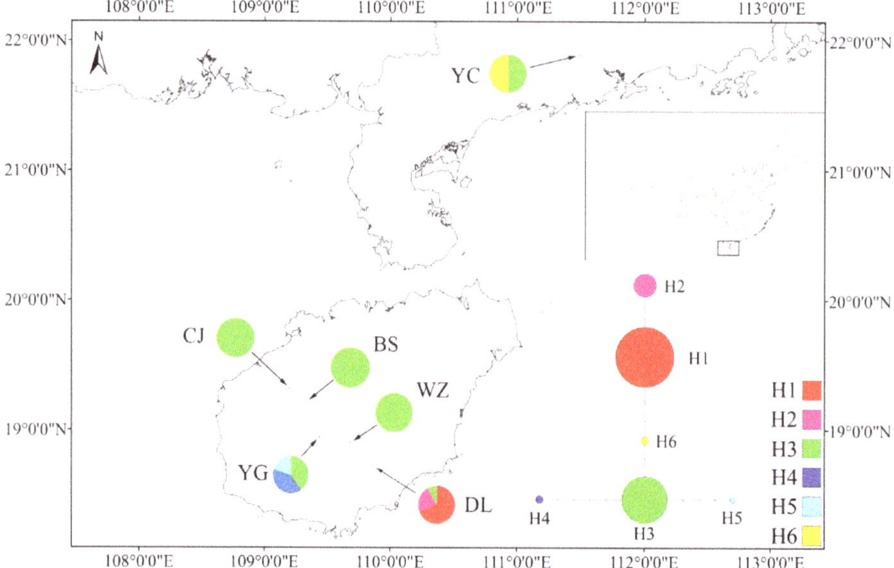

Figure 5. Location of the study populations and chloroplast haplotype relationships and distribution in six accessions of *Michelia shiluensis*. The radius of the pie charts is proportional to the number of individual haplotypes in each population.

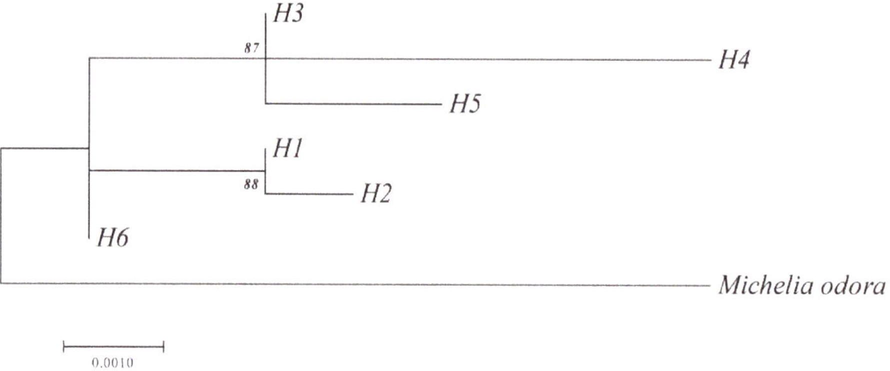

Figure 6. Maximum Likelihood (ML) tree based on six haplotypes of the cpDNA fragment with *Michelia odora* as an outgroup. The numbers are percentage values over 1000 bootstrap replicates. Only bootstrap values over 50% are shown.

AMOVA results (Table 2) show that chloroplast DNA diversity of all groups was 77.56% among populations and 22.44% within populations.

The neutrality test (Table 4) shows that both Tajima's D and Fu's FS were positive, but not significant (p = 0.525 and 0.904, respectively). This result indicates that chloroplast DNA diversity was less affected by selection than nuclear DNA diversity. The analysis of the mismatch distribution using multimodal plots (Figure 7) shows that a recent expansion in population was not supported, indicating that the population may be in dynamic equilibrium. The values of SSR and HRag were 0.059 (p = 0.25) and 0.190 (p = 0.41), respectively.

Table 4. Neutrality test and mismatch distribution from cpDNA fragments of *Michelia shiluensis*.

Hap. pops	Neutrality test		Mismatch distribution	
	Tajima's D(P)	Fu's FS(P)	SSD (PSSD)	HRag (PHRag)
All pops	0.075(0.525)	2.786(0.904)	0.059(0.25)	0.190(0.41)

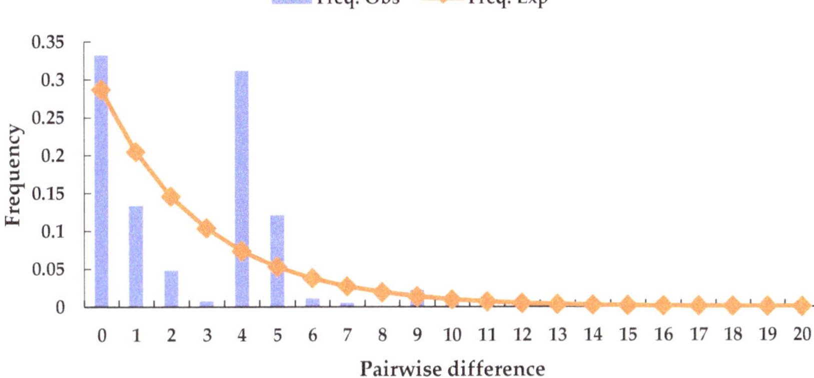

Figure 7. Multimodal plots from mismatch distribution of *Michelia shiluensis* analyzed by DnaSP.

The results of the correlation between geographic distance and genetic distance (Mantel test; Figure 8) show no significant correlation between the genetic distance and geographic distance of the *M. shiluensis* population at the cpDNA level (R = 0.01, p = 0.96), regardless of whether YC is removed (R = 0.63, p = 0.13).

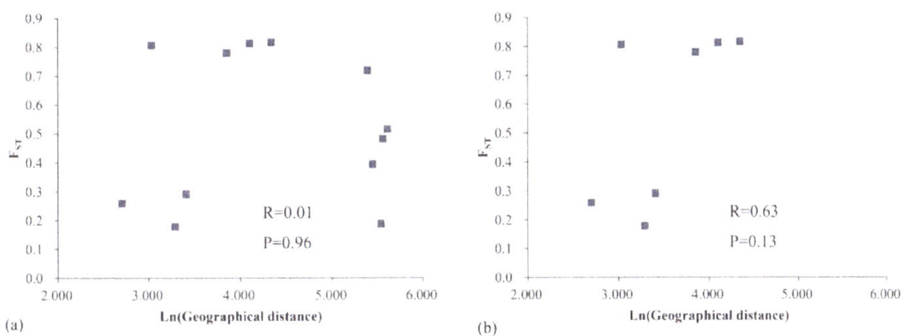

Figure 8. Mantel test among six populations of *Michelia shiluensis* from cpDNA markers. (**a**) Includes YC. (**b**) Excludes YC.

4. Discussion

4.1. High Level of Genetic Diversity

The level of genetic diversity plays an important role in the adaptive evolution and long-term survival of species and populations [13–15]. Generally, higher genetic diversity is found in widely distributed species than in those with restricted ranges [49,50]. However, many studies have shown that high genetic diversity can be maintained in endangered species despite dispersed populations [17,51]. In this study, a relatively high degree of genetic variation was found in the sampled populations of *M. shiluensis*. The nSSR data showed the population-level He to be 0.718, which is higher than that found in *Magnolia tomentosa* (He = 0.675) [52], *Michelia coriacea* (He = 0.47) [53], or *Magnolia wufengensis* (He = 0.184) [54], but lower than that of *Magnolia stellata* (He = 0.773) [9]. It is worth noting that the He found in this study is significantly higher than that found in some narrow-ranging species (He = 0.420) and in some widely distributed species (He = 0.620) [55]. Studies have shown that genetic diversity is positively correlated with population size [56–58]. Surprisingly, in this study, neither Ho (p = 0.953) nor He (p = 0.166) was significantly related to the population size (Figure 1). This outcome has been found in some special cases [59,60]. However, because the population size in this study has remained relatively uniform, this result might require further consideration. The cpDNA data show that the overall nucleotide diversity was relatively low (0.00220), but the haplotype diversity (Hd) was 0.674 (Table 3); this is higher than that of both *M. stellata* (Hd = 0.527) [9] and *Michelia maudiae* (Hd = 0.44) [27], but lower than that of *Michelia formosana* (Hd = 0.953) [61]. High haplotype diversity further indicates that *M. shiluensis* is highly genetically diverse and is not greatly affected by genetic drift [9]. The lower nucleotide diversity may be due to highly conserved sequences and low substitution rates within Magnoliaceae [62].

Although the six sampled *M. shiluensis* groups formed a single large population comprising several smaller subpopulations [63], the genetic diversity in each subpopulation was relatively high. The reasons for high genetic diversity may be as follows: first, we speculate that gene exchange between *M. shiluensis* populations has historically been frequent, and thus, the existing population inherited rich ancestral genetic variation [64]; second, *M. shiluensis* is a perennial plant, and therefore, the recent sharp decline of individuals has not increased the chance of inbreeding, which reduces genetic diversity [65]; third, plants of the *Michelia* genus tend to cross-pollinate, which can reduce the loss of genetic diversity through large gene flows [55,66,67].

4.2. Lack of Genetic Differentiation between Populations

The results of molecular analysis of variance (AMOVA; Table 2) showed that genetic variation among the *M. shiluensis* populations was 10.18% based on nSSR markers, while that based on cpDNA markers was 77.56%. This inconsistency may be related to the differential focus of the markers or high levels of gene flow [68]. Comparative analysis between bi-parental markers and maternally inherited markers can provide comprehensive insights into population dynamics because cpDNA mutations reflect past changes, while nSSR mutations can provide inferences for recent population events [69]. Thus, differences in genetic patterns and rates of evolution often produce large contrasts between nuclear and organelle genetic diversities under conditions of genetic differentiation [28].

According to Wright [70], if Nm is less than 1, genetic drift and differentiation may occur. In this study, Nm reached 1.546 (Table S3), thus, it is very likely that the *M. shiluensis* population has not yet undergone either genetic drift or any apparent degree of genetic differentiation. From the perspective of *M. shiluensis* haplotype distribution (Figure 5), haplotype H3 is distributed in all populations, which greatly reduces the level of differentiation between populations. In addition, the characteristic red seeds of Magnoliaceae are easily seen and eaten by birds [71]. As a result, the seeds can be widely dispersed via long-distance bird flight, thereby increasing the range of gene flow. Based on the above reasons, we can infer that a large amount of gene flow may be the reason for the current low level of genetic differentiation and indistinct geographic structure in the *M. shiluensis* population [72].

Generally, the level of genetic differentiation between populations increases with increasing geographic distance [73]. However, the two markers showed a different pattern when applied to geographically isolated *M. shiluensis* accessions (Figure 4, Figure 8). The results of nSSR showed that the genetic distance increased significantly with the increase in geographic distance (p = 0.04); however, removal of the YC population eliminated significance (p = 0.20). The YC population is far from the main distribution area and isolated by a strait (Figure 5). We speculate that the latter was the source of the significant difference we found. Results based on cpDNA showed no correlation between genetic and geographic distances regardless of whether YC is removed (P=0.96, 0.13). The result of the Mantel test may be due to the fact that, in the primary distribution range of *M. shiluensis* on Hainan Island [23], the maximum distance between groups of less than 80 km is insufficient to limit the range of bird activity; the resulting pattern is consistent with a lack of geographical isolation [9]. In addition to bird-mediated seed dispersal, gravity is another agent of dispersal, which results in a spatially-restricted genetic structure [9].

4.3. Demographic History and Population Structure

Population genetic structure reflects the genetic relationships within and between populations. From the haplotype network diagram, we can infer that *M. shiluensis* is likely to have originated in central Hainan. Colonization in DL fostered the formation of the haplotypes H1 and H2, while haplotypes H4 and H5 were formed after colonization in YG. Based on the large number of individuals and the relatively complete age structure in the DL population, we conclude that H1 and H2 form dominant haplotypes; however, as the number of individuals remaining in YG is small, it is impossible to determine which one of these haplotypes is dominant. Results of nSSR analysis confirm the genetic distinctiveness between DL and other Hainan groups, but suggest that the YG population may be very similar to BS, WZ, and CJ. The large number of private alleles in the DL population may be an important adaptation to an atypical environment [74], or a response to climate change; this result may be related to the local adaptability this species developed in response to the low temperature conditions of the Cenozoic Era [75,76]. It is worth mentioning that *M. shiluensis* has also been found outside Hainan Province. Studies have shown that Hainan Island was connected to Guangxi Province 65 million years ago [77]. In this study, STRUCTURE results of the analysis of nSSR markers indicate that the YC population shares the same gene pool as BS, YG, WZ, and CJ, while the analysis of cpDNA markers indicate that all sampled populations share the H3 haplotype. Furthermore, we speculate that there may be residual *M. shiluensis* in the area around the border between Guangdong and Guangxi.

4.4. Implication for Conservation

One of the main goals in protecting endangered species is to maximize existing genetic variation [17], as genetic variation may largely promote adaptation to a changing environment [78,79]. As all of the sampled *M. shiluensis* groups showed similar levels of genetic diversity, all of these populations make important contributions to the viability and protection of the species. Strengthening in situ conservation is an ideal approach that enables existing genetic resources to continue developing in an existing suitable environment. According to our field investigation, natural reserves are well-established in the DL, WZ, YG, and YC populations. Although no natural reserves are established in BS and CJ, these populations have also received protective support from the local government. The protection of existing genetic resources is only a short-term goal; the medium-term goal is to establish a new generation to improve genetic diversity. Except for DL, the individuals in the remaining populations are mature trees with a high level of genetic diversity, which provides protective value to each individual. One strategy for propagating future generations is to collect seeds in order to cultivate young seedlings and establish breeding in an off-site location [11]. In addition, germplasm resources can be propagated by grafting, and genetically dissimilar individuals can be crossed by artificial pollination to produce highly heterozygous seedlings. Great care needs to be taken to avoid genetic contamination when transplanting seedlings back to their native land [80], as this could allow foreign genes to infiltrate the original population and potentially eliminate it due to a lack of competitiveness. Finally, future research can encompass building a niche model to explore possible habitats of *M. shiluensis*, further expand its distribution range, and generate more genetic variation through local adaptation.

5. Conclusions

In summary, our results from nSSR and cpDNA markers indicate that high levels of population genetic diversity and low levels of genetic differentiation still exist in the endangered plants we studied. Therefore, we infer that the fragmentation and isolation within the *M. shiluensis* population may be due to recent human disturbance rather than to inbreeding or genetic drift. We have proposed some protection strategies for *M. shiluensis* based on these data.

Supplementary Materials: Supplementary materials can be found at http://www.mdpi.com/1999-4907/11/4/469/s1, Table S1. Geographical and climatic information of sampled *Michelia shiluensis* populations; Table S2. Characterization of the eight SSR and two cpDNA primers used in this study; Table S3. Genetic diversity of eight SSR markers within six *Michelia shiluensis* populations; Table S4. F-statistics and Nm at nSSR loci in six *Michelia shiluensis* populations; Table S5. Variable sites of the cpDNA fragment in six *Michelia shiluensis* populations.

Author Contributions: Conceptualization, Y.D.; methodology, Y.D. and T.L.; software, Y.X.; validation, Y.D., T.L., Y.X., Y.W., M.G., Z.X., Y.S., and X.D.; formal analysis, T.L. and Y.X.; resources, Y.D., Y.W., Z.X., Y.S., and X.D.; writing—original draft preparation, Y.D.; writing—review and editing, Y.D.; supervision, X.D.; project administration, X.D.; funding acquisition, X.D. All authors have read and agreed to the published version of the manuscript.

Funding: This study was supported by the Forestry Public Welfare Industry Research of China (Grant Number: 201404116), the National Natural Science Foundation of China (Grant Number: 31670601), and the Forestry Science and Technology Innovation of Guangdong Province grant programs (Grant Numbers: 2014KJCX006, 2017KJCX023).

Acknowledgments: We appreciate the help from Xinsheng Qin and forest rangers in each location during sampling. We are also grateful to Ye Sun and Xinsheng Hu for their invaluable guidance during the manuscript preparation.

Conflicts of Interest: The authors declare no conflicts of interest.

References

1. Tamaki, I.; Setsuko, S.; Tomaru, N. Genetic variation and differentiation in populations of a threatened tree, *Magnolia stellata*: Factors influencing the level of within-population genetic variation. *Heredity* **2008**, *100*, 415–423. [CrossRef] [PubMed]
2. Zhang, Y.B.; Ma, K.P. Geographic distribution patterns and status assessment of threatened plants in China. *Biodivers. Conserv.* **2008**, *17*, 1783–1798. [CrossRef]
3. Caballero, A.; Bravo, I.; Wang, J. Inbreeding load and purging: Implications for the short-term survival and the conservation management of small populations. *Heredity* **2017**, *118*, 177–185. [CrossRef] [PubMed]
4. Lande, R. Anthropogenic, ecological and genetic factors in extinction and conservation. *Popul. Ecol.* **1998**, *40*, 259–269. [CrossRef]
5. Tang, C.Q.; Yang, Y.; Ohsawa, M.; Momohara, A.; Hara, M.; Cheng, S.; Fan, S. Population structure of relict *Metasequoia glyptostroboides* and its habitat fragmentation and degradation in south-central China. *Biol. Conserv.* **2011**, *144*, 279–289. [CrossRef]
6. Hamrick, J.L.; Godt, M.W. Effects of life history traits on genetic diversity in plant species. *Philos. Trans. R. Soc. Lond. Ser. B Biol. Sci.* **1996**, *351*, 1291–1298.
7. Loveless, M.D.; Hamrick, J.L. Ecological determinants of genetic structure in plant populations. *Annu. Rev. Ecol. Syst.* **1984**, *15*, 65–95. [CrossRef]
8. Schoen, D.J.; Morgan, M.T.; Bataillon, T. How does self-pollination evolve? Inferences from floral ecology and molecular genetic variation. *Philos. Trans. R. Soc. Lond. B* **1996**, *351*, 1281–1290.
9. Setsuko, S.; Ishida, K.; Ueno, S.; Tsumura, Y.; Tomaru, N. Population differentiation and gene flow within a metapopulation of a threatened tree, *Magnolia stellata* (Magnoliaceae). *Am. J. Bot.* **2007**, *94*, 128–136. [CrossRef]
10. Hanski, I.; Simberloff, D. The metapopulation approach, its history, conceptual domain, and application to conservation. In *Metapopulation Biology*; Elsevier: Amsterdam, The Netherlands, 1997; pp. 5–26.
11. Ellstrand, N.C.; Elam, D.R. Population genetic consequences of small population size: Implications for plant conservation. *Annu. Rev. Ecol. Syst.* **1993**, *24*, 217–242. [CrossRef]
12. Chen, Y.Y.; Bao, Z.X.; Qu, Y.; Li, W.; Li, Z.Z. Genetic diversity and population structure of the medicinal orchid *Gastrodia elata* revealed by microsatellite analysis. *Biochem. Syst. Ecol.* **2014**, *54*, 182–189. [CrossRef]
13. Frankham, R. Genetics and conservation biology. *C. R. Biol.* **2003**, *326*, 22–29. [CrossRef]
14. Kingston, N.; Waldren, S.; Smyth, N. Conservation genetics and ecology of *Angiopteris chauliodonta* Copel. (Marattiaceae), a critically endangered fern from Pitcairn Island, South Central Pacific Ocean. *Biol. Conserv.* **2004**, *117*, 309–319. [CrossRef]

15. Lande, R.; Barrowdough, G. Effective population size, genetic variation, and their use in population management. In *Viable Populations for Conservation*; Soule, M.E., Ed.; Cambridge University Press: Cambridge, UK, 1987; p. 87.
16. Godt, M.J.W.; Caplow, F.; Hamrick, J. Allozyme diversity in the federally threatened golden paintbrush, *Castilleja levisecta* (Scrophulariaceae). *Conserv. Genet.* **2005**, *6*, 87–99. [CrossRef]
17. Luan, S.; Chiang, T.Y.; Gong, X. High genetic diversity vs. low genetic differentiation in *Nouelia insignis* (Asteraceae), a narrowly distributed and endemic species in China, revealed by ISSR fingerprinting. *Ann. Bot.* **2006**, *98*, 583–589. [CrossRef]
18. Cicuzza, D.; Newton, A.; Oldfield, S. *The Red List of Magnoliaceae*; Botanic Gardens Conservation International: London, UK, 2007.
19. Francisco-Ortega, J.; Wang, F.G.; Wang, Z.S.; Xing, F.W.; Liu, H.; Xu, H.; Xu, W.X.; Luo, Y.B.; Song, X.Q.; Gale, S.; et al. Endemic seed plant species from Hainan Island: A checklist. *Bot. Rev.* **2010**, *76*, 295–345. [CrossRef]
20. Wang, F.; Ye, H.; Zhao, N. Studies on the spermatophytic flora of E' huangzhang Nature Reserve in Yangchun of Guangdong Province. *Guangxi Zhiwu* **2003**, *23*, 495–504.
21. Deng, Y.; Luo, Y.; He, Y.; Qin, X.; Li, C.; Deng, X. Complete chloroplast genome of *Michelia shiluensis* and a comparative analysis with four Magnoliaceae species. *Forests* **2020**, *11*, 267. [CrossRef]
22. Yi, R.; Chen, Y.; Han, J.; Hu, Q.; Li, H.; Wu, H. Identification and biological characteristics of *Diaporthe ueckerae* causing dieback disease on *Michelia shiluensis*. *Sci. Silvae Sin.* **2018**, *54*, 80–88.
23. Wei, Y.; Hong, F.; Yuan, L.; Kong, Y.; Shi, Y. Population distribution and age structure characteristics of *Michelia shiluensis*, an endangered and endemic species in Hainan Island. *Chin. J. Trop. Crops* **2017**, *38*, 2280–2284.
24. Varshney, R.K.; Graner, A.; Sorrells, M.E. Genic microsatellite markers in plants: Features and applications. *Trends Biotechnol.* **2005**, *23*, 48–55. [CrossRef] [PubMed]
25. Sahu, J.; Sen, P.; Choudhury, M.D.; Barooah, M.; Modi, M.K.; Talukdar, A.D. Towards an efficient computational mining approach to identify EST-SSR markers. *Bioinformation* **2012**, *8*, 201–202. [CrossRef] [PubMed]
26. Slavov, G.T.; Howe, G.T.; Gyaourova, A.V.; Birkes, D.S.; Adams, W.T. Estimating pollen flow using SSR markers and paternity exclusion: Accounting for mistyping. *Mol. Ecol.* **2005**, *14*, 3109–3121. [CrossRef]
27. Ueno, S.; Setsuko, S.; Kawahara, T.; Yoshimaru, H. Genetic diversity and differentiation of the endangered Japanese endemic tree *Magnolia stellata* using nuclear and chloroplast microsatellite markers. *Conserv. Genet.* **2005**, *6*, 563–574. [CrossRef]
28. Birky, C.W., Jr.; Fuerst, P.; Maruyama, T. Organelle gene diversity under migration, mutation, and drift: Equilibrium expectations, approach to equilibrium, effects of heteroplasmic cells, and comparison to nuclear genes. *Genetics* **1989**, *121*, 613–627.
29. Birky, C.W., Jr. Uniparental inheritance of mitochondrial and chloroplast genes: Mechanisms and evolution. *Proc. Natl. Acad. Sci. USA* **1995**, *92*, 11331–11338. [CrossRef]
30. Petit, R.J.; Pineau, E.; Demesure, B.; Bacilieri, R.; Ducousso, A.; Kremer, A. Chloroplast DNA footprints of postglacial recolonization by oaks. *Proc. Natl. Acad. Sci. USA* **1997**, *94*, 9996–10001. [CrossRef]
31. Forcioli, D.; Saumitou-Laprade, P.; Valero, M.; Vernet, P.; Cuguen, J. Distribution of chloroplast DNA diversity within and among populations in gynodioecious *Beta vulgaris* ssp. maritima (Chenopodiaceae). *Mol. Ecol.* **1998**, *7*, 1193–1204. [CrossRef]
32. Caron, H.; Dumas, S.; Marque, G.; Messier, C.; Bandou, E.; Petit, R.; Kremer, A. Spatial and temporal distribution of chloroplast DNA polymorphism in a tropical tree species. *Mol. Ecol.* **2000**, *9*, 1089–1098. [CrossRef]
33. Petit, R.J.; El Mousadik, A.; Pons, O. Identifying populations for conservation on the basis of genetic markers. *Conserv. Biol.* **1998**, *12*, 844–855. [CrossRef]
34. Schuelke, M. An economic method for the fluorescent labeling of PCR fragments. *Nat. Biotechnol.* **2000**, *18*, 233–234. [CrossRef]
35. Holland, M.M.; Parson, W. GeneMarker®HID: A reliable software tool for the analysis of forensic STR data. *J. Forensic Sci.* **2011**, *56*, 29–35. [CrossRef]
36. Peakall, R.; Smouse, P.E. GENALEX 6: Genetic analysis in Excel. Population genetic software for teaching and research. *Mol. Ecol. Notes* **2006**, *6*, 288–295. [CrossRef]

37. Liu, K.; Muse, S.V. PowerMarker: An integrated analysis environment for genetic marker analysis. *Bioinformatics* **2005**, *21*, 2128–2129. [CrossRef]
38. Pritchard, J.K.; Wen, W. *Documentation for STRUCTURE Software: Version 2*; University of Chicago: Chicago, IL, USA, 2003.
39. Earl, D.A. STRUCTURE HARVESTER: A website and program for visualizing STRUCTURE output and implementing the Evanno method. *Conserv. Genet. Resour.* **2012**, *4*, 359–361. [CrossRef]
40. Jakobsson, M.; Rosenberg, N.A. CLUMPP: A cluster matching and permutation program for dealing with label switching and multimodality in analysis of population structure. *Bioinformatics* **2007**, *23*, 1801–1806. [CrossRef]
41. Kumar, S.; Stecher, G.; Tamura, K. MEGA7: Molecular evolutionary genetics analysis version 7.0 for bigger datasets. *Mol. Biol. Evol.* **2016**, *33*, 1870–1874. [CrossRef]
42. Letunic, I.; Bork, P. Interactive Tree of Life (iTOL) v4: Recent updates and new developments. *Nucleic Acids Res.* **2019**, *47*, W256–W259. [CrossRef]
43. Yeh, F.; Yang, R.C.; Boyle, T.B.J.; Ye, Z.; Xiyan, J.M.; Yang, R.; Boyle, T.J. *PopGene32, Microsoft Windows-Based Freeware for Population Genetic Analysis. Version 1.32*; University of Alberta: Edmonton, AB, Canada, 2000.
44. Schneider, S.D.; Roessli, D.; Excoffier, L. Arlequin: A software for population genetics data analysis. *User Man. Ver* **2000**, *2*, 2496–2497.
45. Hall, T. BioEdit: An important software for molecular biology. *GERF Bull. Biosci.* **2011**, *2*, 60–61.
46. Rozas, J.; Ferrer-Mata, A.; Sánchez-DelBarrio, J.C.; Guirao-Rico, S.; Librado, P.; Ramos-Onsins, S.E.; Sánchez-Gracia, A. DnaSP 6: DNA sequence polymorphism analysis of large data sets. *Mol. Biol. Evol.* **2017**, *34*, 3299–3302. [CrossRef]
47. Clement, M.; Posada, D.C.K.A.; Crandall, K.A. TCS: A computer program to estimate gene genealogies. *Mol. Ecol.* **2000**, *9*, 1657–1659.
48. Pons, O.; Petit, R. Measuring and testing genetic differentiation with ordered versus unordered alleles. *Genetics* **1996**, *144*, 1237–1245.
49. Verma, S.; Rana, T.S.; Ranade, S.A. Genetic variation and clustering in *Murraya paniculata* complex as revealed by single primer amplification reaction methods. *Curr. Sci. India* **2009**, *96*, 1210–1216.
50. George, S.; Sharma, J.; Yadon, V.L. Genetic diversity of the endangered and narrow endemic *Piperia yadonii* (Orchidaceae) assessed with ISSR polymorphisms. *Am. J. Bot.* **2009**, *96*, 2022–2030. [CrossRef]
51. Qiao, Q.; Zhang, C.Q.; Milne, R.I. Population genetics and breeding system of *Tupistra pingbianensis* (Liliaceae), a naturally rare plant endemic to SW China. *J. Syst. Evol.* **2010**, *48*, 47–57. [CrossRef]
52. Setsuko, S.; Ishida, K.; Tomaru, N. Size distribution and genetic structure in relation to clonal growth within a population of *Magnolia tomentosa* Thunb. (Magnoliaceae). *Mol. Ecol.* **2004**, *13*, 2645–2653. [CrossRef]
53. Zhao, X.; Ma, Y.; Sun, W.; Wen, X.; Milne, R. High genetic diversity and low differentiation of *Michelia coriacea* (Magnoliaceae), a critically endangered endemic in southeast Yunnan, China. *Int. J. Mol. Sci.* **2012**, *13*, 4396–4411. [CrossRef]
54. Wang, L.; Xiao, A.H.; Ma, L.Y.; Chen, F.J.; Sang, Z.Y.; Duan, J. Identification of *Magnolia wufengensis* (Magnoliaceae) cultivars using phenotypic traits, SSR and SRAP markers: Insights into breeding and conservation. *Genet. Mol. Res.* **2017**, *16*. [CrossRef]
55. Nybom, H. Comparison of different nuclear DNA markers for estimating intraspecific genetic diversity in plants. *Mol. Ecol.* **2004**, *13*, 1143–1155. [CrossRef]
56. Lázaro-Nogal, A.; Matesanz, S.; García-Fernández, A.; Traveset, A.; Valladares, F. Population size, center–periphery, and seed dispersers' effects on the genetic diversity and population structure of the Mediterranean relict shrub *Cneorum tricoccon*. *Ecol. Evol.* **2017**, *7*, 7231–7242. [CrossRef]
57. Hewitt, A.; Rymer, P.; Holford, P.; Morris, E.C.; Renshaw, A. Evidence for clonality, breeding system, genetic diversity and genetic structure in large and small populations of *Melaleuca deanei* (Myrtaceae). *Aust. J. Bot.* **2019**, *67*, 36–45. [CrossRef]
58. Tamaki, I.; Setsuko, S.; Tomaru, N. Genetic diversity and structure of remnant *Magnolia stellata* populations affected by anthropogenic pressures and a conservation strategy for maintaining their current genetic diversity. *Conserv. Genet.* **2016**, *17*, 715–725. [CrossRef]
59. Bezemer, N.; Krauss, S.L.; Roberts, D.G.; Hopper, S.D. Conservation of old individual trees and small populations is integral to maintain species' genetic diversity of a historically fragmented woody perennial. *Mol. Ecol.* **2019**, *28*, 3339–3357. [CrossRef]

60. Leimu, R.; Mutikainen, P.; Koricheva, J.; Fischer, M. How general are positive relationships between plant population size, fitness and genetic variation? *J. Ecol.* **2006**, *94*, 942–952. [CrossRef]
61. Lu, S.Y.; Hong, K.H.; Liu, S.L.; Cheng, Y.P.; Wu, W.L.; Chiang, T.Y. Genetic variation and population differentiation of *Michelia formosana* (Magnoliaceae) based on cpDNA variation and RAPD fingerprints: Relevance to post-Pleistocene recolonization. *J. Plant Res.* **2002**, *115*, 203–216. [CrossRef]
62. Azuma, H.; Thien, L.B.; Kawano, S. Molecular phylogeny of *Magnolia* (Magnoliaceae) inferred from cpDNA sequences and evolutionary divergence of the floral scents. *J. Plant Res.* **1999**, *112*, 291–306. [CrossRef]
63. Frankham, R.; Briscoe, D.A.; Ballou, J.D. *Introduction to Conservation Genetics*; Cambridge University Press: Cambridge, UK, 2002.
64. Yu, H.H.; Yang, Z.L.; Sun, B.; Liu, R.N. Genetic diversity and relationship of endangered plant *Magnolia officinalis* (Magnoliaceae) assessed with ISSR polymorphisms. *Biochem. Syst. Ecol.* **2011**, *39*, 71–78. [CrossRef]
65. Young, A.G.; Merriam, H.G.; Warwick, S.I. The effects of forest fragmentation on genetic variation in *Acer saccharum* Marsh. (sugar maple) populations. *Heredity* **1993**, *71*, 277–289. [CrossRef]
66. Chai, Y.; Jin, X.; Cai, M. Pollination biology of *Michelia crassipes* YW Law. In Proceedings of the II International Symposium on Germplasm of Ornamentals, Atlanta, GA, USA, 8–12 August 2016; International Society for Horticultural Science: Leuven, Belgium, 2017; pp. 297–304.
67. Zhao, X.; Sun, W. Abnormalities in sexual development and pollinator limitation in *Michelia coriacea* (Magnoliaceae), a critically endangered endemic to Southeast Yunnan, China. *Flora* **2009**, *204*, 463–470. [CrossRef]
68. Zhang, X.; Shen, S.; Wu, F.; Wang, Y. Inferring genetic variation and demographic history of *Michelia yunnanensis* Franch. (Magnoliaceae) from chloroplast DNA sequences and microsatellite markers. *Front. Plant Sci.* **2017**, *8*, 583. [CrossRef] [PubMed]
69. Han, Q.; Higashi, H.; Mitsui, Y.; Setoguchi, H. Distinct phylogeographic structures of wild radish (*Raphanus sativus* L. var. raphanistroides Makino) in Japan. *PLoS ONE* **2015**, *10*, e0135132. [CrossRef] [PubMed]
70. Wright, S. Evolution in Mendelian populations. *Genetics* **1931**, *16*, 97–159.
71. Tian, K.; Zhang, G.; Cheng, X.; He, S.; Yang, Y.; Yang, Y. The habitat fragility of *Manglietiastrum sinicum*. *Acta Bot. Yunnanica* **2003**, *25*, 551–556.
72. Allendorf, F.W. Isolation, gene flow and genetic differentiation among populations. In *Genetics and Conservation: A Reference for Managing Wild Animal and Plant Populations*; Schoenwald-Cox, C.M., Chambers, S.M., Thomas, W.L., Eds.; Menlo Park/Benjamin Cummings: San Francisco, CA, USA, 1983; pp. 51–65.
73. Moyle, L.C. Correlates of genetic differentiation and isolation by distance in 17 congeneric *Silene* species. *Mol. Ecol.* **2006**, *15*, 1067–1081. [CrossRef]
74. Holsinger, K.E.; Gottlieb, L.; Falk, D.A. Conservation of rare and endangered plants: Principles and prospects. In *Genetics and Conservation of Rare Plants*; Falk, D.A., Holsinger, K.E., Eds.; Oxford University Press: Oxford, UK, 1991; pp. 195–208.
75. Kawecki, T.J.; Ebert, D. Conceptual issues in local adaptation. *Ecol. Lett.* **2004**, *7*, 1225–1241. [CrossRef]
76. Shi, X.; Kohn, B.; Spencer, S.; Guo, X.; Li, Y.; Yang, X.; Shi, H.; Gleadow, A. Cenozoic denudation history of southern Hainan Island, South China Sea: Constraints from low temperature thermochronology. *Tectonophysics* **2011**, *504*, 100–115. [CrossRef]
77. Liang, G. A study of the genesis of Hainan Island. *Geol. China* **2018**, *45*, 693–705.
78. Zhou, T.H.; Dong, S.S.; Li, S.; Zhao, G.F. Genetic structure within and among populations of *Saruma henryi*, an endangered plant endemic to China. *Biochem. Genet.* **2012**, *50*, 146–158. [CrossRef]
79. Ge, X.J.; Zhou, X.L.; Li, Z.C.; Hsu, T.W.; Schaal, B.A.; Chiang, T.Y. Low genetic diversity and significant population structuring in the relict *Amentotaxus argotaenia* complex (Taxaceae) based on ISSR fingerprinting. *J. Plant Res.* **2005**, *118*, 415–422. [CrossRef]
80. Li, J.M.; Jin, Z.X. Genetic structure of endangered *Emmenopterys henryi* Oliv. based on ISSR polymorphism and implications for its conservation. *Genetica* **2008**, *133*, 227–234. [CrossRef]

Article

Comparative Plastome Analyses and Phylogenetic Applications of the *Acer* Section *Platanoidea*

Tao Yu [1], Jian Gao [2], Bing-Hong Huang [3], Buddhi Dayananda [4], Wen-Bao Ma [5], Yu-Yang Zhang [1], Pei-Chun Liao [3,*] and Jun-Qing Li [1,*]

1. Forestry College, Beijing Forestry University, 35 Qinghua East Road, Haidian District, Beijing 100083, China; yutao123@bjfu.edu.cn (T.Y.); yuyangzhang@bjfu.edu.cn (Y.-Y.Z.)
2. Faculty of Resources and Environment, Baotou Teachers' College, No.3, Kexue Rd., BaoTou, Inner Mongolia 014030, China; gaojian5688@163.com
3. School of Life Science, National Taiwan Normal University, No. 88, Sec. 4, TingChow Rd., Wunshan Dist., Taipei 116, Taiwan; sohjiro4321@yahoo.com.tw
4. School of Agriculture and Food Sciences, The University of Queensland, Brisbane QLD 4072, Australia; buddhi6@gmail.com
5. Key Laboratory of National Forestry and Grassland Administration on Sichuan Forest Ecology, Resources and Environment Sichuan Academy of Forestry, Chengdu 610081, China; mawenbao_2000@126.com

* Correspondence: pcliao@ntnu.edu.tw (P.-C.L.); lijq@bjfu.edu.cn (J.-Q.L.); Tel.: +886-2-77346330 (P.-C.L.)

Received: 17 March 2020; Accepted: 16 April 2020; Published: 19 April 2020

Abstract: The *Acer* L. (Sapindaceae) is one of the most diverse and widespread genera in the Northern Hemisphere. Section *Platanoidea* harbours high genetic and morphological diversity and shows the phylogenetic conflict between *A. catalpifolium* and *A. amplum*. Chloroplast (cp) genome sequencing is efficient for the enhancement of the understanding of phylogenetic relationships and taxonomic revision. Here, we report complete cp genomes of five species of *Acer* sect. *Platanoidea*. The length of *Acer* sect. *Platanoidea* cp genomes ranged from 156,262 bp to 157,349 bp and detected the structural variation in the inverted repeats (IRs) boundaries. By conducting a sliding window analysis, we found that five relatively high variable regions (*trnH-psbA*, *psbN-trnD*, *psaA-ycf3*, *petA-psbJ* and *ndhA* intron) had a high potential for developing effective genetic markers. Moreover, with an addition of eight plastomes collected from GenBank, we displayed a robust phylogenetic tree of the *Acer* sect. *Platanoidea*, with high resolutions for nearly all identified nodes, suggests a promising opportunity to resolve infrasectional relationships of the most species-rich section *Platanoidea* of *Acer*.

Keywords: *Acer*; sect. *Platanoidea*; chloroplast genome; sequence divergence; structural variation; phylogenetics

---

1. Introduction

Chloroplasts (cp) are essential organelles in plant cells for the processes of photosynthesis and carbon fixation [1]. They possess uniparental inheritance and their genome has a high conservation structure in most land plants [2]. Generally, cp genomes in most angiosperms are circular DNA molecules composed of four parts, namely two inverted repeats (IRs) at approximately 20–28 kb, a vast single-copy region (LSC) at 80–90 kb and a small single-copy region (SSC) at 16–27 kb [3]. The composition of angiosperm cp genomes is relatively conserved and encodes four ribosomal RNAs (rRNAs), roughly 30 transfer RNAs (tRNAs) and approximately 80 single-copy genes [4]. With the rapid development of next-generation sequencing (NGS) and other methods for obtaining the cp genome sequences, the availability of cp genome sequences has increased dramatically for land plants, offering opportunities for the comprehensive structure comparison, improvement of horticultural plant breeding [5,6] and reconstruction of evolutionary relationships [7,8]. In most angiosperms, the cp genome is inherited from the patrilineal lineage and exhibits little or no recombination [9]. Due to its relatively conserved features, cp sequences are commonly used as DNA barcodes for genetic identification, plant systematic studies, and research into plant biodiversity, biogeography, adaptation, etc. [10,11].

Acer L. (Maple), a diverse genus of family Sapindaceae L., contains more than 124 species [12]. Most extant species of the genus are native to Asia, whereas others occur in North America, Europe and North Africa [12–14]. Most *Acer* species are famous ornamental plants [13], and also another usage for pharmaceutical and chemical products [15]. To date, 17 species belonging to the *Acer* section Platanoidea have been recognised in China, in which the section shows high genetics and morphology diversity. [12]. Among them, four species are widespread in various vegetation regions (*A. amplum*, *A. longipes*, *A. mono* and *A. truncatum*) [16], and some are endangered, such as *A. catalpifolium*, *A. miaotaiense*, and *A. yangjuechi*.

However, some species with diversified leaf morphology have taxonomic controversy due to unresolved phylogenetic relationships (for example, *A. longipes* and *A. amplum*) and require further studies and clarification [17]. The comparative plastome analysis allows detailed insights to affirm the phylogenetic placement of these plants and will be useful for species identification, to verify taxonomic levels and identify phylogenetic relationships [8,18]. Recently, cp genomes of *A. miaotaiense* and *A. truncatum* have been reported, but merely the sequence information was provided without further analyses [19,20]. Thus, a comparative study among these two published cp genomes and five newly generated plastomes of sect. *Platanoidea* is conducted and applied to address phylogenetic and taxonomic validity.

Firstly, we reported newly completed cp genomes of five species of the *Acer* sect. *Platanoidea* (*A. catalpifolium*, *A. amplum*, *A. longipes*, *A. yanjuechi* and *A. mono*). Then, we compared the gene contents and the plastomic organisation with two published cp genomes in the sect. *Platanoidea* to identify variable loci that can apply to the species or population-level studies on *Acer*. The aims of this study are: (i) to deepen our understanding of the genetic and structural diversity within the sect. *Platanoidea*, (ii) to increase our understanding of phylogenetic relationships of species within the sect. *Platanoidea*, and (iii) to reconstruct a phylogenetic tree based on these plastomes. Our study also provides genetic resources for future research in this genus.

2. Materials and Methods

2.1. Sampling and DNA Extraction

Young leaves of five *Acer* species (*A. catalpifolium*, *A. amplum*, *A. longipes*, *A. yanjuechi* and *A. mono*) were collected and dried immediately with silica gel for preservation, for each species we collected the leaves from one healthy plant. Collection information of the plant materials is showed in Table S1. Vouchers taxonomical determination was by the Beijing Forestry University herbarium and deposited at the College of Forestry, Beijing Forestry University, China. We isolated total genomic DNA from silica gel-dried leaves according to the modified CTAB method [21].

2.2. Chloroplast Genome Sequencing, Assembling and Annotation

Purified genomic DNA was sequenced using an Illumina MiSeq sequencer at Shanghai OE Biotech. Co., Ltd. A paired-end library was constructed with an insert size of 300 bp, yielding at least 8 GB of 150 bp paired-end reads for each species. Clean reads were obtained with NGS QC Toolkit v2.3.3 (cut-off read length for HQ = 70%, cut-off quality score = 20, trim reads from 5′ = 3, trim reads from 3′ = 7) [22]. We used MITObim v. 1.8 [23] to assemble the five new *Acer* cp sequences with the reference cp genomes, *A. miaotaiense* (KX098452) [19], *A. davidii* (KU977442) [24] and *A. morrisonense* (KT970611) [25]. Gene functions were annotated using DOGMA [26], and the protein-encoding genes (PCG), tRNAs and rRNA were determined and verified using the BLAST searches (https://blast.ncbi.nlm.nih.gov/Blast.cgi).

2.3. Divergence Hotspot Identification

In order to determine the divergence level, a MAFFT: multiple sequence alignment program [27] was used to align cp sequences of seven *Acer* sect. *Platanoidea* species, and then sliding windows of the nucleotide variability (pi) was conducted using DnaSP 5.0 with 600-bp window length and 200-bp step size [28].

2.4. Phylogenomic Reconstruction

To determine the phylogenetic relationship of the *Acer* sect. *Platanoidea*, we performed phylogenetic analyses using 13 cp genome sequences, which comprised five plastome sequences generated in this study, six plastomes of the *Acer* species collected from GenBank, and two of the *Dipteronia* species as an outgroup (Table S2). Consequently, a total length of 160,886 bp was aligned using MAFFT [27]. The best-fitting substitution model (GTR + I + G) was inferred using Modeltest 3.7 [29]. Finally, phylogenomic relationships were reconstructed with Bayesian Inference (BI) and Maximum Likelihood (ML) using MrBayes 3.2 [30] and phyML v3.0 [31], respectively. For the BI tree, 10 million generations were simulated using two parallel Markov Chain Monte Carlo (MCMC) simulations, sampled every 1000 generations. The first 25% of the simulations were discarded (burn-in) to generate a consensus tree. For the ML tree, 1000 bootstrap replicates were conducted to evaluate the supporting values of each node.

3. Results and Discussion

3.1. Chloroplast Genome Organisation of the Acer sect. Platanoidea

The nucleotide sequences of the seven *Acer* sect. *Platanoidea* cp genomes ranged from 156,262 bp to 157,349 bp in length (Figure 1, Table 1), which are similar to other reported cp genomes of *Acer* [19,25]. The variation of chloroplast genome length is mainly caused by the change of LSC region length. The quadripartite structures of these cp genomes were identical to most angiosperms containing a LSC region, SSC region and two inverted repeat regions (IRa and IRb) [32]. The overall GC content of these cp genomes accounts for 37.9% and the GC content of IR regions accounts for 42.8% higher than the LSC (35.9%) and the SSC (32.3%). The new sequences possess 117 genes, including four unique rRNAs, 31 tRNAs, and 82 PCGs, respectively (Table 2). Among them, most cp genes were single copy, while 23 genes exhibited double copies, including four rRNA (4.5S, 5S, 16S and 23S rRNA), nine tRNA (*trnA-UGC*, *trnI-CAU*, *trnI-GAU*, *trnL-CAA*, *trnM-CAU*, *trnN-GUU*, *trnR-ACG*, *trnT-GGU* and *trnV-GAC*) and 10 PCGs (*ndhB*, *rpl2*, *rps12*, *rpl23*, *rps19*, *rps7*, *ycf1*, *orf42*, *ycf2* and *ycf15*). Additionally, a total of 18 genes harboured introns, and three genes (*ycf3*, *clpP* and *rps12*) contained two introns.

Table 1. General features of the *Acer* sect. *Platanoidea* chloroplast genomes compared in this study.

Name	Total (bp)	LSC (bp)	SSC (bp)	IR (bp)	GC%
A. catalpifolium	157,349	85,745	18,066	26,769	37.9
A. miaotaiense	156,595	86,327	18,068	26,100	37.9
A. longipes	157,137	85,531	18,068	26,769	37.9
A. amplum	156,995	85,386	18,077	26,766	37.9
A. mono	156,985	85,378	18,069	26,769	37.9
A. yangjuechi	157,088	85,483	18,069	26,768	37.9
A. truncatum	156,262	86,019	18,073	26,085	37.9

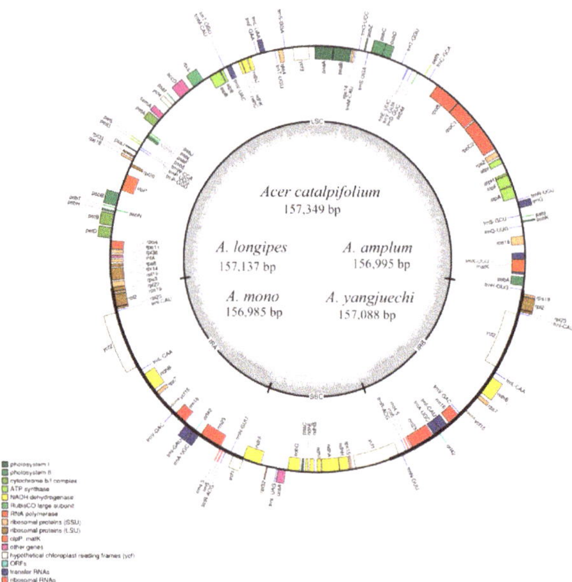

Figure 1. Merged gene map of the complete chloroplast genomes of five *Acer* sect. *Platanoidea* species. Genes belonging to different functional groups are colour-coded. The genes drawn inside the circle are transcribed clockwise, while those outside are transcribed counter-clockwise. Darker grey in the inner circle corresponds to the GC content of the chloroplast genome.

Table 2. Genes presented in the *Acer* sect. *Platanoidea* chloroplast genome.

	Group of Gene	Genes Name
1	Photostsyem I	*psaA, psaB, psaC, psaI, psaJ*
2	Photostsyem II	*psbA, psbB, psbC, psbD, psbE, psbF, psbh, psbI, psbJ, psbK, psbL, psbM, psbN, psbT, psbZ*
3	Cytochrome b/f complex	*petA, petB*, petD*, petG, petL, petN*
4	ATP synthase	*atpA, atpB, atpE, atpF*, atpH, atpI*
5	NADH dehydrogenase	*ndhA*, ndhB*, ndhC, ndhD, ndhE, ndhF, ndhG, ndhH, ndhI, ndhJ, ndhK*
6	RubisCO large subunit	*rbcL*
7	RNA polymerase	*ropA, ropB, ropC1*, ropC2*
8	Ribosomal proteins(SSU)	*rps2, rps3, rps4, rps7, rps8, rps11, rps12**, rps14, rps15, rps16*, rps18, rps19*
9	Ribosomal proteins (LSU)	*rpl2*, rpl14, rpl16*, rpl20, rpl22, rpl23(×2), rpl32, rpl33, rpl36*
10	Other gene	*clpP **, matK, accD, ccsA, infA, cemA*
11	Proteins of unknown function	*ycf1, ycf2, ycf3**, ycf4, ycf15*
12	ORFs	*Orf42*
13	Transfer RNAs	31 tRNAs(six contain a single intron)
14	Ribosomal RNAs	rrn4.5, rrn5, rrn16, rrn23

Note: a single asterisk (*) following after gene names indicate intron-containing genes, and double asterisks (**) following after gene names indicate two introns in the gene.

3.2. Comparative Analysis of the Genomic Structure

Contraction and expansion of the IRs, LSC and SSC are important to the evolution of cp genomes [33,34], which is the leading cause of gene order and the size changes of the cp genome [35]. Detailed structure comparisons among the 13 cp genomes of the *Acer* species were presented in Figure 2. *rpl22* in the LSC/IR boundary and *rps19* was the last gene in most *Acer* sect. *Platanoidea* cp genomes included *A. mono, A. amplum, A. catalpifolium, A. yangjuechi, A. longipes, A. morrisonense, A. davidii* and *A. griseum*. However, the different structures of *rps19* in the LSC/IR boundary and the last gene *rpl2* were found in *A. truncatum, A. miaotaiense* and *A. buergerianum*. Among the Arecoideae species, *rpl22* and *rps19* gene order changes in the IRA/LSC borders were also observed and has become the most varied rearrangement in this section [36]. Similarly, the rearrangement has been reported in the Apiales species, which also has two structure types of *rpl23* and *rps19* in the LSC/IR boundary [37]. Length variations of the *Acer* cp genomes were found by contraction and expansion of the LSC and IRs, the length of the LSC varies from 85,379 bp to 86,327 bp, and the length of the IR varies from 26,085 bp to 26,769 bp (Figure 2). Moreover, the type of unique structural borders of the cp genome (*A. truncatum*, *A. miaotaiense*, and *A. buergerianum*) also show a contraction of IRs, which indicates variations in boundary genes may be caused by variations in length.

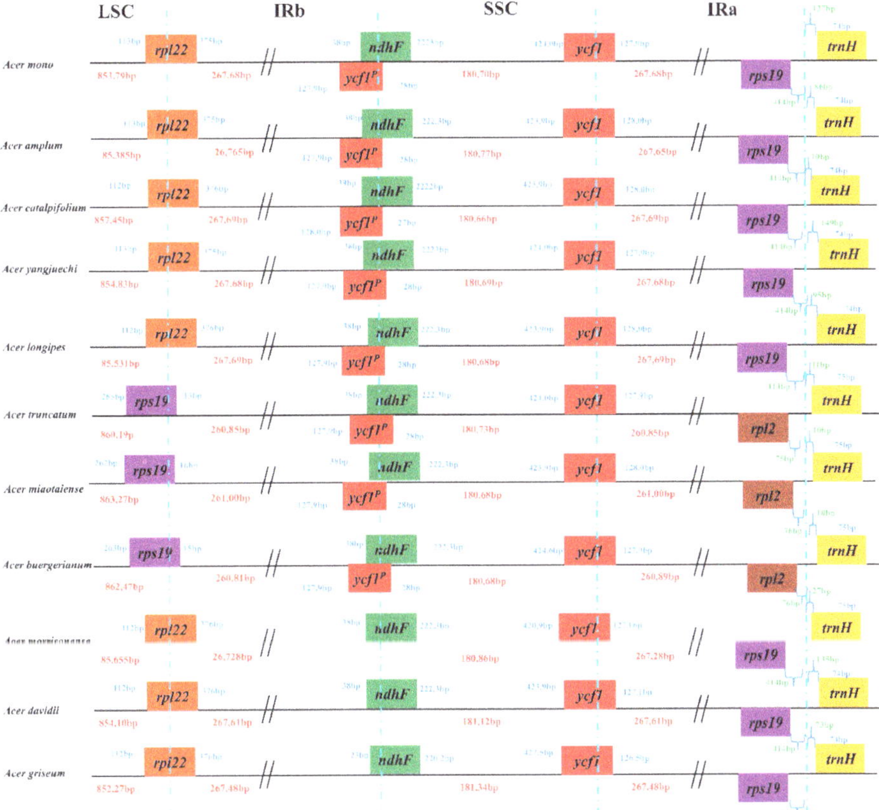

Figure 2. Comparison of the junction sites of the LSC, IRs and SSC regions among 13 Acer sect. Platanoidea species chloroplast genomes. Different colour boxes indicate specific genes and those above the genome lines indicate their transcriptions in a forward direction, while the under-line boxes are in the reverse direction. The length of the LSC, SSC, and IR regions is shown in red, the length of the gene distance from the boundary is shown in green, the length of the gene is shown in blue, and the P represents the pseudogene.

3.3. Divergence Hotspot of the Acer sect. Platanoidea Species

Sliding window analysis of the whole cp genome was performed to identify hotspots of the *Acer* sect. *Platanoidea* species. In Figure 3, it is apparent that the *trnH-psbA*, *psbN-trnD*, *psaA-ycf3*, *petA-psbJ* and *ndhA* intron nucleotide variability was higher than other regions. Most divergent hotspot loci are located in the LSC region, which allows for the proper design of the genetic markers. Only one hotspot *ndhA* intron was located in the SSC region. The IR regions were much more conserved. This result was similar to other cp genomes [37,38]. The general barcode *trnH-psbA* has demonstrated extreme variation in plant groups [39,40]. Thus, the highest variation *trnH-psbA* has the potential to be used in DNA barcoding in the *Acer* sect. *Platanoidea*. Additionally, the evolutionary history of *A. mono* has been inferred by using *psbA-trnH*, *trnL-trnF* and an intron of *rpl16* [16]. The regions of the *psaA-ycf3*, *petA-psbJ* and *ndhA* intron have been indicated as high variations in previous studies. In witch-hazel (genus *Hamamelis* L., Hamamelidaceae), appropriate variations of *psaA-ycf3* were used to reconstruct phylogenetic relationships [41]. In *Scutellaria*, *petA-psbJ* was one of six fast-evolving DNA sequences in the cp genome [42], while a systematic study shows that Muhlenbergiinae has high variation at the *ndhA* intron [43].

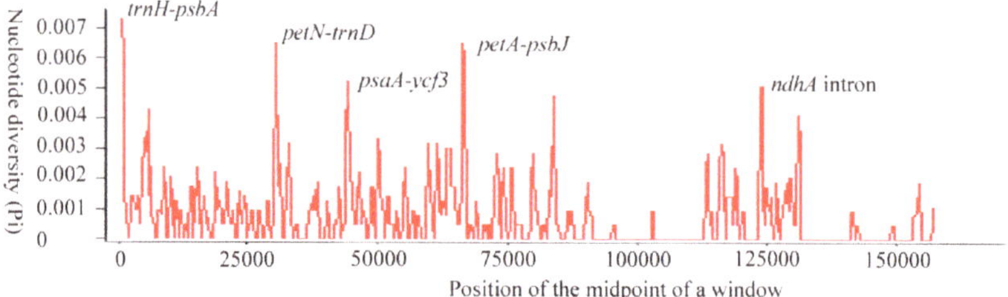

Figure 3. Sliding-window analysis on the cp genomes for seven *Acer* sect. *Platanoidea* species.

The endangered plants, *A. catalpifolium*, *A. yangjuechi* and *A. miaotaiense* in the *Acer* sect. *Platanoidea*, have small population sizes [44]. Population genetics studies of these species are relatively weak, and with limited conservation goals [45,46]. These high variability regions can provide alternative sites for subsequent studies and will contribute to the conservation of endangered plants.

3.4. Phylogenetic Analysis

The phylogenomic inference based on cp genomes shows high bootstrap supports in most nodes that provided robust evolutionary placement and relationship of the *Acer* species (Figure 4). The results showed that seven sampled species of the *Acer* sect. *Platanoidea* formed a single clade, which is consistent with previous studies of the *Acer* phylogeny [14,47]. In this clade, *A. catalpifolium*, *A. mono*, *A. miaotaiense and A. yangjuechi* had the closest phylogenetic relationship and formed a subclade, while *A. truncatum*, *A. longipes*, *A. amplum* formed another one. Previous phylogenetic inference of the *Acer* did not contain species with a small population size in the sect. *Platanoidea* (such as *A. catalpifolium* and *A. yangjuechi*) [14,47]. Phylogenetic analysis exhibited in this study for the issue of *A. longipes A. amplum* complex in the sect. Platanoidea raised earlier [17], and the phylogenetic position of *A. catalpifolium* was also redefined.

It is somewhat surprising that *A. mono* is not a sister with *A. truncatum* but with *A. yangjuechi*, which differs from some published studies [13]. Since *A. mono* is widespread in Asia and comprised of multiple local varieties, we cannot rule out that the possibility of the grouping of *A. mono* and *A. yangjuechi* due to adjacent sampling localities of *A. mono* and *A. yangjuechi* in Lin'an, Zhejiang province of China, the only extant habitat of *A. yangjuechi* [48]. Liu et al. [49] showed that the population composition of the Lin'an population is significantly different from the neighbouring populations of *A. mono*. The clustering of *A. mono* and *A. yangjuechi* inferred in this study may not only reflect the truth of geographic divergence in the genetics of *A. mono* but also implies the chloroplast capture by ancient hybridisation events between the two species in Lin'an.

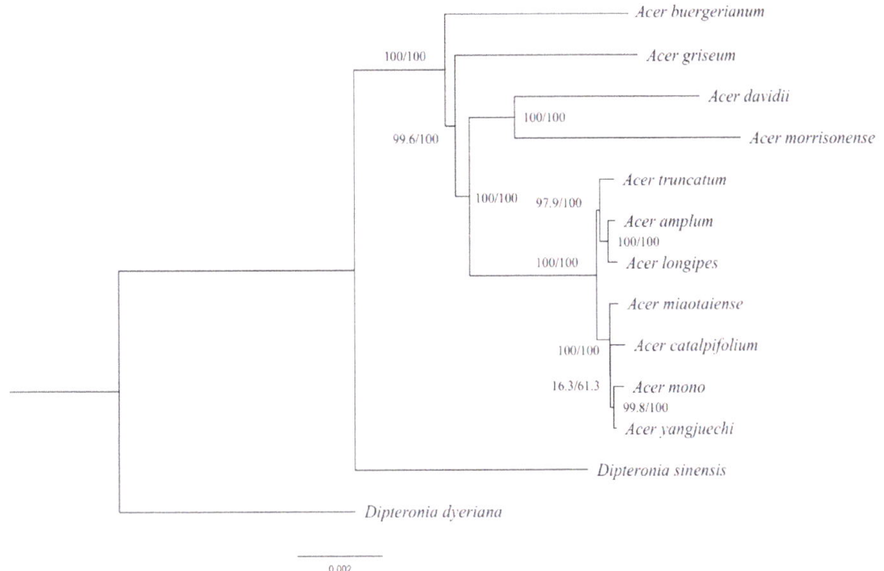

Figure 4. Phylogenetic tree of 13 *Acer* species inferred by Maximum Likelihood (ML) and Bayesian Inference (BI) methods, based on the whole cp genome sequences. The numbers above the branches are the bootstrap values of ML methods and the posterior probabilities of BI.

4. Conclusions

In this study, we firstly reported complete cp genomes of five *Acer* sect. *Platanoidea* species (*A. catalpifolium, A. amplum, A. longipes, A. yanjuechi* and *A. mono*) using the NGS technology. In comparison with other published *Acer* species from NCBI, we found that the *Acer* species have similar cp genome structure and gene content. The divergence hotspots identified in the cp genome of the *Acer* sect. *Platanoidea* could be applied to develop molecular markers for further population genetics studies. The high variation at the IR/LSC and IR/SSC boundaries were also reported. The phylogenetic analysis strongly supported that *A. catalpifolium* has the closest relationship with *A. miaotaiense*, followed by *A. mono*, and *A. yanjuechi*, which confirms the species-complex relationship of *A. longipes* and *A. amplum*. The available genomic data presented in this paper provides a basis for further research on the evolutionary history and conservation genetics of endangered species of genus *Acer*.

Supplementary Materials: The following are available online at http://www.mdpi.com/1999-4907/11/4/462/s1, Table S1: General features of the *Acer* sect. *Platanoidea* chloroplast genomes compared in this study. Table S2: Acer taxa sampled in this study.

Author Contributions: T.Y. and J.G. conceived and designed the work; J.G. and Y.-Y.Z. collected samples; T.Y., J.G. and B.-H.H. performed the experiments and analysed the data; T.Y. and J.G. wrote the manuscript; P.-C.L., W.-B.M., B.D. and J.-Q.L. critically reviewed the manuscript. All authors have read and agreed to the published version of the manuscript.

Funding: This research was financially supported by the program "Reintroduction Technologies and Demonstration of Extremely Rare Wild Plant Population" of National Key Research and Development Program (2016YFC0503106) to J.Q.L., the Ministry of Science and Technology of Taiwan (grant number: 108-2628-B-003-001) and National Taiwan Normal University (NTNU) to P.C.L. and "The biogeographical feature and competitive hybridization of Maple (*Acer* L.) in East Asia" of National Natural Science Foundation of China (41901063) to J.G.

Conflicts of Interest: The authors declare no conflict of interest.

References

1. Neuhaus, H.E.; Emes, M.J. Nonphotosynthetic metabolism in plastids. *Annu. Rev. Plant Biol.* **2000**, *51*, 111–140. [CrossRef] [PubMed]
2. Armbrust, E.V. Uniparental inheritance of chloroplast genomes. *Adv. Photosynth. Respir.* **1998**, *7*, 93–113.
3. Choi, K.S.; Chung, M.G.; Park, S.J. The complete chloroplast genome sequences of three veroniceae species (plantaginaceae): Comparative analysis and highly divergent regions. *Front. Plant Sci.* **2016**, *7*. [CrossRef] [PubMed]
4. Yu, T.; Lv, J.; Li, J.; Du, F.K.; Yin, K. The complete chloroplast genome of the dove tree Davidia involucrata (Nyssaceae), a relict species endemic to China. *Conserv. Genet. Resour.* **2016**, *8*, 1–4. [CrossRef]
5. Sonah, H.; Deshmukh, R.K.; Singh, V.P.; Gupta, D.K.; Singh, N.K.; Sharma, T.R. Genomic resources in horticultural crops: Status, utility and challenges. *Biotechnol. Adv.* **2011**, *29*, 199–209. [CrossRef]
6. Xiong, J.S.; Ding, J.; Li, Y. Genome-editing technologies and their potential application in horticultural crop breeding. *Hortic. Res.* **2015**, *2*. [CrossRef]
7. Cai, J.; Ma, P.F.; Li, H.T.; Li, D.Z. Complete plastid genome sequencing of fourtiliaspecies (malvaceae): A comparative analysis and phylogenetic implications. *PLoS ONE* **2015**, *10*. [CrossRef]
8. Ruhsam, M.; Rai, H.S.; Mathews, S.; Ross, T.G.; Graham, S.W.; Raubeson, L.A.; Mei, W.; Thomas, P.I.; Gardner, M.F.; Ennos, R.A.; et al. Does complete plastid genome sequencing improve species discrimination and phylogenetic resolution in Araucaria? *Mol. Ecol. Resour.* **2015**, *15*, 1067–1078. [CrossRef]
9. Yi, X.; Gao, L.; Wang, B.; Su, Y.J.; Wang, T. The complete chloroplast genome sequence of Cephalotaxus oliveri (Cephalotaxaceae): Evolutionary comparison of cephalotaxus chloroplast DNAs and insights into the loss of inverted repeat copies in gymnosperms. *Genome Biol. Evol.* **2013**, *5*, 688–698. [CrossRef]
10. Wambugu, P.W.; Brozynska, M.; Furtado, A.; Waters, D.L.; Henry, R.J. Relationships of wild and domesticated rices (Oryza AA genome species) based upon whole chloroplast genome sequences. *Sci. Rep.* **2015**, *5*. [CrossRef]
11. Brozynska, M.; Furtado, A.; Henry, R.J. Genomics of crop wild relatives: Expanding the gene pool for crop improvement. *Plant Biotechnol. J.* **2016**, *14*, 1070–1085. [CrossRef] [PubMed]
12. Xu, T.Z.; Chen, Y.S.; de Jong, P.C.; Oterdoom, H.J. Aceraceae. In *Flora of China*; Wu, Z.Y., Raven, P.H., Hong, D.Y., Eds.; Science press: Beijing, China, 2008; Volume 11, pp. 515–553.
13. Van Gelderen, D.M.; De Jong, P.C.; Oterdoom, H.J. *Maples of the World*; Timber Press: Portland, OR, USA, 1994; ISBN 0881920002.
14. Renner, S.S.; Grimm, G.W.; Schneeweiss, G.M.; Stuessy, T.F.; Ricklefs, R.E. Rooting and dating maples (Acer) with an uncorrelated-rates molecular clock: Implications for north American/Asian disjunctions. *Syst. Biol.* **2008**, *57*, 795–808. [CrossRef] [PubMed]
15. Rosado, A.; Vera-Vélez, R.; Cota-Sánchez, J.H. Floral morphology and reproductive biology in selected maple (Acer, L.) species (Sapindaceae). *Braz. J. Bot.* **2018**, *41*, 1–14. [CrossRef]
16. Guo, X.; Wang, H.; Bao, L.; Wang, T.; Bai, W.; Ye, J.; Ge, J. Evolutionary history of a widespread tree species Acer mono in East Asia. *Ecol. Evol.* **2014**, *4*, 4332–4345. [PubMed]
17. Grimm, G.W.; Denk, T. The Colchic region as refuge for relict tree lineages: Cryptic speciation in field maples. *Turk. J. Bot.* **2014**, *38*, 1050–1066. [CrossRef]
18. Kumar, S.; Hahn, F.M.; McMahan, C.M.; Cornish, K.; Whalen, M.C. Comparative analysis of the complete sequence of the plastid genome of Parthenium argentatum and identification of DNA barcodes to differentiate Parthenium species and lines. *BMC Plant Biol.* **2009**, *9*, 131. [CrossRef]
19. Zhang, Y.; Li, B.; Chen, H.; Wang, Y. Characterization of the complete chloroplast genome of Acer miaotaiense (Sapindales: Aceraceae), a rare and vulnerable tree species endemic to China. *Conserv. Genet. Resour.* **2016**, *8*, 1–3. [CrossRef]
20. Wang, R.; Fan, J.; Chang, P.; Zhu, L.; Zhao, M.; Li, L. Genome survey sequencing of acer truncatum bunge to identify genomic information, simple sequence repeat (ssr) markers and complete chloroplast genome. *Forests* **2019**, *10*, 87. [CrossRef]
21. Doyle, J.J. A rapid DNA isolation procedure for small amounts of fresh leaf tissue. *Phytochem. Bull.* **1987**, *19*, 11–15.

22. Dai, M.; Thompson, R.C.; Maher, C.; Contrerasgalindo, R.; Kaplan, M.H.; Markovitz, D.M.; Omenn, G.; Meng, F. NGSQC: Cross-platform quality analysis pipeline for deep sequencing data. *BMC Genom.* **2010**, *11*. [CrossRef]
23. Hahn, C.; Bachmann, L.; Chevreux, B. Reconstructing mitochondrial genomes directly from genomic next-generation sequencing reads–A baiting and iterative mapping approach. *Nucleic Acids Res.* **2013**, *41*, e129. [CrossRef] [PubMed]
24. Jia, Y.; Yang, J.; He, Y.L.; He, Y.; Niu, C.; Gong, L.L.; Li, Z.H. Characterization of the whole chloroplast genome sequence of Acer davidii Franch (Aceraceae). *Conserv. Genet. Resour.* **2016**, *8*, 1–3. [CrossRef]
25. Li, Z.-H.; Xie, Y.-S.; Zhou, T.; Jia, Y.; He, Y.-L.; Yang, J. The complete chloroplast genome sequence of Acer morrisonense (Aceraceae). *Mitochondrial DNA Part A* **2015**, *28*, 309–310. [CrossRef] [PubMed]
26. Wyman, S.K.; Jansen, R.K.; Boore, J.L. Automatic annotation of organellar genomes with DOGMA. *Bioinformatics* **2004**, *20*, 3252–3255. [CrossRef]
27. Katoh, K.; Kuma, K.; Toh, H.; Miyata, T. MAFFT version 5: Improvement in accuracy of multiple sequence alignment. *Nucleic Acids Res.* **2005**, *33*, 511–518. [CrossRef]
28. Librado, P.; Rozas, J. DnaSP v5: A software for comprehensive analysis of DNA polymorphism data. *Bioinformatics* **2009**, *25*, 1451–1452. [CrossRef]
29. Posada, D.; Buckley, T.R. Model selection and model averaging in phylogenetics: Advantages of akaike information criterion and bayesian approaches over likelihood ratio tests. *Syst. Biol.* **2004**, *53*, 793–808. [CrossRef]
30. Huelsenbeck, J.P.; Ronquist, F. MRBAYES: Bayesian inference of phylogenetic trees. *Bioinformatics* **2001**, *17*, 754–755. [CrossRef]
31. Guindon, S.; Lethiec, F.; Duroux, P.; Gascuel, O. PHYML Online–A web server for fast maximum likelihood-based phylogenetic inference. *Nucleic Acids Res.* **2005**, *33*, 557–559. [CrossRef]
32. Daniell, H.; Lin, C.-S.; Yu, M.; Chang, W.-J. Chloroplast genomes: Diversity, evolution, and applications in genetic engineering. *Genome Biol.* **2016**, *17*, 134. [CrossRef]
33. Guo, H.; Liul, J.; Luo, L.; Wei, X.; Zhang, J.; Qi, Y.; Zhang, B.; Liu, H.; Xiao, P. Complete chloroplast genome sequences of Schisandra chinensis: Genome structure, comparative analysis, and phylogenetic relationship of basal angiosperms. *Sci. China Life Sci.* **2017**, *60*, 1286–1290. [CrossRef] [PubMed]
34. Curci, P.L.; Paola, D.D.; Danzi, D.; Vendramin, G.G.; Sonnante, G. Complete chloroplast genome of the multifunctional crop globe artichoke and comparison with other asteraceae. *PLoS ONE* **2015**, *10*. [CrossRef] [PubMed]
35. Qian, J.; Song, J.; Gao, H.; Zhu, Y.; Xu, J.; Pang, X.; Yao, H. The complete chloroplast genome sequence of the medicinal plant salvia miltiorrhiza. *PLoS ONE* **2013**, *8*. [CrossRef] [PubMed]
36. Xie, D.-F.; Yu, Y.; Deng, Y.-Q.; Li, J.; Liu, H.-Y.; Zhou, S.-D.; He, X.-J. Comparative analysis of the chloroplast genomes of the Chinese endemic genus Urophysa and their contribution to chloroplast phylogeny and adaptive evolution. *Int. J. Mol. Sci.* **2018**, *19*, 1847. [CrossRef]
37. He, L.; Qian, J.; Li, X.; Sun, Z.; Xu, X.; Chen, S. Complete chloroplast genome of medicinal plant Lonicera japonica: Genome rearrangement, intron gain and loss, and implications for phylogenetic studies. *Molecules* **2017**, *22*, 249. [CrossRef]
38. Han, Y.W.; Duan, D.; Ma, X.F.; Jia, Y.; Liu, Z.L.; Zhao, G.F.; Li, Z.H. Efficient identification of the forest tree species in aceraceae using DNA barcodes. *Front. Plant Sci.* **2016**, *7*. [CrossRef]
39. Kress, W.J.; Erickson, D.L. A two-locus global DNA barcode for land plants: The coding rbcL gene complements the non-coding trnH-psbA spacer region. *PLoS ONE* **2007**, *2*. [CrossRef]
40. Pang, X.; Chang, L.; Shi, L.; Rui, L.; Dong, L.; Li, H.; Cherny, S.S.; Chen, S. Utility of the trnh–Psba intergenic spacer region and its combinations as plant dna barcodes: a meta-analysis. *PLoS ONE* **2012**, *7*. [CrossRef]
41. Lahaye, R.R.Y.; Savolainen, V.; Duthoit, S.; Maurin, O.; Bank, M. A test of psbK-psbI and atpF-atpH as potential plant DNA barcodes using the flora of the Kruger National Park (South Africa) as a model system. *Nat. Preced.* **2008**. [CrossRef]
42. Timme, R.E.; Kuehl, J.V.; Boore, J.L.; Jansen, R.K. A comparative analysis of the Lactuca and Helianthus (Asteraceae) plastid genomes: Identification of divergent regions and categorization of shared repeats. *Am. J. Bot.* **2007**, *94*, 302–312. [CrossRef]

43. Falchi, A.; Paolini, J.; Desjobert, J.M.; Melis, A.; Costa, J.; Varesi, L. Phylogeography of Cistus creticus L. on Corsica and Sardinia inferred by the TRNL-F and RPL32-TRNL sequences of cpDNA. *Mol. Phylogenetics Evol.* **2009**, *52*, 538–543. [CrossRef] [PubMed]
44. Baillie, J.E.M.; Hilton, C.; Stuart, S.N. *IUCN Red List of Threatened Species: A Global Species Assessment*; IUCN-The World Conservation Union: Grand, Switzerland, 2004.
45. Zhang, Y.; Yu, T.; Ma, W.; Tian, C.; Sha, Z.; Li, J. Morphological and physiological response of Acer catalpifolium Rehd. Seedlings to water and light stresses. *Glob. Ecol. Conserv.* **2019**, *19*, 10. [CrossRef]
46. Zhao, J.; Yao, X.; Xi, L.; Yang, J.; Chen, H.; Zhang, J. Characterization of the chloroplast genome sequence of acer miaotaiense: comparative and phylogenetic analyses. *Molecules* **2018**, *23*, 1740. [CrossRef] [PubMed]
47. Li, J.H.; Yue, J.P.; Shoup, S. Phylogenetics of Acer (Aceroideae, Sapindaceae) based on nucleotide sequences of two chloroplast non-coding regions. *Harv. Pap. Bot.* **2006**, *11*, 101–115. [CrossRef]
48. Li, Q.; Liu, X.; Su, J.; Cong, Y. Extraction of genomic DNA from Acer yangjuechi and system optimization of SRAP-PCR. *Jiangsu J. Agric. Sci.* **2009**, *25*, 538–541.
49. Liu, C.; Tsuda, Y.; Shen, H.; Hu, L.; Saito, Y.; Ide, Y. Genetic structure and hierarchical population divergence history of Acer mono var. mono in South and Northeast China. *PLoS ONE* **2014**, *9*. [CrossRef]

Article

Complete Chloroplast Genome of *Michelia shiluensis* and a Comparative Analysis with Four Magnoliaceae Species

Yanwen Deng [1,2], Yiyang Luo [1,2], Yu He [1,2], Xinsheng Qin [2], Chonggao Li [3] and Xiaomei Deng [1,2,*]

1 Guangdong Key Laboratory for Innovative Development and Utilization of Forest Plant Germplasm, Guangzhou 510642, China
2 College of Forestry and Landscape Architecture, South China Agricultural University, Guangzhou 510642, China
3 Guangzhou Hengsheng Construction Engineering Co., Ltd., Guangzhou 510000, China
* Correspondence: xmdeng_scau@163.com; Tel.: +86-137-1116-8219

Received: 7 February 2020; Accepted: 25 February 2020; Published: 27 February 2020

Abstract: *Michelia shiluensis* is a rare and endangered magnolia species found in South China. This species produces beautiful flowers and is thus widely used in landscape gardening. Additionally, its timber is also used for furniture production. As a result of low rates of natural reproduction and increasing levels of human impact, wild *M. shiluensis* populations have become fragmented. This species is now classified as endangered by the IUCN. In the present study, we characterized the complete chloroplast genome of *M. shiluensis* and found it to be 160,075 bp in length with two inverted repeat regions (26,587 bp each), a large single-copy region (88,105 bp), and a small copy region (18,796 bp). The genome contained 131 genes, including 86 protein-coding genes, 37 tRNAs, and 8 rRNAs. The guanine-cytosine content represented 39.26% of the overall genome. Comparative analysis revealed high similarity between the *M. shiluensis* chloroplast genome and those of four closely related species: *Michelia odora*, *Magnolia laevifolia*, *Magnolia insignis*, and *Magnolia cathcartii*. Phylogenetic analysis shows that *M. shiluensis* is most closely related to *M. odora*. The genomic information presented in this study is valuable for further classification, phylogenetic studies, and to support ongoing conservation efforts.

Keywords: Hainan Province; endemic species; conservation; codon usage; sequence divergence; phylogeny

1. Introduction

Michelia shiluensis Chun and Y. F. Wu (Magnoliaceae) is an endangered flowering plant that is sparsely distributed throughout Hainan Province, China [1]. It is characterized by leafy branches and beautiful flowers, and is, therefore, widely used in landscape gardening [2]. This species is also a source of excellent quality wood which is in demand for furniture production [3]. In recent decades, there has been a serious decline in wild populations of this species as a result of the illegal harvesting to supply both the timber and horticultural markets [4]. Moreover, this species naturally has a low seeding rate and its wild populations are declining [5]. Consequently, *M. shiluensis* is categorized as a Class II National Key Protected Species in China [6] and is considered endangered (EN) by the International Union for Conservation of Nature [7]. Currently, most studies on *M. shiluensis* have focused on its use in landscape gardening and its protection in China [5]; however, there remains a lack of evolutionary and phylogenetic research.

The chloroplast is an important organelle in plants with its own genome (hereafter, cp genome) and participates in photosynthesis and other functions [8]. The cp genome of most land plants has a circular structure, including four segments: A large single-copy (LSC), a small single-copy (SSC), and two invert repeats (IRs) [9]. Although the cp genome is generally conserved, it has undergone intra- and inter-species rearrangement during evolution [10,11], including IR expansion and contraction. The information obtained from sequence rearrangements can be applied in phylogenetic analyses to solve taxonomic problems, such as low-level classifications, using genome comparison [12–17]. In the section *Michelia*, complete cp genomes have been reported for only *Magnolia alba* (NC037005), *Magnolia laevifolia* (NC035956), and *Michelia odora* (NC023239). Therefore, analysis of the cp genomes of other *Michelia* plants is necessary because of the similarity of morphology among Magnoliaceae species [18].

In the present study, we characterized the cp genome of *M. shiluensis* and compared its sequence features with four closely related species (*M. odora*, *M. laevifolia*, *Magnolia insignis*, and *Magnolia cathcartii*). The phylogenetic relationships among 28 Magnoliaceae species were constructed based on 79 protein-coding gene (PCG) sequences and show that *M. shiluensis* is most closely related to *M. odora*.

2. Materials and Methods

2.1. Plant Material and DNA Extraction

Fresh leaves of *M. shiluensis* were collected in the South China Botanical Garden (113°21′ E, 23°10′ N), China and transported to the laboratory at the South China Agricultural University. Total genomic DNA was isolated from the leaves using the CTAB method [19].

2.2. Genome Sequencing and Annotation

An Illumina shotgun library was established according to the manufacturer's protocol, and high-throughput sequencing was conducted using the HiSeq X TEN platform (Illumina, San Diego, CA, USA). After filtration using SOAPnuke [20], 4.93 GB of clean data were generated. Filtered reads were assembled de novo using SPAdes (version 3.10.1) [21] by referencing them against the cp genome sequence of *M. odora* (NC037005.1) using BLAST v2.2.30 (National Center for Biotechnology Information, Bethesda, MD, USA). Gene annotation was performed using GeSeq [22]. The cp genome map was generated using Organellar Genome DRAW (version 1.2) [23]. The annotated sequence was submitted to GenBank (accession number MN418056).

2.3. Sequence and Repeat Analysis

We used the Editseq v7.1.0 [24] software to calculate the guanine-cytosine (GC) content. MEGA v7.0.26 [25] was used to generate the relative synonymous codon usage (RSCU) values based on 79 PCGs. RNA editing sites in PCGs were predicted using the PREP suite [26] with the default settings.

The REPuter [27] online service was used to identify repeats (forward, reverse, complement, and palindromic) in the cp genome with default parameters. Chloroplast simple sequence repeats (cpSSRs) were identified using MISA-web [28] with minimal repeat numbers of 8, 5, 4, 3, 3, and 3 for mono-, di-, tri-, tetra-, penta-, and hexanucleotide repeats, respectively.

2.4. Genome Comparison and Sequence Divergence

Comparisons between five Magnoliaceae cp genomes were visualized using online mVISTA software [29] with the annotation of *M. shiluensis* as the reference in Shuffle-LAGAN mode. The borders of four different regions among the five cp genomes of Magnoliaceae were visualized using IRscope [30]. The nucleotide diversity (Pi), the rate of nonsynonymous (dN) substitutions, the rate of synonymous (dS) substitutions were determined using DNAsp v6.12.03 [31] to investigate the nucleotide diversity of sequences and genes that are considered to be under selection pressure.

2.5. Phylogenetic Analysis

To research the phylogenetic relationships and allow for comparisons among Magnoliaceae species, a maximum likelihood tree was constructed using RAxML [32], with 1000 bootstrap replicates, based on the PCG sequences found in 28 Magnoliaceae cp genomes. All 28 Magnoliaceae cp genome sequences were downloaded from the NCBI nucleotide database.

3. Result

3.1. Structures and Features of M. shiluensis Chloroplast Genome

The complete cp genome of *M. shiluensis* was 160,075 bp in length and comprised two IR regions of 26,587 bp each, separated by an LSC region of 88,105 bp, and an SSC region of 18,796 bp. The cp genome had the following base proportions: Adenine (A), 29.99%; thymine (T), 30.75%; cytosine (C), 19.98%; and guanine (G), 19.28%. Therefore, the total GC content was 39.26%. The GC content of the LSC, SSC, and IR regions were 37.95%, 34.28%, and 43.20%, respectively (Table 1).

Table 1. Summary of the chloroplast genomes of *Michelia shiluensis* and four closely related species (*Michelia odora*, *Magnolia laevifolia*, *Magnolia insignis*, and *Magnolia cathcartii*).

	M. shiluensis	*M. odora*	*M. laevifolia*	*M. insignis*	*M. cathcartii*
Accession	MN418056	NC023239	NC035956	NC035657	NC023234
			Genome		
Length (bp)	160,075	160,070	160,120	160,117	159,950
GC (%)	39.26	39.26	39.24	39.24	39.22
			LSC		
length (bp)	88,105	88,098	88,145	88,195	88,142
GC (%)	37.95	37.95	37.9	37.92	37.91
Length (%)	55.04	55.04	55.05	55.08	55.11
			SSC		
length (bp)	18,796	18,800	18,799	18,782	18,790
GC (%)	34.28	34.28	34.32	34.25	34.13
Length (%)	11.74	11.74	11.74	11.73	11.75
			IR		
length (bp)	26,587	26,586	26,588	26,570	26,509
GC (%)	43.2	43.2	43.2	43.19	43.2
Length (%)	16.61	16.61	16.61	16.59	16.57
			No. of Genes (duplicated in IR)		
Genes	131(18)	131(18)	131(18)	131(18)	131(18)
PCGs	86(7)	86(7)	86(7)	86(7)	86(7)
tRNA	37(7)	37(7)	37(7)	37(7)	37(7)
rRNA	8(4)	8(4)	8(4)	8(4)	8(4)
With introns	16(5)	16(5)	16(5)	16(5)	16(5)

Notes: GC: Guanine-cytosine; LSC: Large single-copy; IR: Invert repeat; SSC: Small single-copy; PCG: Protein-coding gene.

The cp genome of *M. shiluensis* contained 113 unique genes, including 79 PCGs, 30 tRNAs, and four rRNAs (Table 1, Figure 1). A total of 58 genes were found to be involved in self-replication, 12 genes encoded small ribosomal subunit proteins, eight genes encoded large ribosomal subunit proteins, 30 genes encoded tRNA, and four genes encoded RNA polymerase subunits. A total of 44 genes were found to be involved in photosynthesis, including six genes for ATP synthase, 11 genes for NADH dehydrogenase, six genes for the cytochrome b/f complex, five genes for photosystem I, 15 genes for photosystem II, and one gene for the large chain of Rubisco. In total, 18 genes were duplicated in the cp genome of *M. shiluensis*, including seven PCGs, seven tRNA genes, and four rRNA genes, all of which were located in the IR region (Table 2). None of the genes contained stop codons in coding sequences, therefore, no pseudogenes were detected.

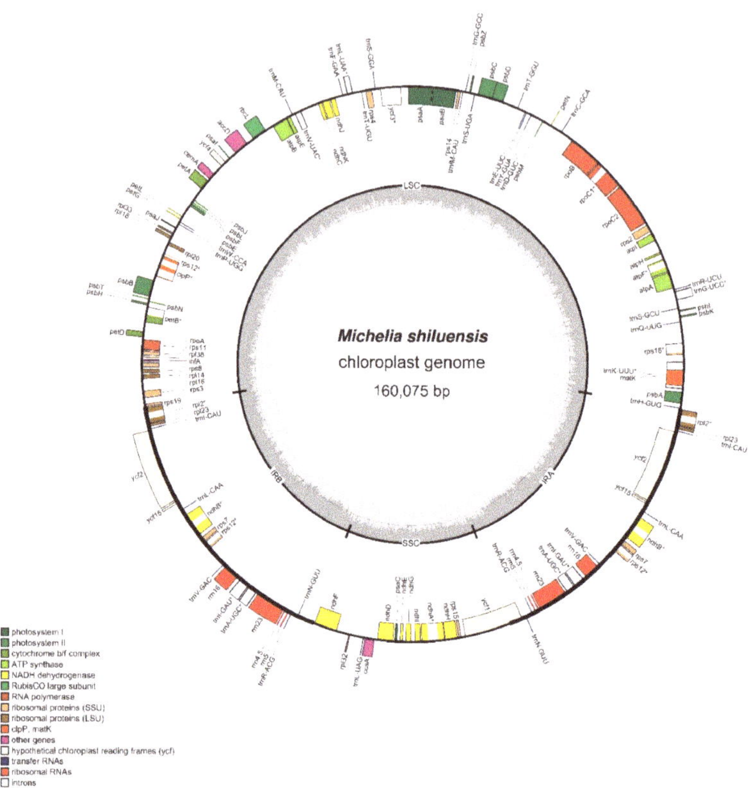

Figure 1. Gene map of the *Michelia shiluensis* chloroplast genome. Genes on the outside of the circle are transcribed counter-clockwise, while genes on the inside are transcribed clockwise. Different colors represent different kinds of functional genes. The guanine-cytosine content is indicated by darker gray and the adenine-thymine content is indicated by light gray.

A total of 16 genes were found to have introns, including 10 PCGs and six tRNA genes. Of these genes, *clpP*, *trnA-UGC*, *trnI-GAU*, and *ycf3* had two introns, whereas *atpF*, *ndhA*, *ndhB*, *petB*, *rpl2*, *rpoC1*, *rps12*, *rps16*, *trnG-UCC*, *trnK-UUU*, *trnL-UAA*, and *trnV-UAC* had one intron. The *rps12* gene which encodes the 40S ribosomal protein S12, was trans-spliced, with one exon located in the LSC region, and the other two exons located in the IR region. The largest intron was located in the *trnK* gene (2490 bp) with the *matK* gene inside; *trnL-UAA* had the smallest intron (491 bp) (Table 3).

We compared the basic cp genome features of *M. shiluensis* with four Magnoliaceae species. The cp genome lengths of *M. laevifolia* and *M. insignis* were 45 and 42 bp longer than that of *M. shiluensis*, respectively, while the cp genome lengths of *M. odora* and *M. cathcartii* were five and 125 bp shorter, respectively. Compared with *M. shiluensis*, the variation in the lengths of the LSC, SSC, and IR regions ranged from 7 to 90, 3 to 14, and 1 to 78 bp, respectively. In addition, the GC content of the whole genome and of each region of *M. shiluensis* were highly similar to those of the other four species. Moreover, there was no variation with respect to the total number of genes, PCGs, tRNA genes, rRNA genes, and genes with introns (Table 1).

Table 2. List of the annotated genes in the *Michelia shiluensis* chloroplast genome.

Category of Genes	Subcategory of Genes	Gene Names			
Self-replication	**rRNA genes**	***rrn 4.5* ***	***rrn5* ***	***rrn16* ***	***rrn23* ***
	tRNA genes	*trnA-UGC* *	*trnC-ACA*	*trnD-GUC*	*trnE-UUC*
		trnF-GAA	*trnfM-CAU*	*trnG-GCC*	*trnG-UCC*
		trnH-GUG	*trnI-CAU*	*trnI-GAU* *	*trnK-UUU*
		trnL-CAA *	*trnL-UAA*	*trnL-UAG*	*trnM-CAU* *
		trnN-GUU *	*trnP-UGG*	*trnQ-UUG*	*trnR-ACG* *
		trnR-UCU	*trnS-GCU*	*trnS-GGA*	*trnS-UGA*
		trnT-GGU	*trnT-UGU*	*trnV-GAC* *	*trnV-UAC*
		trnW-CCA	*trnY-GUA*		
	Small subunit of ribosome	*rps2*	*rps3*	*rps4*	*rps7* *
		rps8	*rps11*	*rps12* *	*rps14*
		rps15	*rps16*	*rps18*	*rps19*
	Large subunit of ribosome	*rpl2* *	*rpl14*	*rpl16*	*rpl20*
		rpl23 *	*rpl32*	*rpl33*	*rpl36*
	RNA polymerase subunits	*rpoA*	*rpoB*	*rpoC1*	*rpoC2*
Photosynthesis	ATP synthase gene	*atpA*	*atpB*	*atpE*	*atpF*
		atpH	*atpI*		
	NADH dehydrogenase	*ndhA*	*ndhB* *	*ndhC*	*ndhD*
		ndhE	*ndhF*	*ndhG*	*ndhH*
		ndhI	*ndhJ*	*ndhK*	
	Cytochrome b/f complex	*petA*	*petB*	*petD*	*petG*
		petL	*petN*		
	Photosystem I	*psaA*	*psaB*	*psaC*	*psaI*
		psaJ			
	Photosystem II	*psbA*	*psbB*	*psbC*	*psbD*
		psbE	*psbF*	*psbH*	*psbI*
		psbJ	*psbK*	*psbL*	*psbM*
		psbN	*psbT*	*psbZ*	
	Large chain of Rubisco	*rbcL*			
Other genes	ATP-dependent protease	*clpP*			
	Cytochrome c biogenesis	*ccsA*			
	Acetyl-CoA carboxylase	*accD*			
	Translation initiation factor IF-1	*infA*			
	Membrane protein	*cemA*			
	Maturase	*matK*			
Unknown function	Hypothetical chloroplast reading frame	*ycf1*	*ycf2* *	*ycf3*	*ycf4*
		ycf15 *			

Note: "*" indicates duplicated genes.

Table 3. Characteristics of the genes that contain introns in the cp genome of *Michelia shiluensis*.

Gene	Location	Exon I (bp)	Intron I (bp)	Exon II (bp)	Intron II (bp)	Extron III (bp)
trnK-UUU	LSC	35	2490	37		
rps16	LSC	217	825	44		
trnG-UCC	LSC	24	768	48		
atpF	LSC	411	706	144		
rpoC1	LSC	1624	722	434		
ycf3	LSC	154	729	227	739	126
trnL-UAA	LSC	35	491	50		
trnV-UAC	LSC	37	584	56	565	39
clpP	LSC	246	630	291	781	69
petB	LSC	5	786	641		
rpl2	IR	432	658	387		
rps12	IR/LSC	114	-	25	536	232
ndhB	IR	756	700	777		
trnI-GAU	IR	42	936	35		
trnA-UGC	IR	38	799	35		
ndhA	SSC	541	1078	551		

3.2. Codon Usage and RNA Analysis

Based on the PCGs, 22,791 codons were detected (excluding the stop codons) (Table 4). The three most abundant amino acids were leucine (2423 codons), isoleucine (2085 codons), and serine (1719 codons), and the three least abundant amino acids were cysteine (314 codons), tryptophan (427 codons), and methionine (602 codons) (Figure S1). Of the 30 most frequent codons (RSCU > 1), most of them end with A or U, and only the UUG codon ends with G. In contrast, most of the 32 least frequent codons (RSCU < 1) end with C or G. In addition, two codons, AUG and UGG, have no codon bias (RSCU = 1).

PREP suite was used to edit predictions in the genome of *M. shiluensis* by manipulating the first codon position of the first nucleotide (Table S1). A total of 106 RNA editing sites were detected from the PCGs in *M. shiluensis*; with the majority of the amino acid conversions involving the conversion of serine to leucine. Most of the RNA editing sites were located on the *ndhB* gene (14 sites), followed by *ndhD* (11 sites), and *matK* (nine sites). Most of the conversions changed from a polar group to a nonpolar group, while only two sites changed from a nonpolar group to a polar group (proline to serine); one of these was located on the *psbE* gene while the other was located on the *ccsA* gene.

Table 4. Relative synonymous codon usage (RSCU) in the chloroplast genome of *Michelia shiluensis*.

Codon	Amino Acid	Count	RSCU	Codon	Amino Acid	Count	RSCU
UUU	Phe	714	1.1	UAU	Tyr	600	1.47
UUC	Phe	586	0.9	UAC	Tyr	214	0.53
UUA	Leu	671	1.66	UAA	*	147	0.98
UUG	Leu	510	1.26	UAG	*	137	0.91
CUU	Leu	439	1.09	CAU	His	429	1.41
CUC	Leu	206	0.51	CAC	His	178	0.59
CUA	Leu	400	0.99	CAA	Gln	560	1.41
CUG	Leu	197	0.49	CAG	Gln	237	0.59
AUU	Ile	893	1.28	AAU	Asn	712	1.44
AUC	Ile	545	0.78	AAC	Asn	278	0.56
AUA	Ile	647	0.93	AAA	Lys	777	1.37
AUG	Met	602	1	AAG	Lys	359	0.63
GUU	Val	474	1.37	GAU	Asp	631	1.54
GUC	Val	202	0.58	GAC	Asp	189	0.46
GUA	Val	487	1.41	GAA	Glu	780	1.42
GUG	Val	223	0.64	GAG	Glu	318	0.58
UCU	Ser	457	1.6	UGU	Cys	214	1.36
UCC	Ser	266	0.93	UGC	Cys	100	0.64
UCA	Ser	373	1.3	UGA	*	166	1.11
UCG	Ser	192	0.67	UGG	Trp	427	1
CCU	Pro	345	1.41	CGU	Arg	256	1.15
CCC	Pro	213	0.87	CGC	Arg	72	0.32
CCA	Pro	307	1.25	CGA	Arg	289	1.3
CCG	Pro	117	0.48	CGG	Arg	129	0.58
ACU	Thr	426	1.49	AGU	Ser	324	1.13
ACC	Thr	237	0.83	AGC	Ser	107	0.37
ACA	Thr	358	1.26	AGA	Arg	410	1.84
ACG	Thr	120	0.42	AGG	Arg	179	0.8
GCU	Ala	537	1.82	GGU	Gly	542	1.33
GCC	Ala	185	0.63	GGC	Gly	171	0.42
GCA	Ala	335	1.13	GGA	Gly	628	1.54
GCG	Ala	126	0.43	GGG	Gly	291	0.71

Note: "*" indicates the stop codon.

3.3. *Repeat Sequence Analysis*

The REPuter results show that the *M. shiluensis* cp genome contains a total of 49 repeats: 23 palindromic, 18 forward, and eight reverse repeats (Figure 2). The repeat size ranged from 18 to 33 bp. The most abundant repeats were 18 bp (12 sites) followed by 20 bp (10 sites) (Figure S2).

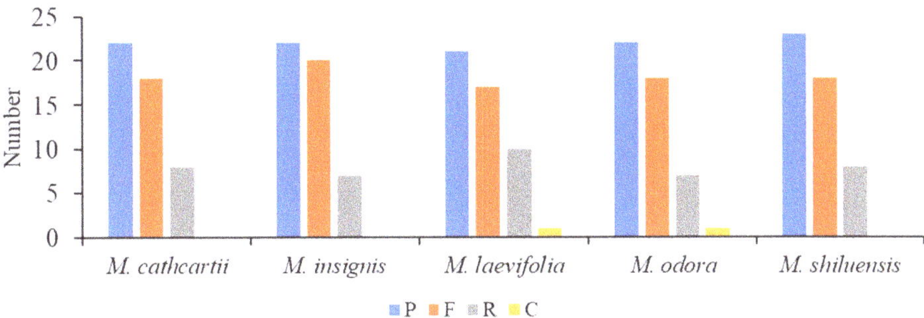

Figure 2. Comparison of repeats among five Magnoliaceae species: *Michelia shiluensis*, *Michelia odora*, *Magnolia laevifolia*, *Magnolia insignis*, and *Magnolia cathcartii*. (F: Forward; R: Reverse; C: Complement; and P: Palindromic).

In the first location, 46.9% of repeats were detected in the intergenic spacers (IGSs), while 34.7% were in the PCGs, and 18.4% were in the tRNA genes. Of all the PCGs, the *ycf2* gene had five forward repeats and four palindromic repeats and was the gene with the most repeats (Table S2). Comparison of the repeat types with the other four species revealed no substantial variation among the five species (Figure 2). *Michelia shiluensis* had the highest frequency of palindromic repeats (23), while *M. laevifolia* had the lowest (21). *Magnolia cathcartii*, *M. odora*, and *M. shiluensis* had the same number of forward repeats (18), while *M. shiluensis* and *M. cathcartii* had eight reverse repeats. In addition, only one complement repeat was found in the genomes of *M. laevifolia* and *M. odora*, whereas no complements were identified in the cp genomes of *M. shiluensis* and *M. insignis*. A total of 141 cpSSRs were found in the cp genome of *M. shiluensis* (Table S3). The majority of them were mononucleotide repeats (118), followed by tetranucleotide repeats (9), and dinucleotide repeats (8) (Figure 3). No pentanucleotide repeats were detected in the cp genome of *M. shiluensis*. The longest repeat was 18 bp while the shortest was 8 bp. Noncoding regions, including IGSs (97) and introns (19), contained most of the SSRs while only 25 repeats were located in coding regions, including *cemA*, *ndhD*, *ndhF*, *psbC*, *rpoB*, *rpoC1*, *rpoC2*, *rps19*, *rps3*, *ycf1*, *ycf2*, and *ycf4* (Table 5). The cpSSRs were mainly distributed in the LSC region (72.34%), followed by the SSC region (17.73%), with just 4.96% in the IR. The cpSSRs in *M. shiluensis* had base bias towards A-T bases. In total, 113 SSRs had A or T bases, accounting for 80.14% of the total SSRs (Figure 3). Comparison among the five species of Magnoliaceae show high similarity in SSR type and distribution. The variation in the total amount, mono-, di-, tri-, tetra-, penta-, and hexanucleotide repeats among the five species was 5, 4, 0, 2, 0, 1, and 1, respectively (Figure 3). The number of SSRs in the IR were the same among the five species while the counts in different locations and regions were highly conserved.

Table 5. Distribution of single sequence repeats in different locations and regions among five Magnoliaceae species: *Michelia shiluensis*, *Michelia odora*, *Magnolia laevifolia*, *Magnolia insignis*, and *Magnolia cathcartii*.

Species	Number	Location			Regions		
		LSC	IR	SSC	CDS	Intron	IGS
M. cathcartii	143	103	7	26	25	18	100
M. insignis	141	102	7	25	26	15	100
M. laevifolia	141	101	7	26	27	18	96
M. odora	138	99	7	25	24	19	95
M. shiluensis	141	102	7	25	25	19	97

Notes: LSC: Large single copy; IR: Invert region; SSC: Small single copy; CDS: Coding sequence; IGS: Intergenic spacer.

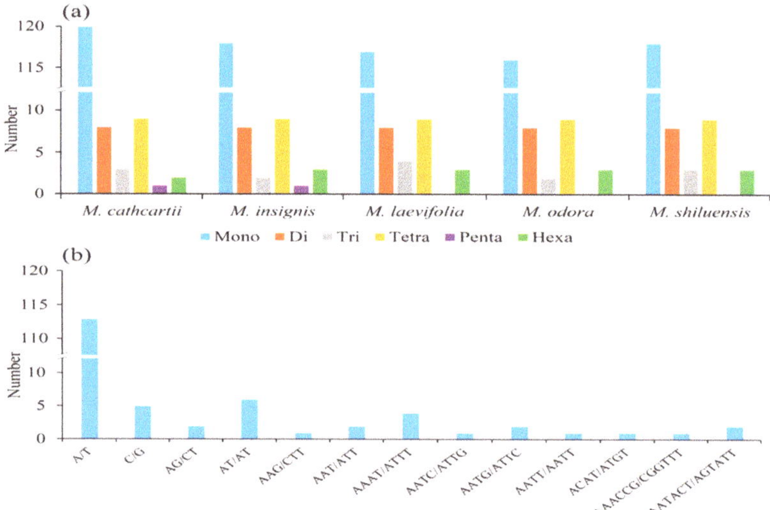

Figure 3. Single sequence repeats (SSRs) in the chloroplast genome of *Michelia shiluensis*. (**a**) Comparison of the SSRs among five Magnoliaceae species (*M. shiluensis*, *Michelia odora*, *Magnolia laevifolia*, *Magnolia insignis*, and *Magnolia cathcartii*); (**b**) base composition of SSRs in the cp genome of *M. shiluensis*.

3.4. Genome Comparison and Sequence Divergence

The mVISTA online software was used to compare the variation in the whole cp genome among the five species (Figure 4). The alignments indicated that the whole cp genome of the five species was highly conserved, especially in the IR region. The noncoding sequences had relatively more divergence than the coding sequences. The noncoding sequences that contained high levels of divergence were *rps16-trnQ*, *atpH-atpI*, *trnT-psbD*, *petA-psbJ*, and *ndhF-trnL*. In the coding sequences, only *ycf1* show relatively more variation than the other genes. No obvious insertions were found among the five species.

The four junctions in the regions of the cp genomes of the five species were shown using IRscope (Figure 5). There was a conserved structure on each border, and slight distance differences among the five species. Gene *rps19* was fully located in the LSC at a distance of 1–6 bp from the LSC/IRb border, while gene *rpl2* was fully located in the IRb. The *ndhF* gene was found in the SSC region and was 61 bp away from the IRb/SSC border in *M. odora*, *M. laevifolia*, and *M. insignis*, while it was 21 bp longer in *M. shiluensis*, and 7 bp shorter in *M. cathcartii*. The SSC/IRa border was inside the *ycf1* gene in all five species. Compared to *M. shiluensis* and *M. odora*, the part of the *ycf1* gene in the SSC region of *M. laevifolia* and *M. insignis* was almost 20 bp longer, and this resulted in the differences in gene length. However, the *ycf1* gene of *M. cathcartii* was almost 30 bp shorter on both sides of the SSC/IRa border; thus, the *ycf1* gene of *M. cathcartii* was almost 60 bp shorter than those of the other four species. The distance from the *trnH* gene to the IRa/LSC border was 11 bp in *M. shiluensis*, *M. odora*, and *M. laevifolia*, while it was 16 bp in *M. insignis* and 9 bp in *M. cathcartii*. Due to the short length in the IR region, the whole length of the *M. cathcartii* cp genome was significantly shorter than those of the other four species.

To detect the selective pressures on the PCGs in the *M. shiluensis* cp genome, the rate of nonsynonymous (dN) substitutions, the rate of synonymous (dS) substitutions, and their ratio (dN/dS) were calculated based on the 79 PCG sequences of the five Magnoliaceae species (Figure 6). Only four genes had a dN/dS ratio greater than 1 (*accD* in *M. insignis* vs. *M. cathcartii*, score 1.14; *ndhD* in *M. shiluensis* vs. *M. cathcartii*, score 1.29; *ndhF* in *M. odora* vs. *M. cathcartii*, score 1.89; and *rpoC2* in *M. laevifolia* vs. *M. cathcartii*, score 2.50), which indicates that most genes are under the influence of negative selection, while only a few genes are under the influence of positive selection.

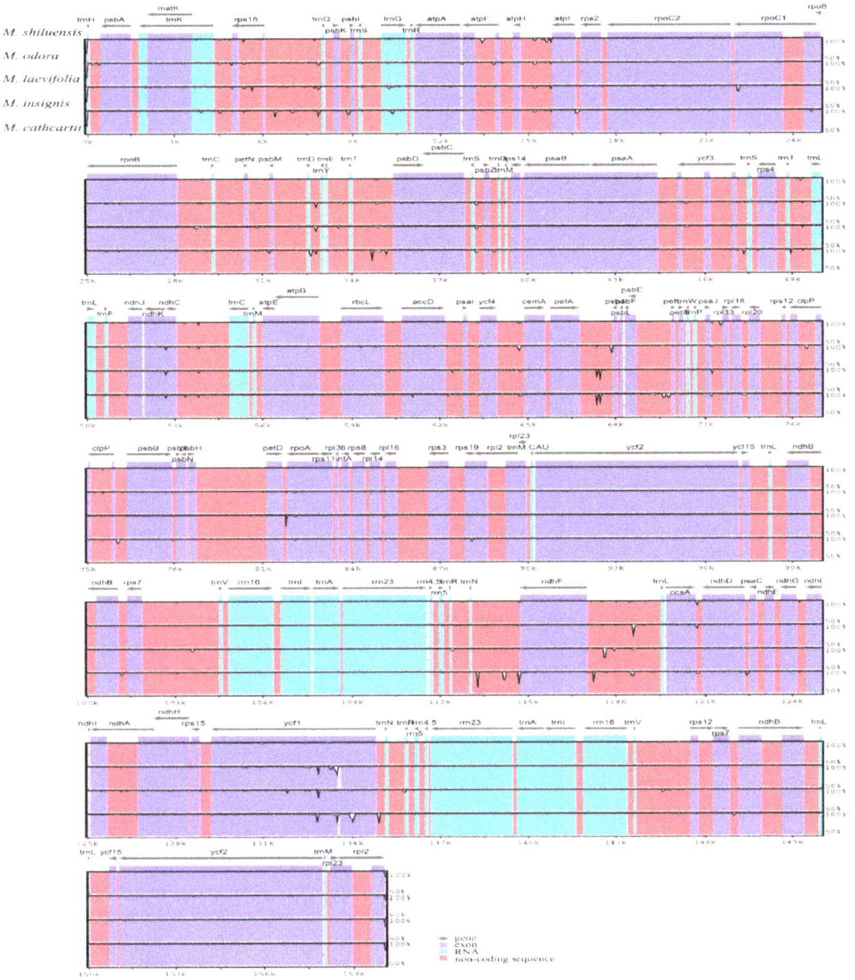

Figure 4. Sequence alignment of five whole chloroplast genomes in Magnoliaceae (*Michelia shiluensis*, *Michelia odora*, *Magnolia laevifolia*, *Magnolia insignis*, and *Magnolia cathcartii*) using *M. shiluensis* as a reference in *mVISTA*.

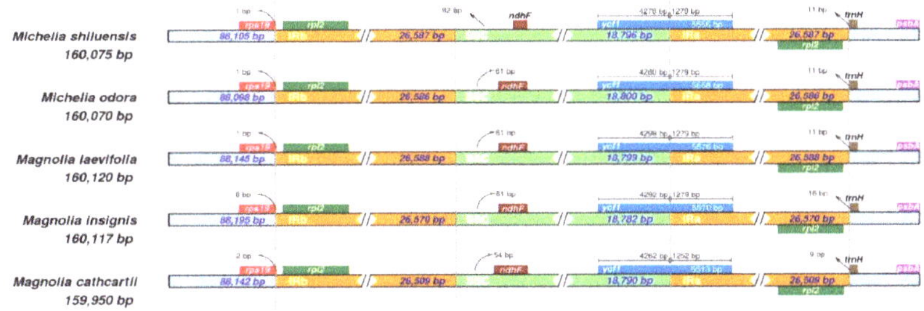

Figure 5. The four junctions of the regions in the chloroplast genomes of the five *Magnoliaceae* species, determined using *IRscope*.

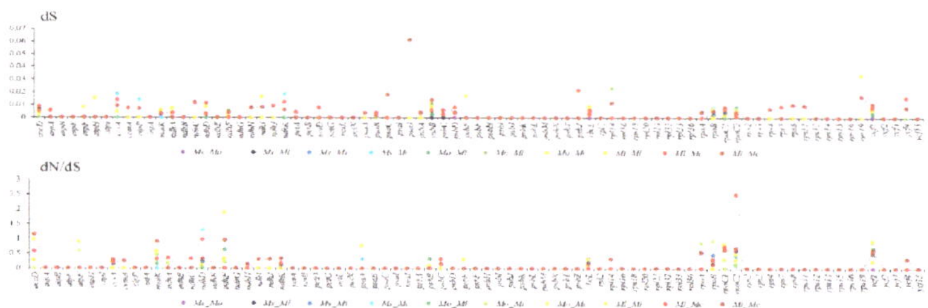

Figure 6. The synonymous (dS) and nonsynonymous substitutions (dN)/dS ratio values of 79 protein-coding genes from five Magnoliaceae chloroplast genomes (Ms: *Michelia shiluensis*; Mo: *Michelia odora*; Ml: *Magnolia laevifolia*; Mi: *Magnolia insignis*; Mc: *Magnolia cathcartii*).

In the coding region, the mean Pi in the PCGs was 0.00117 (ranging from 0 to 0.00606); and the mean values of Pi in the LSC, IR, and SSC regions were 0.001192, 0.000186, and 0.001634, respectively (Figure 7). Meanwhile, in the IGSs, the mean Pi value was 0.00295 (ranging from 0 to 0.02416); and the mean Pi value in the LSC, IR, and SSC regions were 0.0297, 0.00045, and 0.00731, respectively. This result indicates that the Pi value in the coding region is lower than that in the IGSs. The results also demonstrate that the IR region is the most conserved region among the five species, followed by the LSC and SSC regions. In total, 20 mutation sites (Pi > 0.005) were identified, including 19 sites in IGSs and one site in a PCG. The mutation sites in IGSs were as follows: *trnH-psbA*, *psbK-psbI*, *atpA-atpF*, *rps2-rpoC2*, *trnT-psbD*, *ycf3-trnS*, *ndhJ-ndhK*, *ndhK-ndhC*, *accD-psaI*, *psbL-psbF*, *petL-petG*, *trnW-trnP*, *trnP-psaJ*, *rpl18-rpl20*, *ndhF-trnL*, *ccsA-ndhD*, *ndhD-psaC*, *ndhG-ndhI*, and *ndhI-ndhA*. One gene, *psaJ*, was unique and had a Pi value greater than 0.005.

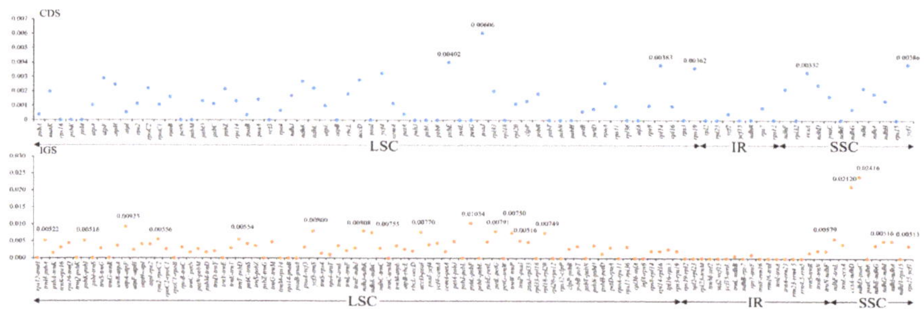

Figure 7. Nucleotide diversity (Pi) in the chloroplast genome of five Magnoliaceae species (*Michelia shiluensis*; *Michelia odora*; *Magnolia laevifolia*; *Magnolia insignis*; and *Magnolia cathcartii*).

3.5. Phylogenetic Analysis

To reveal the evolutionary relationships between the investigated species and to enable comparison with traditional phylogenies, a maximum likelihood phylogenetic tree was constructed using RAxML (with 1000 bootstrap replicates) based on the PCG sequences found in 28 Magnoliaceae cp genomes (Figure 8). The phylogenetic tree generated 25 nodes; most of which had 100% bootstrap support. This result strongly supports the notion that *M. shiluensis* is most closely related to *M. odora*.

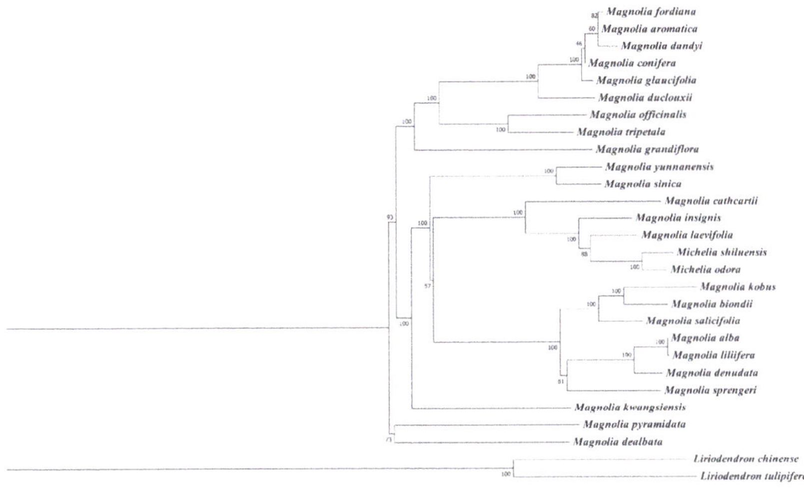

Figure 8. Maximum likelihood tree with 1000 bootstrap replicates constructed using RAxML based on chloroplast genomes of 28 Magnoliaceae species (26 species of *Magnolia* and *Michelia*, and two species of *Liriodendron* as outgroups). Bootstrap values (%) are shown above branches. Accession numbers: *Magnolia alba* NC037005, *Magnolia liliiflora* NC023238, *Magnolia denudata* NC018357, *Magnolia sprengeri* NC023242, *Magnolia salicifolia* NC023240, *Magnolia biondii* KY085894, *Magnolia kobus* NC023237, *Michelia odora* NC023239, *Michelia shiluensis* MN418056, *Magnolia laevifolia* NC035956, *Magnolia insignis* NC035657, *Magnolia cathcartii* NC023234, *Magnolia yunnanensis* NC024545, *Magnolia sinica* NC023241, *Magnolia kwangsiensis* NC015892, *Magnolia conifera* NC037001, *Magnolia dandyi* NC037004, *Magnolia aromatica* NC037000, *Magnolia fordiana* MF990562, *Magnolia glaucifolia* NC037003, *Magnolia duclouxii* NC037002, *Magnolia tripetala* NC024027, *Magnolia officinalis* NC020316, *Magnolia grandiflora* NC020318, *Magnolia pyramidata* NC023236, *Magnolia dealbata* NC023235, *Liriodendron chinense* NC030504, and *Liriodendron tulipifera* NC008326.

4. Discussion

In this study, we characterized the complete cp genome of *M. shiluensis*, an endangered and valuable species (Figure 1). By comparing five closely related species, we found that gene content, order, structure, and other features were highly conserved among them (Figures 3–5). *Michelia shiluensis* was shown to be most closely related to *M. odora* (Figure 8). This finding can help to further our understanding of the characterization of the *M. shiluensis* cp genome and reveal information concerning the evolution, population genetics, and phylogeny of this species.

Normally, the length of the cp genome in higher plants is in the range of about 120–160 kb, with a stable structure and conserved sequence [8,33]. The cp genome of *M. shiluensis* displayed a typical quadripartite structure, with an LSC and an SSC which were separated by two IR regions (Figure 1). The whole length of this genome was 160,075 bp, with 39.26% GC content, and containing 113 unique genes and 16 genes with one or two introns (Tables 1–3). Among the 26 Magnoliaceae species, the length of the cp genome ranged from 158,177 to 160,183 bp, the GC content ranged from 39.15% to 39.30%, and they collectively contained 112 common genes, including 79 PCGs, 29 tRNA, and four rRNA genes; also, one or two introns were found among these 16 genes. The results for the *M. shiluensis* cp genome were consistent with a previous analysis of 26 Magnoliaceae species [34], except for the number of genes, one additional tRNA gene (*trnV-GAC*) was detected in *M. shiluensis*. Similar to other angiosperms, a high GC content was detected in the IR region of *M. shiluensis*, which may be a result of the existence of high-GC rRNA sequences [9,35–37]. Introns play a vital role in selective gene splicing [38]. However, introns have been lost among some species during their evolution [39,40].

In this study, no introns were lost in the cp genome of *M. shiluensis* during evolution, which reflects the fact that the cp genome is highly conserved in Magnoliaceae [34].

In total, 22,791 codons were found in the cp genome of *M. shiluensis* (Table 4), among which, the codons for leucine were the most abundant (10.63%). This result has also been observed in *Ailanthus altissima* [41] and *Justicia flava* [42]. Among the preferred codons (RSCU > 1), we found that most of them ended with A or U, except UUG. This is not unique to the *M. shiluensis* cp genome as similar findings have been observed in *Papaver rhoeas* and *P. orientale* [36], *Ageratina adenophora* [43], and *Oryza sativa* [44]. Of the PCGs in the *M. shiluensis* cp genome, 106 possible sites for RNA editing were detected (Table S1). The majority of the amino acid conversion was from serine to leucine, and the *ndhB* gene accounted for a high number of editable sites (14 of the total 106 sites). Similar results have been obtained for *Forsythia suspensa* [45] and *Sanionia uncinata* [46].

Repeat sequences play an important role in genomic structural variation, expansion, and rearrangement [8,40]. Previous research has indicated that most of the repeat sequences are located in the IGS regions followed by the coding regions [47,48]. A similar result was found in this study, with 46.9% of repeats detected in the IGS regions, followed by 34.7% in the coding regions, and the remainder in the tRNAs (Table S2). The cpSSR is an effective marker [49,50] that is widely used in population genetics, biogeographic studies, and phylogenetic evaluation [51,52]. In the cp genome of *M. shiluensis*, over 80% of the SSRs consisted of A or T bases, and over 80% were mononucleotide repeats. Similar results have been observed in other studies [48,50,53]. The majority of SSRs are found in the SSC and LSC regions [50] and, in this respect, *M. shiluensis* is no exception (Table 5). These two regions accounted for 90.07% of the SSRs, and only seven SSRs were found in the IR region.

Although the cp genome of angiosperms is relatively conserved in structure and size [54,55], the expansion and contraction of the IR region, as caused by evolutionary events, has resulted in minor changes in the IR boundary and size of the genome [39,56], thus increasing the chloroplast genetic diversity of angiosperms [57,58]. In this study, comparative analysis of five Magnoliaceae species revealed that the IR lengths were similar in *M. shiluensis, M. odora*, and *M. laevifolia* (Figure 5). However, the IR region of *M. insignis* had completely lost 11 bp in the *rps12-trnV*, *rrn23-rrn4.5* IGSs, while *M. cathcartii* had lost 5 bp in the *rpl2* intron, 6 bp in the *rps12* intron, 26 bp in *ycf1*, and 41 bp in the *trnN-ndhF* IGS. The losses of these bases resulted in differences in the lengths of the IR regions among the five species.

DNA barcoding is a technique that is widely applied in plant identification studies [59,60]. However, only a few regions have been used for the DNA barcoding of Magnoliaceae [61–63]. We used mVISTA to compare the genomes of five Magnoliaceae species and revealed that the IR region is more highly conserved than the LSC and SSC regions, and that the coding region was more highly conserved than the noncoding region (Figure 4), consistent with other angiosperms [9,38]. Five regions in the *M. shiluensis* cp genome had high levels of variation (four on IGSs and one on a PCG). The Pi value was also investigated among the 79 PCGs and 125 IGSs (Figure 7), and only 20 regions were found to have a Pi value greater than 0.005; which confirmed the low base substitution rate in Magnoliaceae [64]. Regions with a high degree of variation can be used to develop high resolution DNA barcoding for identification.

Due to the high morphological similarity among Magnoliaceae species [18,65], there have been some difficulties with respect to the classification of the family. Thus, the classification of Magnoliaceae has always been controversial [66–72]. The cp genome contains sufficient information and has been shown to be more effective than cpDNA fragments for clarifying low level phylogenetic relationships in plants [53,73]. In this study, the phylogenetic results of 28 Magnoliaceae plants based on PCG sequences revealed that *M. shiluensis* is most closely related to *M. odora* (Figure 8), which is consistent with phylogenetic results based on the *ndhF* sequence [70]. According to traditional morphological classification, *M. insignis* has been placed in the subgenera *Manglietia*, and *M. alba* has been placed in the section *Michelia* [69,74]. However, the phylogenetic relationship results based on the cp genome show that *M. insignis* is located in the section *Michelia* clade and *M. alba* is located in the subgenera

Yulania clade. This result differs from that of traditional morphological classification [69,74] and the results of three nuclear gene sequences [75]. These findings confirm that not even a complete cp genome can distinguish species in young evolutionary lineages, and that phylogenetic conclusions may require consideration of certain features in the nuclear genome [76].

5. Conclusions

The complete cp genome provided by this study can be used for in-depth genetic research on *M. shiluensis* and Magnoliaceae species in general, and may also play an important role in the development of new conservation and management strategies to ultimately aid species conservation efforts.

Supplementary Materials: Supplementary materials can be found at http://www.mdpi.com/1999-4907/11/3/267/s1. Figure S1, Amino acid frequency among 79 protein-coding genes in the *Michelia shiluensis* chloroplast genome; Table S1, Possible RNA editing sites in the chloroplast genome of *Michelia shiluensis*; Table S2, Repeats in the chloroplast genome of *Michelia shiluensis*; Figure S2, Different lengths of repeats in the *Michelia shiluensis* cp genome; Table S3, Single sequence repeats in the chloroplast genome of *Michelia shiluensis*.

Author Contributions: Conceptualization, Y.D.; methodology, Y.D. and Y.H.; software, Y.L. and C.L.; validation, Y.D., Y.L., Y.H., C.L., and X.D.; formal analysis, Y.L. and Y.H.; resources, X.Q. and X.D.; writing—original draft preparation, Y.D.; writing—review and editing, Y.D.; supervision, X.D.; project administration, X.D.; funding acquisition, X.D. All authors have read and agreed to the published version of the manuscript.

Funding: This study was supported by the Forestry Public Welfare Industry Research of China (Grant Number: 201404116), the National Natural Science Foundation of China (Grant Number: 31670601), and the Forestry Science and Technology Innovation of Guangdong Province grant programs (Grant Numbers: 2014KJCX006 and 2017KJCX023).

Acknowledgments: We are sincerely grateful to Xiaomei Deng (supervisor to Yanwen Deng) for her support during this study, which included assistance with funding, materials, resources, and consultations with other academics. We are also grateful to Xinsheng Hu for his invaluable guidance during the manuscript preparation.

Conflicts of Interest: The authors declare no conflict of interest.

References

1. Francisco-Ortega, J.; Wang, F.-G.; Wang, Z.-S.; Xing, F.-W.; Liu, H.; Xu, H.; Xu, W.-X.; Luo, Y.-B.; Song, X.-Q.; Gale, S. Endemic seed plant species from Hainan Island: A checklist. *Bot. Rev.* **2010**, *76*, 295–345. [CrossRef]
2. Bao, S.; Zhou, S.; Yu, Y. Comparative anatomy of the leaves in *Michelia*. *Guangxi Zhi Wu* **2002**, *22*, 140–144.
3. Xiong, J.; Wang, L.-J.; Qian, J.; Wang, P.-P.; Wang, X.-J.; Ma, G.-L.; Zeng, H.; Li, J.; Hu, J.-F. Structurally diverse sesquiterpenoids from the endangered ornamental plant *Michelia shiluensis*. *J. Nat. Prod.* **2018**, *81*, 2195–2204. [CrossRef]
4. Chen, B. Forest wild biospecies diversity and its preservation in China. *Chin. J. Ecol.* **1993**, *12*, 39–43. [CrossRef]
5. Wei, Y.; Hong, F.; Yuan, L.; Kong, Y.; Shi, Y. Population distribution and age structure characteristics of *Michelia shiluensis*, an endangered and endemic species in Hainan Island. *Chin. J. Trop. Crops* **2017**, *38*, 2280–2284. [CrossRef]
6. State Forestry Administration of China. List of national key protected wild plants (first batch). *State Counc. Bull.* **1999**, *13*, 6.
7. Bhandari, M. International Union for Conservation of Nature. In *The Wiley-Blackwell Encyclopedia of Globalization*; Ritzer, G., Ed.; John Wiley & Sons: Hoboken, NJ, USA, 2012. [CrossRef]
8. Wicke, S.; Schneeweiss, G.M.; dePamphilis, C.W.; Müller, K.F.; Quandt, D. The evolution of the plastid chromosome in land plants: Gene content, gene order, gene function. *Plant Mol. Biol.* **2011**, *76*, 273–297. [CrossRef] [PubMed]
9. Chen, Y.; Hu, N.; Wu, H. Analyzing and characterizing the chloroplast genome of *Salix wilsonii*. *BioMed Res. Int.* **2019**, *2019*, 5190425. [CrossRef] [PubMed]
10. Tangphatsornruang, S.; Uthaipaisanwong, P.; Sangsrakru, D.; Chanprasert, J.; Yoocha, T.; Jomchai, N.; Tragoonrung, S. Characterization of the complete chloroplast genome of *Hevea brasiliensis* reveals genome rearrangement, RNA editing sites and phylogenetic relationships. *Gene* **2011**, *475*, 104–112. [CrossRef] [PubMed]
11. Walker, J.F.; Jansen, R.K.; Zanis, M.J.; Emery, N.C. Sources of inversion variation in the small single copy (SSC) region of chloroplast genomes. *Am. J. Bot.* **2015**, *102*, 1751–1752. [CrossRef]

12. Johansson, J.T. There large inversions in the chloroplast genomes and one loss of the chloroplast generps16 suggest an early evolutionary split in the genusAdonis (Ranunculaceae). *Plant Syst. Evol.* **1999**, *218*, 133–143. [CrossRef]
13. Lee, H.-L.; Jansen, R.K.; Chumley, T.W.; Kim, K.-J. Gene relocations within chloroplast genomes of Jasminum and Menodora (Oleaceae) are due to multiple, overlapping inversions. *Mol. Biol. Evol.* **2007**, *24*, 1161–1180. [CrossRef] [PubMed]
14. Jansen, R.K.; Wojciechowski, M.F.; Sanniyasi, E.; Lee, S.-B.; Daniell, H. Complete plastid genome sequence of the chickpea (*Cicer arietinum*) and the phylogenetic distribution of rps12 and clpP intron losses among legumes (Leguminosae). *Mol. Phylogenet. Evol.* **2008**, *48*, 1204–1217. [CrossRef] [PubMed]
15. Shaw, J.; Lickey, E.B.; Schilling, E.E.; Small, R.L. Comparison of whole chloroplast genome sequences to choose noncoding regions for phylogenetic studies in angiosperms: The tortoise and the hare III. *Am. J. Bot.* **2007**, *94*, 275–288. [CrossRef]
16. Mardanov, A.V.; Ravin, N.V.; Kuznetsov, B.B.; Samigullin, T.H.; Antonov, A.S.; Kolganova, T.V.; Skyabin, K.G. Complete sequence of the duckweed (*Lemna minor*) chloroplast genome: Structural organization and phylogenetic relationships to other angiosperms. *J. Mol. Evol.* **2008**, *66*, 555–564. [CrossRef]
17. Park, I.; Kim, W.-J.; Yang, S.; Yeo, S.-M.; Li, H.; Moon, B.C. The complete chloroplast genome sequence of *Aconitum coreanum* and *Aconitum carmichaelii* and comparative analysis with other *Aconitum* species. *PLoS ONE* **2017**, *12*, e0184257. [CrossRef]
18. Li, J.; Conran, J. Phylogenetic relationships in Magnoliaceae subfam. Magnolioideae: A morphological cladistic analysis. *Plant Syst. Evol.* **2003**, *242*, 33–47. [CrossRef]
19. Doyle, J.J.; Doyle, J.L. A rapid DNA isolation procedure for small quantities of fresh leaf tissue. *Phytochem. Bull.* **1987**, *19*, 11–15.
20. Chen, Y.; Shi, C.; Huang, Z.; Zhang, Y.; Li, S.; Li, Y.; Ye, J.; Yu, C.; Li, Z. SOAPnuke: A MapReduce acceleration-supported software for integrated quality control and preprocessing of high-throughput sequencing data. *Gigascience* **2017**, *7*, gix120. [CrossRef]
21. Bankevich, A.; Nurk, S.; Antipov, D.; Gurevich, A.A.; Dvorkin, M.; Kulikov, A.S.; Lesin, V.M.; Nikolenko, S.I.; Pham, S.; Prjibelski, A.D.; et al. SPAdes: A new genome assembly algorithm and its applications to single-cell sequencing. *J. Comput. Biol.* **2012**, *19*, 455–477. [CrossRef]
22. Tillich, M.; Lehwark, P.; Pellizzer, T.; Ulbricht-Jones, E.S.; Fischer, A.; Bock, R.; Greiner, S. GeSeq-versatile and accurate annotation of organelle genomes. *Nucleic Acids Res.* **2017**, *45*, W6–W11. [CrossRef] [PubMed]
23. Lohse, M.; Drechsel, O.; Kahlau, S.; Bock, R. OrganellarGenomeDRAW—A suite of tools for generating physical maps of plastid and mitochondrial genomes and visualizing expression data sets. *Nucleic Acids Res.* **2013**, *41*, W575–W581. [CrossRef] [PubMed]
24. Burland, T.G. DNASTAR's Lasergene sequence analysis software. In *Bioinformatics Methods and Protocols*; Misener, S., Krawetz, A.K., Eds.; Springer: Basel, Switzerland, 2000; pp. 71–91.
25. Kumar, S.; Stecher, G.; Tamura, K. MEGA7: Molecular evolutionary genetics analysis version 7.0 for bigger datasets. *Mol. Biol. Evol.* **2016**, *33*, 1870–1874. [CrossRef] [PubMed]
26. Mower, J.P. The PREP suite: Predictive RNA editors for plant mitochondrial genes, chloroplast genes and user-defined alignments. *Nucleic Acids Res.* **2009**, *37*, W253–W259. [CrossRef]
27. Kurtz, S.; Choudhuri, J.V.; Ohlebusch, E.; Schleiermacher, C.; Stoye, J.; Giegerich, R. REPuter: The manifold applications of repeat analysis on a genomic scale. *Nucleic Acids Res.* **2001**, *29*, 4633–4642. [CrossRef]
28. Beier, S.; Thiel, T.; Münch, T.; Scholz, U.; Mascher, M. MISA-web: A web server for microsatellite prediction. *Bioinformatics* **2017**, *33*, 2583–2585. [CrossRef]
29. Frazer, K.A.; Pachter, L.; Poliakov, A.; Rubin, E.M.; Dubchak, I. VISTA: Computational tools for comparative genomics. *Nucleic Acids Res.* **2004**, *32*, W273–W279. [CrossRef]
30. Amiryousefi, A.; Hyvönen, J.; Poczai, P. IRscope: An online program to visualize the junction sites of chloroplast genomes. *Bioinformatics* **2018**, *34*, 3030–3031. [CrossRef]
31. Rozas, J.; Ferrer-Mata, A.; Sánchez-DelBarrio, J.C.; Guirao-Rico, S.; Librado, P.; Ramos-Onsins, S.E.; Sánchez-Gracia, A. DnaSP 6: DNA sequence polymorphism analysis of large data sets. *Mol. Biol. Evol.* **2017**, *34*, 3299–3302. [CrossRef]
32. Stamatakis, A. RAxML version 8: A tool for phylogenetic analysis and post-analysis of large phylogenies. *Bioinformatics* **2014**, *30*, 1312–1313. [CrossRef]

33. Palmer, J.D. Comparative organization of chloroplast genomes. *Annu. Rev. Genet.* **1985**, *19*, 325–354. [CrossRef] [PubMed]
34. Shen, Y.; Chen, K.; Gu, C.; Zheng, S.; Ma, L. Comparative and phylogenetic analyses of 26 Magnoliaceae species based on complete chloroplast genome sequences. *Can. J. For. Res.* **2018**, *48*, 1456–1469. [CrossRef]
35. Asaf, S.; Khan, A.L.; Khan, A.R.; Waqas, M.; Kang, S.-M.; Khan, M.A.; Lee, S.-M.; Lee, I.-J. Complete chloroplast genome of *Nicotiana otophora* and its comparison with related species. *Front. Plant Sci.* **2016**, *7*, 843. [CrossRef]
36. Zhou, J.; Cui, Y.; Chen, X.; Li, Y.; Xu, Z.; Duan, B.; Li, Y.; Song, J.; Yao, H. Complete chloroplast genomes of *Papaver rhoeas* and *Papaver orientale*: Molecular structures, comparative analysis, and phylogenetic analysis. *Molecules* **2018**, *23*, 437. [CrossRef] [PubMed]
37. Meng, J.; Li, X.; Li, H.; Yang, J.; Wang, H.; He, J. Comparative analysis of the complete chloroplast genomes of four Aconitum Medicinal Species. *Molecules* **2018**, *23*, 1015. [CrossRef] [PubMed]
38. Zhong, Q.; Yang, S.; Sun, X.; Wang, L.; Li, Y. The complete chloroplast genome of the Jerusalem artichoke (*Helianthus tuberosus* L.) and an adaptive evolutionary analysis of the ycf2 gene. *PeerJ* **2019**, *7*, e7596. [CrossRef]
39. He, L.; Qian, J.; Li, X.; Sun, Z.; Xu, X.; Chen, S. Complete chloroplast genome of medicinal plant *Lonicera japonica*: Genome rearrangement, intron gain and loss, and implications for phylogenetic studies. *Molecules* **2017**, *22*, 249. [CrossRef]
40. Maréchal, A.; Brisson, N. Recombination and the maintenance of plant organelle genome stability. *New Phytol.* **2010**, *186*, 299–317. [CrossRef]
41. Saina, J.; Li, Z.-Z.; Gichira, A.; Liao, Y.-Y. The complete chloroplast genome sequence of tree of heaven (*Ailanthus altissima* (Mill.) (Sapindales: Simaroubaceae), an important pantropical tree. *Int. J. Mol. Sci.* **2018**, *19*, 929. [CrossRef]
42. Yaradua, S.S.; Alzahrani, D.A.; Albokhary, E.J.; Abba, A.; Bello, A. Complete chloroplast genome sequence of *Justicia flava*: Genome comparative analysis and phylogenetic relationships among Acanthaceae. *BioMed Res. Int.* **2019**, *2019*, 4370258. [CrossRef]
43. Nie, X.; Lv, S.; Zhang, Y.; Du, X.; Wang, L.; Biradar, S.S.; Tan, X.; Wan, F.; Weining, S. Complete chloroplast genome sequence of a major invasive species, crofton weed (*Ageratina adenophora*). *PLoS ONE* **2012**, *7*, e36869. [CrossRef]
44. Liu, Q.; Xue, Q. Comparative studies on codon usage pattern of chloroplasts and their host nuclear genes in four plant species. *J. Genet.* **2005**, *84*, 55–62. [CrossRef] [PubMed]
45. Wang, W.; Yu, H.; Wang, J.; Lei, W.; Gao, J.; Qiu, X.; Wang, J. The complete chloroplast genome sequences of the medicinal plant *Forsythia suspensa* (Oleaceae). *Int. J. Mol. Sci.* **2017**, *18*, 2288. [CrossRef] [PubMed]
46. Park, M.; Park, H.; Lee, H.; Lee, B.-h.; Lee, J. The complete plastome sequence of an Antarctic bryophyte *Sanionia uncinata* (Hedw.) Loeske. *Int. J. Mol. Sci.* **2018**, *19*, 709. [CrossRef] [PubMed]
47. Yang, Y.; Zhou, T.; Duan, D.; Yang, J.; Feng, L.; Zhao, G. Comparative analysis of the complete chloroplast genomes of five *Quercus* species. *Front. Plant Sci.* **2016**, *7*, 959. [CrossRef] [PubMed]
48. Zhou, T.; Chen, C.; Wei, Y.; Chang, Y.; Bai, G.; Li, Z.; Kanwal, N.; Zhao, G. Comparative transcriptome and chloroplast genome analyses of two related *Dipteronia* species. *Front. Plant Sci.* **2016**, *7*, 1512. [CrossRef]
49. Provan, J.; Powell, W.; Hollingsworth, P.M. Chloroplast microsatellites: New tools for studies in plant ecology and evolution. *Trends Ecol. Evol.* **2001**, *16*, 142–147. [CrossRef]
50. Ebert, D.; Peakall, R. Chloroplast simple sequence repeats (cpSSRs): Technical resources and recommendations for expanding cpSSR discovery and applications to a wide array of plant species. *Mol. Ecol. Res.* **2009**, *9*, 673–690. [CrossRef]
51. Mohammad-Panah, N.; Shabanian, N.; Khadivi, A.; Rahmani, M.-S.; Emami, A. Genetic structure of gall oak (*Quercus infectoria*) characterized by nuclear and chloroplast SSR markers. *Tree Genet. Genomes* **2017**, *13*, 70. [CrossRef]
52. Zeng, J.; Chen, X.; Wu, X.F.; Jiao, F.C.; Xiao, B.G.; Li, Y.P.; Tong, Z.J. Genetic diversity analysis of genus *Nicotiana* based on SSR markers in chloroplast genome and mitochondria genome. *Acta Tabacaria Sin.* **2016**, *22*, 89–97. [CrossRef]
53. Dong, W.; Xu, C.; Li, W.; Xie, X.; Lu, Y.; Liu, Y.; Jin, X.; Suo, Z. Phylogenetic resolution in *Juglans* based on complete chloroplast genomes and nuclear DNA sequences. *Front. Plant Sci.* **2017**, *8*, 1148. [CrossRef] [PubMed]
54. Liu, L.-X.; Li, R.; Worth, J.R.; Li, X.; Li, P.; Cameron, K.M.; Fu, C.-X. The complete chloroplast genome of Chinese bayberry (*Morella rubra*, Myricaceae): Implications for understanding the evolution of Fagales. *Front. Plant Sci.* **2017**, *8*, 968. [CrossRef] [PubMed]

55. Lu, R.-S.; Li, P.; Qiu, Y.-X. The complete chloroplast genomes of three *Cardiocrinum* (Liliaceae) species: Comparative genomic and phylogenetic analyses. *Front. Plant Sci.* **2017**, *7*, 2054. [CrossRef] [PubMed]
56. Raubeson, L.A.; Peery, R.; Chumley, T.W.; Dziubek, C.; Fourcade, H.M.; Boore, J.L.; Jansen, R.K. Comparative chloroplast genomics: Analyses including new sequences from the angiosperms *Nuphar advena* and *Ranunculus macranthus*. *BMC Genom.* **2007**, *8*, 174. [CrossRef]
57. Kim, K.-J.; Lee, H.-L. Complete chloroplast genome sequences from Korean ginseng (*Panax schinseng* Nees) and comparative analysis of sequence evolution among 17 vascular plants. *DNA Res.* **2004**, *11*, 247–261. [CrossRef]
58. Huang, H.; Shi, C.; Liu, Y.; Mao, S.-Y.; Gao, L.-Z. Thirteen *Camellia* chloroplast genome sequences determined by high-throughput sequencing: Genome structure and phylogenetic relationships. *BMC Evol. Biol.* **2014**, *14*, 151. [CrossRef]
59. Zhang, W.; Fan, X.; Zhu, S.; Zhao, H.; Fu, L. Species-specific identification from incomplete sampling: Applying DNA barcodes to monitoring invasive *Solanum* plants. *PLoS ONE* **2013**, *8*, e55927. [CrossRef]
60. Dick, C.W.; Webb, C.O. Plant DNA barcodes, taxonomic management, and species discovery in tropical forests. In *DNA Barcodes. Methods in Molecular Biology (Methods and Protocols)*; Kress, W., Webb, C.O., Eds.; Humana Press: Totowa, NJ, USA, 2012; Volume 858, pp. 379–393.
61. Yu, H.; Wu, K.; Song, J.; Zhu, Y.; Yao, H.; Luo, K.; Dai, Y.; Xu, S.; Lin, Y. Expedient identification of Magnoliaceae species by DNA barcoding. *Plant Omics* **2014**, *7*, 47.
62. Huan, H.V.; Trang, H.M.; Toan, N.V. Identification of DNA barcode sequence and genetic relationship among some species of *Magnolia* Family. *Asian J. Plant Sci.* **2018**, *17*, 56–64. [CrossRef]
63. Huan, H.V.; Nguyen, L.T.T.; Quang, N.M. To create DNA barcode data of *Magnolia chevalieri* (Dandy) V.S. Kumar for identification species and researching genetic diversity. *J. For. Sci. Technol.* **2019**, *18*, 3–9.
64. Azuma, H.; Thien, L.B.; Kawano, S. Molecular phylogeny of *Magnolia* (Magnoliaceae) inferred from cpDNA sequences and evolutionary divergence of the floral scents. *J. Plant Res.* **1999**, *112*, 291–306. [CrossRef]
65. Bentley, R.A. *Manual of Botany: Including the Structure, Classification, Properties, Uses, and Functions of Plants*; J. & A. Churchill: London, UK, 1882.
66. Azuma, H.; García-Franco, J.G.; Rico-Gray, V.; Thien, L.B. Molecular phylogeny of the Magnoliaceae: The biogeography of tropical and temperate disjunctions. *Am. J. Bot.* **2001**, *88*, 2275–2285. [CrossRef] [PubMed]
67. Dandy, J.E. A revised survey of the genus Magnolia together with *Manglietia* and *Michelia*. In *Magnolias*; Treseder, N.G., Ed.; Faber & Faber: London, UK, 1978; pp. 27–37.
68. Liang, C.B.; Nooteboom, H.P. Notes on Magnoliaceae III: The Magnoliaceae of China. *Ann. Mo. Bot. Gard.* **1993**, *80*, 999–1104. [CrossRef]
69. Figlar, R.B.; Nooteboom, H.P. Notes on Magnoliaceae IV. *Blumea* **2004**, *49*, 87–100. [CrossRef]
70. Kim, S.; Park, C.W.; Kim, Y.D.; Suh, Y. Phylogenetic relationships in family Magnoliaceae inferred from ndhF sequences. *Am. J. Bot.* **2001**, *88*, 717–728. [CrossRef] [PubMed]
71. Kim, S.; Nooteboom, H.P.; Park, C.-W.; Suh, Y. Taxonomic revision of *Magnolia* section *Maingola* (Magnoliaceae). *Blumea* **2002**, *47*, 319–339.
72. Nooteboom, H. Different looks at the classification of the Magnoliaceae. In Proceedings of the International Symposium on the Family Magnoliaceae, Guangzhou, China, 18–22 May 1998; Science Press: Beijing, China, 2000; pp. 26–37.
73. Zhao, M.-L.; Song, Y.; Ni, J.; Yao, X.; Tan, Y.-H.; Xu, Z.-F. Comparative chloroplast genomics and phylogenetics of nine *Lindera* species (Lauraceae). *Sci. Rep.* **2018**, *8*, 8844. [CrossRef]
74. Noteboom, H. Notes on Magnoliaceae with a revision of *Pachylarnax* and *Elmerrillia* and the Malasian species of *Manglietia* and *Michelia*. *Blumea* **1985**, *31*, 65–121.
75. Nie, Z.-L.; Wen, J.; Azuma, H.; Qiu, Y.-L.; Sun, H.; Meng, Y.; Sun, W.-B.; Zimmer, E.A. Phylogenetic and biogeographic complexity of Magnoliaceae in the Northern Hemisphere inferred from three nuclear data sets. *Mol. Phylogenet. Evol.* **2008**, *48*, 1027–1040. [CrossRef]
76. Ruhsam, M.; Rai, H.S.; Mathews, S.; Ross, T.G.; Graham, S.W.; Raubeson, L.A.; Mei, W.; Thomas, P.I.; Gardner, M.F.; Ennos, R.A. Does complete plastid genome sequencing improve species discrimination and phylogenetic resolution in *Araucaria*? *Mol. Ecol. Resour.* **2015**, *15*, 1067–1078. [CrossRef]

Article

Constitutive and Cold Acclimation-Regulated Protein Expression Profiles of Scots Pine Seedlings Reveal Potential for Adaptive Capacity of Geographically Distant Populations

Danas Baniulis [1,*], Monika Sirgėdienė [2], Perttu Haimi [1], Inga Tamošiūnė [1] and Darius Danusevičius [2]

[1] Lithuanian Research Centre for Agriculture and Forestry, Babtai, 54333 Kaunas, Lithuania; perttu.haimi@gmail.com (P.H.); inga.tamosiune@lammc.lt (I.T.)

[2] Vytautas Magnus University Agriculture Academy, Institute of Forest Biology and Silviculture, Akademija, 53361 Kaunas, Lithuania; m.raskauskaite@asu.lt (M.S.); darius.danusevicius@asu.lt (D.D.)

* Correspondence: danas.baniulis@lammc.lt; Tel.: +370-37-555-253

Received: 5 December 2019; Accepted: 8 January 2020; Published: 10 January 2020

Abstract: Geographically distant Scots pine (*Pinus sylvestris* L.) populations are adapted to specific photoperiods and temperature gradients, and markedly vary in the timing of growth patterns and adaptive traits. To understand the variability of adaptive capacity within species, molecular mechanisms that govern the physiological aspects of phenotypic plasticity should be addressed. Protein expression analysis is capable of depicting molecular events closely linked to phenotype formation. Therefore, in this study, we used comparative proteomics analysis to differentiate Scots pine genotypes originating from geographically distant populations in Europe, which show distinct growth and cold adaptation phenotypes. Needles were collected from 3-month-old seedlings originating from populations in Spain, Lithuania and Finland. Under active growth-promoting conditions and upon acclimation treatment, 65 and 53 differentially expressed proteins were identified, respectively. Constitutive protein expression differences detected during active growth were associated with cell metabolism and stress response, and conveyed a population-specific adaptation to the distinct climatic conditions. Acclimation-induced protein expression patterns suggested the presence of a similar cold adaptation mechanism among the populations. Variation of adaptive capacity among the genotypes was potentially represented by a constitutive low level of expression of the Ser/Thr-protein phosphatase, the negative regulator of the adaptive response. Also, overall less pronounced acclimation-induced response in seedlings from the Spanish population was observed. Thus, our study demonstrates that comparative proteomic analysis of young conifer seedlings is capable of providing insights into adaptation processes at the cellular level, which could help to infer variability of adaptive capacity within the plant species.

Keywords: conifer adaptation; phenotypic plasticity; comparative proteomics; stress response

---

1. Introduction

Sustainability of future forests is threatened by anthropogenic impacts, mainly due to pollution and climate change [1,2]. Therefore, special attention is being given to understanding of the molecular basis of adaptive traits in tree species (e.g., [3,4]). High genetic diversity, phenotypic plasticity and epigenetic mechanisms are the key factors of the adaptive capacity of forest tree populations (e.g., [5,6]). Trees have a low evolutionary rate in terms of nucleotide substitution and diversification rates as compared to herbaceous species [7]. On the other hand, tree populations are larger, less genetically structured and contain larger genetic diversity [7,8]. Most forest tree species show substantial variability in adaptive traits at levels much higher than is observed at neutral genetic markers and form a collection of highly differentiated populations with contrasting adaptation to local climates [9]. Intra-specific genetic variation of tree species is linked to the high adaptability under constantly changing environments [3] providing diversity and plasticity of genetic variants for evolution, where phenotypic plasticity is a property of a genotype to produce different phenotypes in response to particular environmental conditions that vary both temporally and geographically. Among the main evolutionary forces influencing the variability patterns, natural selection primarily acts to increase genetic differentiation among populations, while gene flow and phenotypic plasticity tend to reduce it [10]. In wide-spread wind pollinated conifers such as Scots pine, the effects of genetic drift and mutations are weak. Although natural variation in epigenetic marks and the relation to phenotypic traits in trees is still an under-explored area, epigenetic variation has been suggested to contribute to the phenotypic plasticity and adaptive potential of forest trees [11].

Gene expression is a dynamic process and depends on the specific tissue, developmental stage and environmental signals. Gene expression analysis at both transcriptome and proteome level could provide details about population variability, and the mechanisms involved in evolution and adaptation [12,13]. Protein expression is the final step in gene expression process, thus, proteomic analysis bridges the gap between the genotype and phenotype, providing a depiction of molecular events that are most directly linked to the phenotype at cellular or organism level. Quantitative variation of protein abundance is the result of a complex network of regulatory mechanisms that integrate genetic variation and interaction with the environment of individual plants [14,15]. A study on *Arabidopsis* ecotypes showed that the proteomics approach could constitute a powerful tool to mine the biodiversity between ecotypes of a single plant species [16]. Proteomics was used to study phenotypic and molecular response to water deficit of two *Eucalyptus* genotypes [17] and adaptive variation in *Picea abies* (L.) H.Karst. ecotypes [18,19].

Scots pine (*Pinus sylvestris* L.) is the most widespread coniferous tree species occupying an extensive area in Eurasia. The complex biogeographic history of Scots pine in Europe has been well studied [20–22]. At the neutral part of the genome, in open pollinated wide-spread conifers such as Scots pine, the phylogenic structure is weak and population differentiation is low (e.g., [23,24]). Based on the maternally inherited mitochondrial DNA markers, the European range of Scots pine is grouped into three major haplotypes: south western (refugia in Pyrenees), central (refugia in Balkan and Italy) and south eastern (refugium in the southern edge of the Ural mountains) [20,25–27]. Gene flow and isolation by distance were other major factors affecting population differentiation in the Scots pine that were defined using neutral DNA markers [28]. Numerous studies based on indoor and field trials revealed a significant clinal variation among the Scots pine populations over latitude gradients in adaptive traits such as phenology [29,30].

Timing of cold acclimation is of adaptive significance for northerly conifers such as the Scots pine [31–33]. The seedlings cease active growth and enter the endodormancy stage in response to elongating nights towards the autumn [34], followed by shoot hardiness development in response to chilling temperatures in the early autumn [35,36]. In trees, cold acclimation treatment results in complex structural, biochemical and genetic shifts from elongating meristems to completely dormant and frost hardy tissues [37]. In brief, it causes protoplasmic dehydration, concentration of solutes, reduction of osmotic potential, upregulation of abscisic acid, a change in carbohydrate metabolism along with changes in membrane permeability and fatty acid composition [38,39]. The most prominent changes are induced in the chloroplasts and manifest as structural reorganization of the organelar membranes [40,41]. At protein level, the increased amount of antioxidant enzymes [42–44] and accumulation of dehydrins mostly localized in chloroplasts and the mitochondria membrane system was reported during the acclimation [45]. Gene expression studies showed upregulation of enzymes involved with synthesis of sugars and sugar derivatives, dehydration stress and other stress tolerance enzymes, such as heat-shock proteins [46].

Owing to the photoperiod and temperature-triggered adaptation in the northern regions, conifers have population-specific requirement for critical night length to budset and chilling temperatures to induce variable levels of shoot frost hardiness in early autumn [36]. When transferred to a single location, the critical night length for bud set and shoot hardiness development vary markedly between geographically distant populations within a latitudinal gradient [29,32,47]. Therefore, after a cold acclimation treatment, markedly different gene expression profiles could be expected in seedlings from the geographically distant Scots pine populations.

To probe an adaptive capacity within the Scots pine species, our study assessed phenotypic traits and quantitative and qualitative protein expression patterns in the Scots pine seedlings originating from the geographically distant populations of northern and southern Europe. Protein expression patterns were assessed by two-dimensional protein electrophoresis. Firstly, differences among the three populations were assessed during the active growth period. Furthermore, an acclimation treatment of the seedlings was used to extend the analysis of protein expression profiles with specific emphasis on cold adaptation processes in the distinct Scots pine genotypes.

2. Materials and Methods

2.1. Seedling Production and Cold Acclimation Treatment

We used seeds of Scots pine collected from the three autochthonous populations representing the Mediterranean, temporal and boreal climatic zones in Europe: central Spain (Valsain Forest mountains, Sierra de Guadarrama province, 40°49′ lat., 3°57′ long., 1828 m.a.s.l. alt.), central Lithuania (Kazlu Ruda forest tact, 54°46′ lat., 23°32′ long., 79 m.a.s.l. alt.) and northern Finland (Suomussalmi forest, 65°07′ lat., 28°49′ long., 233 m.a.s.l. alt.), respectively (Figure 1). The seeds were collected in native Scots pine stands naturally-born over generations in large forest tracts, where the influence of non-autochthonous sources is negligible. These populations mainly represent a geographically large latitudinal gradient of genepools adapted to distinct photoperiodic and temperate requirements to initiate dormancy and cold hardiness processes that have a major effect on timing of cold acclimation of the seedlings. The seed lots were mixtures from more than 20 trees within a forest stand, which was considered as representative of a population.

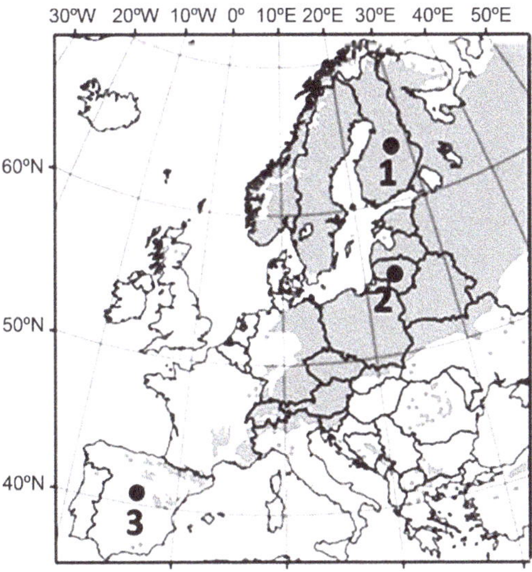

Figure 1. Location of the Finnish (1), Lithuanian (2) and Spanish (3) populations used in the study.

Seeds were soaked overnight in sterile deionized water and surface sterilized for 30 min with 30% hydrogen peroxide. Floating seeds were discarded, and the remaining seeds were sown in sterile plastic containers (length 50 cm × width 17 cm × height 15 cm) on sterilized peat treated with chalk and soaked with sterile nutritive solution. Seeds were germinated and seedlings were grown in the climatic growth chamber LT-36VL (Percival) during a 3-month period. Constant temperature of 24 °C, 100% humidity and a long-day (16 h, 30 μmol m^{-2} s^{-1}; fluorescent light 5500 K photoperiod) was used to promote active growth. Seedlings were watered at a two-day interval and fertilizer containing 44 g/L nitrogen, 11 g/L P_2O_5, 33 g/L K_2O, 2.2–3.3 g/L MgO, was used every second week. Two containers for each population were used, and the containers were randomly relocated within the chamber once per week to reduce undesired external effects within in the chamber. After germination, seedlings were thinned out to maintain comparable plant density (approx. 100 seedlings per container). In 3 months after germination, samples for the active growth experimental group were collected, and the remaining seedlings were exposed to the 2-week cold acclimation treatment in the same growth chamber as follows: for one week, the plants were maintained at 10 °C 100% humidity and 8 h photoperiod, followed by one week of darkness with the same temperature and humidity conditions.

2.2. Seedling Morphology Measurements

Before the cold acclimation treatment, morphometric traits of 40 seedlings per population were measured to the nearest millimeter. To assess growth cessation during the acclimation treatment, the needle weight of pooled sample representing four repeats for the active growth and acclimated experimental groups of each population was measured. The pooled samples were prepared by combining equal amount of material from first crowns of needles from 25 seedlings.

2.3. Two-Dimensional Gel Electrophoresis and Protein Identification

Four biological repeats representing the active growth and acclimated experimental groups of each population were prepared as pooled needle samples as described in the previous section, and were frozen in liquid nitrogen and stored at −70 °C. Protein samples were prepared using phenol extraction and ammonium acetate precipitation, as described previously [48]. Briefly, samples were ground in liquid nitrogen and suspended in 500 μL of the extraction buffer (0.5 M Tris-HCl, pH −7.5, 0.1 M KCl, 0.05 M ethylenediaminetetraacetic acid, 0.7 M sucrose, 2% polyvinylpolypyrrolidone, 2% β-mercaptoethanol, 1 mM phenylmethanesulfonylfluoride). Equal volume of Tris-buffered phenol (pH −7.5) was added and incubated with shaking for 30 min at 4 °C. The tubes were centrifugated for 15 min at 15,000× *g* at 4 °C, the upper phase was recovered, and the phenol extraction was repeated. Proteins were precipitated by addition of equal volume of 3M ammonium acetate in methanol, incubation at −20 °C overnight and centrifugation as before. Protein pellets were washed twice by resuspending in cold methanol and once in cold acetone and centrifugation for 5 min as before. After aspiration of acetone, protein pellets were dried for few minutes and solubilised in two-dimensional electrophoresis lysis buffer (6 M urea, 2 M thiourea, 4% 3-cholamidopropyl dimethylammonio 1-propanesulfonate, 10 mM Tris-HCl, pH−8.5). Protein concentrations were measured using a Bradford assay [49]. Four biological repeats were prepared for each treatment. Internal standards were prepared from a pooled mixture of all protein extracts.

Protein separation and detection was performed using a differential gel electrophoresis procedure as described previously [50]. Briefly, sample aliquots of 50 mg were used for labeling with Cy3 and Cy5 fluorescent dyes, and the internal standard was labeled with Cy2 dye (Lumiprobe). For the preparative gel, 500 mg of unlabeled internal standard was mixed with 50 mg of the labeled internal standard. Proteins were applied to 24 cm immobilized pH gradient strips with a linear gradient of pH 4–7 and isoelectric focusing was performed with an Ettan IPGphor (GE Healthcare, Chicago, IL, USA). Further, the proteins were separated on 1-mm thick 12.5% polyacrylamide gels with Ettan DALTsix electrophoresis apparatus (GE Healthcare) and gels were scanned using a FLA 9000 fluorescence scanner

(GE Healthcare). Relative protein quantifications were performed using DeCyder 2-D Differential Analysis Software, v.7.0 (GE Healthcare).

Preparative gel was fixed in 50% methanol and 10% acetic acid. Protein spots were excised manually and subjected to protein digestion with trypsin, according to a method described previously [51]. Protein digests were loaded and desalted on a 100 mm × 20 mm Acclaim PepMap C18 trap column and separated on a 75 mm × 150 mm Acclaim PepMap C18 column using an Ultimate 3000 rapid separation liquid chromatography system (Thermo-Scientific, Waltham, MA, USA), coupled to a Maxis G4 Q-TOF mass spectrometer detector with a Captive Spray nano-electrospray ionization source (Bruker Daltonics, Bremen, Germany). Peptide identification was performed using the MASCOT server (Matrix Science, Boston, MA, USA) against the *Pinus taeda* L. genome database [52]. The mass error tolerance for peptide matching was limited to 5 ppm. The threshold value for the identification of proteins was a Mascot score >50 and at least two peptides. Where more than one protein was identified from the same spot, the exponentially modified protein abundance index (emPAI) [53] was employed for the relative quantitation of the proteins. Spots were designated as protein mixture if the emPAI indexes were similar for several of the identified proteins.

2.4. Data Analysis

A *t*-test was used to assess to assess the differences in the seedling morphology traits among the populations (40 seedlings per population).

The Biological Variation Analysis module of the DeCyder software was used to match protein spots of the four biological repeats across different gels and to perform statistical analysis of differential protein expression. A two-way analysis of variance (ANOVA) test was used to assess statistically significant ($p \leq 0.01$) population and cold acclimation treatment effects. The threshold value of at least 2-fold difference in protein abundance was used.

To find patterns in the expression profiles, a cluster analysis was performed using Euclidian distance matrix and complete linkage method at the Extended Data Analysis module of the DeCyder software.

Blast2GO software [54] was used for the annotation and gene ontology analysis of the identified protein sequences. The identified *Pinus taeda* sequences were searched with Basic Local Alignment Search Tool (BLAST) against Swissprot database and protein sequences and annotated with Gene Ontology (GO) terms.

3. Results

3.1. Seedling Morphology

After 3-month cultivation under growth-promoting conditions (16 h light, high humidity and constant temperature of 24 °C), the seedlings of the Finnish population had the shortest length of stem and cotyledons (Table 1). There was no significant difference in the seedling height between the Lithuanian and Spanish populations. Seedlings of the Spanish population had more cotyledons and produced the largest needle biomass, as estimated by fresh weight of the sampled needles. During the acclimation treatment, there was no significant change of needle fresh weight for the seedlings from the Spanish population, indicating growth cessation. Meanwhile, the needle mass increased approximately 1,5-fold for the two northern populations.

Table 1. Quantitative traits of pine seedlings before and after the cold acclimation treatment.

Trait	Population		
	Spanish	Lithuanian	Finnish
Number of cotyledons (4 weeks after germination)	7.0 ± 0.12 [a]	6.0 ± 0.13 [b]	6.0 ± 0.13 [b]
Seedling height before the acclimation treatment (12 weeks after seeding), mm	58.6 ± 0.23 [a]	63.6 ± 0.23 [a]	31.9 ± 0.11 [b]
Length of hypocotyl before the acclimation treatment (12 weeks after seeding), mm	30.2 ± 0.16 [a]	30.8 ± 0.16 [a]	17.0 ± 0.06 [b]
Needle weight before the acclimation treatment (12 weeks after seeding), g	4.2 ± 0.51 [a]	3.2 ± 0.26 [b]	1.5 ± 0.06 [c]
Needle weight at the end of acclimation treatment (14 weeks after seeding), g	4.5 ± 0.16 [a]	4.7 ± 0.14 [a]	2.2 ± 0.34 [b]

Seedling height and cotyledon number were measured on 40 seedlings per population. Needle weight represents fresh weight of four pooled samples of 25 seedlings per population. The data are presented as mean and standard error of the mean. Means followed by the same letter are not significantly different ($p < 0.05$ by Tukey least significant difference (LSD) test).

3.2. Differentially Expressed Proteins

The average number of detected protein spots was 1464 ± 145 per gel after alignment (Figure S1). Analysis of biological variance revealed statistically significant and >2-fold variation in the abundance of 76 proteoforms among the six experimental groups (non-aclimated and acclimated seedlings from the Spanish, Lithuanian and Finnish populations) (Figures 2 and 3). The largest number of unique differentially expressed protein spots were a characteristic of the Finnish population (25 and 20 in the non-acclimated and acclimated experimental groups, respectively), followed by the Spanish population (15 and 19). The Lithuanian population had the lowest number of unique spots (2 and 1) (Figure 2a,b). The number of proteoforms differentially expressed between populations was similar for the non-acclimated and acclimated experimental group. The abundance of 53 proteoforms changed after cold acclimation treatment (Figure 2c), and approx. 72% of the detected changes were independent of the seedling origin.

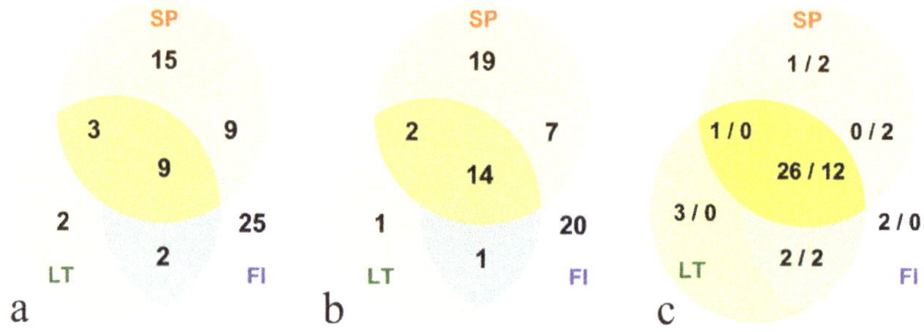

Figure 2. Quantitative distribution of differentially expressed protein spots identified by two-dimensional gel electrophoresis analysis of needle protein samples isolated from seedlings originating from the Spanish (SP), Finish (FI) and Lithuanian (LT) Scots pine populations. Number of differentially expressed protein spots between the populations under growth promoting conditions (**a**) and upon acclimation (**b**) is shown. Number of upregulated and downregulated protein spots upon acclimation treatment are indicated by numbers separated by slash (**c**). The portion of the cycles overlapping between two or all three populations indicate a number of protein spots that are differentially expressed between the two or among all three populations, respectively.

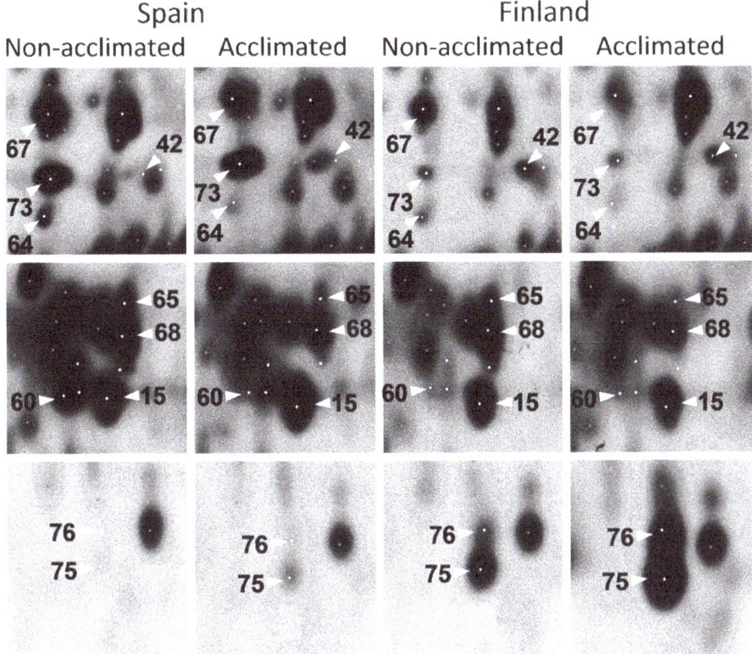

Figure 3. Quantitative differences in protein abundance between non-aclimated and acclimated Scots pine seedlings originating from the geographically separated populations. Representative images of ten identified spots are shown. Complete gel image is presented in Figure S1.

Through liquid chromatography-tandem mass spectrometry fingerprinting of trypsin-digested peptides, 26 proteoforms were unequivocally identified and annotated (Table A1). Among the identified proteins, 14 were associated with metabolic process (0044237), 17 were associated with a related term of cellular metabolic process (0008152) and 15 were linked to response to stress (0006950). At a lower GO hierarchy level, the terms included small molecule metabolic process (0044281), cellular protein metabolic process (0044267), cellular component organization (0016043), regulation of biological process (0050789), oxidation-reduction process (0055114), defense response (0006952) and response to salt stress (0009651), as indicated in Figure 4. Three proteins annotated as thaumatin-like protein (PITA_000066768-RA), chloroplastic TIC 62 protein (PITA_000059572-RA) and NAD-binding Rossmann-fold superfamily protein (PITA_000059572-RA) (Table S1) had no GO terms assigned but they are known to be involved in stress response processes (Albrecht and Bowman, 2012; Dutta et al., 2009; Rahjam et al., 2007). Another 5 proteoforms were identified as proteins coded by the genes of unknown function, and 29 spots contained a mixture of 2–4 proteins (Table S1).

3.3. Protein Expression Patterns

Cluster analysis of variation of abundance of the 76 differentially expressed protein spots revealed protein expression patterns associated with pine seedling origin and/or their response to acclimation treatment (Figure 4). Cold acclimation treatment had little or no significant effect on the expression of the proteins assigned to the clusters 4–5, 7, 9 and 11–12 and these mostly reflected population-depended protein expression differences. The most prominent difference among the three populations was observed for the two proteoforms assigned to the cluster 12. The lactoylglutathione lyase had approximately 5- to 10-fold higher abundance in the seedlings from the Spanish population as compared to the other two populations. Four proteoforms included in the cluster 11 had higher abundance in the Spanish population (none of them could be identified). Furthermore, six proteoforms (including tubulin beta chain-like and tau class glutathione s-transferase (GST) were assigned to the cluster 9 and had higher abundance in the Spanish and Lithuanian populations.

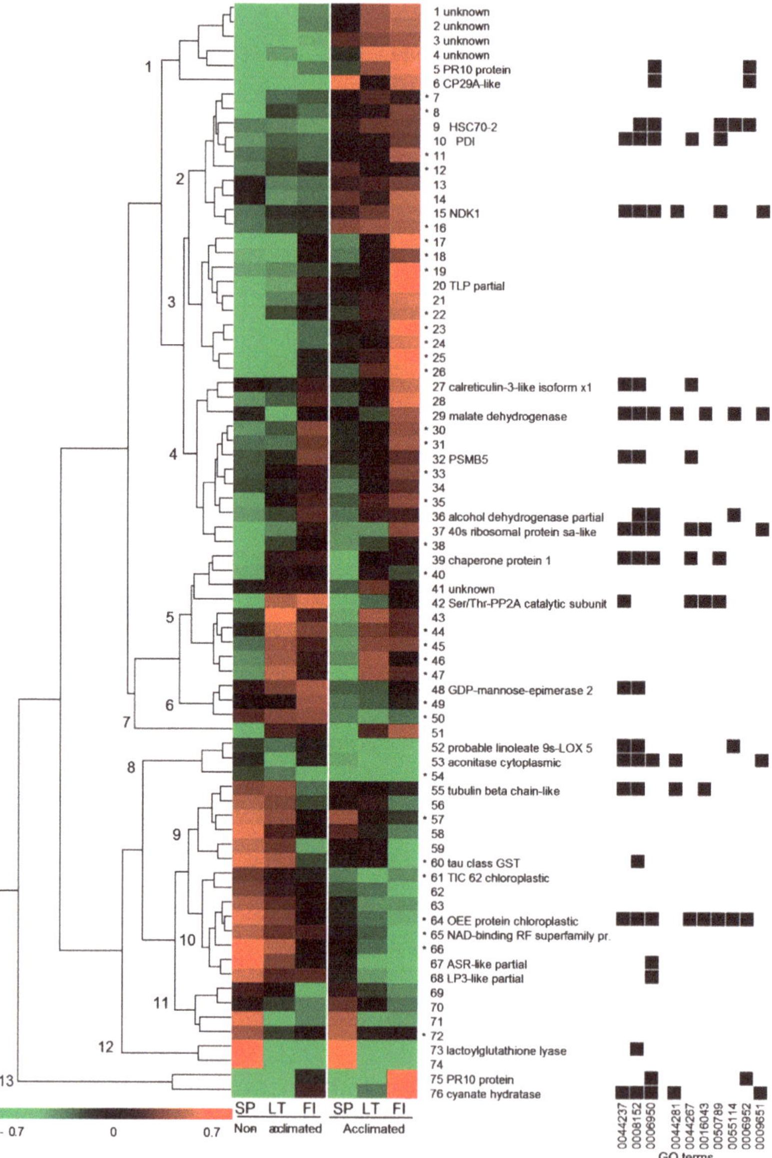

Figure 4. Cluster analysis of abundance of the 76 protein spots differentially expressed in needles of the Scots pine seedlings among the three populations and/or under cold acclimation treatment. The standardized abundance scale is show at the lower left corner. Thirteen major clusters are marked by numbers. Star indicates spots that were identified as mixtures of 2–4 proteins (listed in Table S1). GO terms associated with the identified unique protein sequences are shown on the right (the terms corresponding to the indicated numbers are listed in the Results section).

In contrast, a group of 22 proteins (including calreticulin-3-like isoform x1, malate dehydrogenase (MDH), proteasome subunit beta type-5 (PSMB5), alcohol dehydrogenase, 40 s ribosomal protein sa-like, chaperone protein 1, Ser/Thr-protein phosphatase 2A (PP2A) catalytic subunit) had higher abundance either in the Finnish population or both northern populations and were assigned to the clusters 4, 5 and 7.

Expression pattern of the remaining clusters 1–3, 6, 8, 10 and 13 reflected population- and/or acclimation-dependent differences (Figure 4). Before the acclimation, the majority of the 18 proteoforms separated into the clusters 1, 3 and 13 (including CP29-like, thaumatine-like protein (TLP), cyanate hydratase and two proteoforms of pathogenesis-related 10 (PR-10) were significantly more abundant in the seedlings of the Finnish or both northern populations. These proteins were significantly upregulated in the acclimated seedlings from all the three populations, however, for cluster 3, the response was more prominent in the Finnish and/or in both northern populations. Cyanate hydratase (cluster 13) was upregulated approximately 10-fold in the acclimated seedlings of the Finnish population compared to the non-acclimated control, representing the strongest response to the acclimation treatment.

Among the 10 proteoforms included in cluster 2, only two significant differences (unidentified proteoforms 7 and 8) were detected among the populations before the acclimation treatment. All proteins of this cluster were upregulated by the acclimation treatment, and the response was more prominent in the seedlings for the northern populations, leading to increase of significant differences among the populations (protein disulfide-isomerase (PDI), nucleoside diphosphate kinase 1 (NDK1) and three unidentified proteoforms).

Furthermore, an abundance of 14 proteoforms included in the clusters 6, 8 and 10 was markedly decreased upon the acclimation treatment. The response to acclimation had similar effect on protein expression for all three populations, however, the protein abundance pattern was different. The northern populations had higher abundance of the guanosine-diphosphate (GDP)-mannose epimerase 2 and proteoform 49 among the three proteoforms of the cluster 6. Meanwhile the remaining proteins (clusters 8 and 10) including probable linoleate 9s-LOX 5, aconitase, TIC 62, oxygen-evolving enhancer protein (OEE), nicotinamide adenine dinucleotide (NAD)-binding RF superfamily protein, abscisic acid water deficit stress and ripening-like (ASR-like) and water deficit inducible lipoprotein 3 (LP3)-like, were more abundant in the samples from the Spanish population.

Notably, only a few proteoforms (including four identified proteins) showed a strictly population-specific regulation upon the acclimation treatment. Malate dehydrogenase was upregulated in both northern populations, although the acclimation-induced expression level in the Lithuanian population was still below the constitutive expression level in the Spanish population. Calreticulin-3-like isoform and alcohol dehydrogenase were upregulated in the Finnish and Lithuanian populations, respectively. Meanwhile, the catalytic subunit of Ser/Thr-PP2A was downregulated in the northern populations, and the protein abundance was reduced to the level that was comparable to the constitutive expression observed in the Spanish population.

4. Discussion

An adaptive capacity of species in response to specific environmental cues could be inferred from mechanistic information at molecular level and its integration with ecologically relevant phenotypes [55]. A previous study with conifer tree species by [19] has demonstrated that proteomic analysis of young seedlings imparts a variation of gene expression that could be of importance for adaptation to contrasting altitude conditions in the closely related Norway spruce ecotypes, and the adaptive capacity could be further explored by heat stress treatment of the seedlings [18]. Our study revealed a similar quantitative variation of protein expression profiles in the needles of young Scots pine seedlings from geographically distant populations, although it reflects a variation of different physiological traits of potentially adaptive significance that could be species specific. Conifer seedlings develop an acclimation capacity in response to a short photoperiod and low temperatures at an early stage of of development [56], therefore cold acclimation treatment of the seedlings have stimulated additional differences in protein expression profiles that reflect a variation in cold adaptation capacity of the pine genotypes.

4.1. Stress Response Proteins Upregulated in the Northern Populations under Active Growth Conditions

Under active growth promoting conditions, both northern populations and in particular the Finnish population had higher abundance of proteins involved in protein metabolism (synthesis or degradation), such as heat shock protein (HSP), calreticulin, structural constituents of 40S ribosomes and 26S proteasome (Figure 4), that are also known as constituents of the stress tolerance mechanism. The chaperone protein 1 is a 100 kDa HSP that is critical for thermotolerance [57] and plant development [58]. Studies with *Arabidopsis* mutants revealed that constitutive expression of Hsp100 leads to preformed adaptation that ensures effective protection under heat stress [59]. Calreticulin is another chaperone protein which folds newly synthesized proteins and is also involved in regulation of Ca^{2+} homeostasis in the endoplasmic reticulum [60]. Upregulation of calreticulin has been observed under salt and osmotic stress conditions in *Arabidopsis* [61], and a function of the signaling component involved in response to cold stress in rice was proposed for the protein [62]. The proteasome subunit beta type-5 is a component of the 26S proteasome that mediates ubiquitin-dependent proteolysis of proteins and plays an important role in the stress-induced degradation of misfolded proteins, as well as the regulation of stress response signaling [63]. The 40s ribosomal protein sa-like is a structural constituent of ribosomes that regulates protein synthesis. A function of the p40 protein homologue is crucial for the active tissue growth [64] and its transcriptional upregulation under salt stress has been described in *Arabidopsis* [65].

Furthermore, upregulation of the TLP and PR-10, has been unique to the Finnish population. TLP proteins belong to the PR5 protein family and are induced under biotic, osmotic or cold stress in a variety of plants [66–68]. It is also well established that the PR-10 proteins are involved in biotic and abiotic stress response [69]. In perennial plants, expression of the PR-10 is induced during winter dormancy in mulbery and it provides either cryoprotectection for freeze-labile enzymes or serves as a nitrogen-storage protein [70].

In addition, the seedlings from the Finnish population have had higher abundance of alcohol dehydrogenase, malate dehydrogenase, GDP-mannose-epimerase 2, and cyanate hydratase enzymes that are involved in the primary cell metabolism, but also play role in the stress tolerance. Upregulation of alcohol dehydrogenase is essential in anaerobic respiration under cold stress conditions [71,72]. Malate is important intermediate and regulator of metabolic processes. Plants have several malate dehydrogenases that catalyze conversion of malate and oxaloacetate coupled to reduction or oxidation of the NAD pool, and this function has been linked to an abiotic stress tolerance in alfalfa and apple [73,74]. Upregulation of the cyanate hydratase upon salt stress was observed in tomato [75] and *Suaeda aegyptiaca* (Hasselq.) Zohary plants [76]. An accumulation of cyanide compounds is associated with elevated levels of ethylene synthesis as a consequence of plant stress response, therefore cyanate hydratase is considered a key enzyme responsible for the cyanide detoxification [77,78]. GDP-mannose-epimerase converts GDP-D-mannose to GDP-L-galactose, a precursor of ascorbate and cell wall polysaccharides [79]. This branching metabolic pathway plays a dual role in neutralization of the reactive oxygen species [80] as well as plant development [81].

The seedlings of the northern populations showed shorter phenotype and accumulated less needle biomass as compared to the Spanish population under active growth conditions (Table 1). It could be presumed that the reduced growth and upregulation of the stress response proteins might be an indicator of onset of the cold acclimation processes triggered by photoperiod used for the active growth experiments (16 h light and 8 h dark), owing to the short night length critical for growth cessation for the northern populations [34]. However, the seedlings had a relatively high level of the catalytic subunit of PP2A enzyme under active growth conditions that was effectively downregulated by the acclimation treatment. The PP2A is involved in cold stress signaling as a negative modulator of adaptive responses [82], and transcript mRNA levels of the PP2A has been reduced in tomato and alfalfa plants upon cold stress treatment [83,84]. This would suggest that the seedlings of the northern populations maintained active growth under the conditions used in the experiment, and the observed growth and protein expression differences were rather a consequence of physiological differences among the genotypes. Higher abundance of the proteins involved in stress tolerance could be linked to a constitutive cold adaptation mechanism characteristic of the northern genotypes.

4.2. Stress Response Proteins Upregulated in the Lithuanian and/or Spanish Population under Active Growth Conditions

The clusters 9 through 12 shown in Figure 4 include 20 proteoforms that have been upregulated mostly in the Lithuanian and/or Spanish populations under active growth conditions. Lactoylglutathione lyase, also known as glyoxalase I, is highly expressed in the Spanish population (approx. 10 and 5-fold compared to the Lithuanian and Finnish populations, respectively). The enzyme detoxifies methylglyoxal, a cytotoxic compound that is produced through glycolysis intermediates and generates free radicals [85]. Levels of methylglyoxal have been shown to rise under salinity stress and are regulated by glutathione and glyoxalase I [86].

As it could be expected due to geographical origin, the Lithuanian population tends to follow a similar protein expression pattern as the Finnish population located further north. However, expression of the proteins assigned to the clusters 9 and 10 (including tubulin beta-chain like, tau class GST, chloroplastic OEE, ASR- and LP3-like) in the Lithuanian population is similar to the Spanish population. The tau class GST, is involved in the reactive oxygen species (ROS) detoxification reactions that conjugate reduced gluthatione with a variety of target compounds. Expression of GSTs in plants is responsive to both biotic and abiotic stresses [87] and overexpression of the enzyme leads to chilling and osmotic stress tolerance in tobacco seedlings [88]. Beta-tubulin is one of the two structural components of cell microtubules and it is intimately involved in regulation of cell morphogenesis. In addition, it has been shown that the ROS signaling-mediated rearrangement in tubulin cytoskeleton plays crucial part in cell adaptation to stress [89]. Further, the abscisic acid induced ASR-like protein plays an important role in drought and salinity stress [90], and overexpression of *ASR* genes results in an increased cold tolerance [91,92]. LP3-like protein is a water-deficit-induced protein, which is homologous to the ASR and accumulates in loblolly pine roots upon water-deficit stress [93]. OEE is involved in the assembly of the Photosystem II and has a protective effect on the manganese cluster [94]. Increased expression of OEE1 was shown under salt or water deficiency stress [95,96].

The observed variation of stress-related protein expression among the populations from the southern and northern boundaries of the continent indicates presence of the constitutive stress tolerance mechanisms that are pertinent to the climatic conditions of the geographically distant habitats. The Lithuanian population combines a mixture of the traits.

4.3. Effect of Acclimation Treatment on Protein Expression

Cold acclimation treatment revealed additional protein expression differences that could be separated into three major groups that include proteins either (i) upregulated or (ii) downregulated in seedlings of all the three populations, or (iii) regulated in a population-specific manner. These protein expression differences suggest the presence of distinct elements of the cold acclimation mechanism and/or variation of their regulation among the geographically distant pine populations.

The function of the first group of proteins was linked to stress response. HSC70-2 is a protein folding catalyst involved in protein stabilization under abiotic stress conditions [97]. PDI belongs to another well-known family of chaperones that mediates formation, isomerization and reduction/oxidation of disulfide bonds during protein synthesis [98]. Ribonucleoprotein chloroplastic-like is required for normal chloroplast development under cold stress conditions by stabilizing transcripts of mRNAs [99]. Nucleoside diphosphate kinase 1 is a cytosolic enzyme that is essential in homeostasis of cellular nucleoside triphosphate pools [100], and its role under cold stress conditions has been described [101]. The stress response related proteins PR-10, TLP and cyanate hydratase have been also highly upregulated upon acclimation treatment, however, their abundance in the Spanish and Lithuanian populations have not exceed the level observed in the Finnish population under active growth conditions.

Acclimation treatment downregulates chloroplastic Tic62, a component of Tic translocon that mediated translocation of nuclear-encoded precursor proteins across the inner envelope of chloroplasts, and chloroplastic OEE were downregulated upon the acclimation treatment. Further downregulation of several enzymes, including cytoplasmic aconitase involved in the Krebs cycle, probable linoleate 9S-lipoxygenase 5 of the oxylipin synthesis pathway and GDP-mannose-epimerase 2 enzyme involved in the ascorbate and cell wall polysaccharide synthesis might be the result of reduced metabolic and photosynthetic activity in the seedlings maintained under the acclimation conditions of low temperature and reduced light. However, seasonal changes in chloroplast ultrastructure [102] and depression of photosynthesis [103] is also a part of the winter adaptation process in conifers.

Abundance of ASR-like and homologous LP3-like protein have been also downregulated in the seedling needle samples. This may appear to contradict the previous studies demonstrating that the ASP and LP3 play essential role in stress response [90,91,93]. However, the study by Padmanabhan et al. [93] has shown that the transcript of the LP3 protein is preferentially induced in roots with a constitutive basal level of expression observed in stems and needles, which would explain low expression levels in our experimental set up.

Among the proteins regulated in a population-specific manner, PP2A, the negative modulator of adaptive response [82], has been downregulated, and MDH and calreticulin-3-like isoforms have been upregulated in the northern populations upon the acclimation treatment. In the Spanish population, the response to acclimation treatment has a distinctive reduced intensity compared to the northern populations.

5. Conclusions

Our study demonstrates that comparative proteomic analysis of young Scots pine seedlings is capable of reflecting on adaptive capacity among the geographically distant populations. Under active growth conditions, quantitative protein expression differences were related to biological processes of cell metabolism and stress response and could represent elements of the adaptive mechanism tailored to the specific growth environment. Furthermore, acclimation treatment revealed an additional stratum of protein expression differences that could provide insight into molecular basis of the cold adaptation capacity among the genotypes. As could be expected, the acclimation-induced protein expression patterns confirmed the presence of a similar cold acclimation mechanism among the pine populations, however the analysis also revealed a discrete variation of the adaptive traits. The most prominent differences were substantiated by the constitutive low levels of the PP2A enzyme expression and the overall less-pronounced response to the acclimation treatment of seedlings from the Spanish population. The identified proteins involved in the constitutive or acclimation-induced expression profiles presented interesting candidates for further studies on cold adaptation physiology in pine.

Supplementary Materials: The following are available online at http://www.mdpi.com/1999-4907/11/1/89/s1: Figure S1: Two-dimensional electrophoresis gel of the needle protein samples isolated from Scots pine seedlings; Table S1: Proteins differentially expressed in seedlings of the different populations of Scots pine under active growth conditions and upon acclimation treatment that were identified as unknown protein sequences of the *Pinus taeda* genome or contained several protein mixtures.

Author Contributions: Conceptualization, D.B. and D.D.; investigation, D.B., M.S., P.H., I.T.; writing—original draft preparation, D.B., D.D.; writing—review and editing, M.S., P.H., I.T.; funding acquisition, D.D. All authors have read and agreed to the published version of the manuscript.

Funding: This research was funded by the Ministry of Science and Education of Lithuania, grant number VP1-3.1-ŠMM-08-K-01-025.

Conflicts of Interest: The authors declare no conflict of interest.

Appendix A

Table A1. Proteins differentially expressed in seedlings of the different populations of Scots pine under active growth conditions (white bars) and upon acclimation treatment (grey bars).

Spot No.	Accession	Mascot Score /p. no. /SC, %	Thr. M.W. /pI	Sequence Description	Protein Expression Profile
5	PITA_000033197-RA [a]	413 /7 /35.6	17.8 /6.2	pathogenesis-related 10 (PR-10) protein	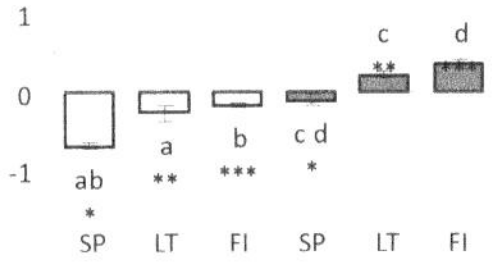
6	PITA_000003453-RA [a]	725 /8 /28.3	32.3 /4.8	29 kda ribonucleoprotein (CP29) chloroplastic-like	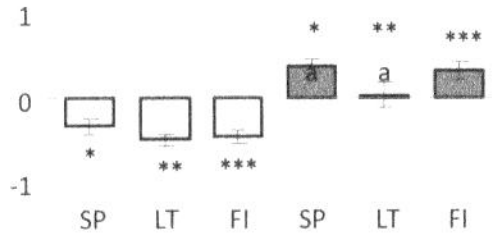
9	PITA_000000569-RA	1258 /19 /32.8	64.4 /5.2	heat shock cognate 70 kda protein 2 (HSP 70-2)	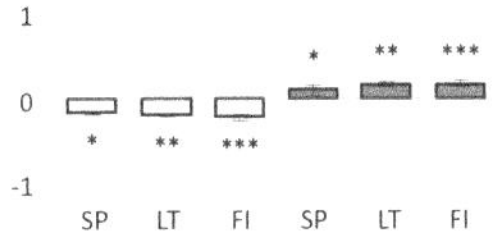
10	PITA_000048817-RA	1513 /23 /42.6	55.8 /4.8	protein disulfide-isomerase (PDI)	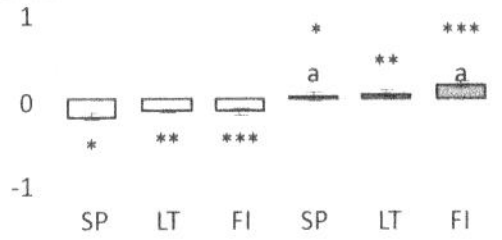
15	PITA_000069617-RA [a]	231 /5 /35.8	16.3 /5.9	nucleoside diphosphate kinase 1 (NDK 1)	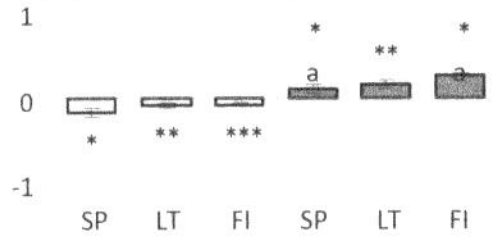
20	PITA_000066768-RA [a]	348 /5 /33.6	25.1 /5.0	thaumatin like partial	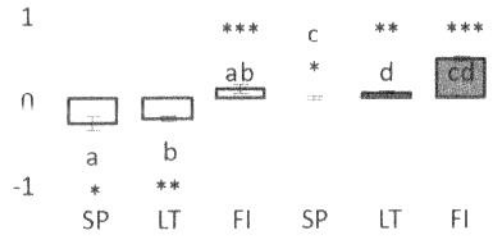
27	PITA_000060113-RA [a]	479 /9 /25.9	49.8 /5.7	calreticulin-3-like isoform x1	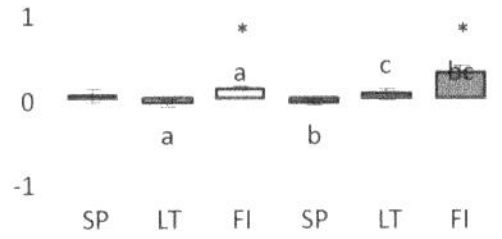
29	PITA_000042977-RA [a]	828 /12 /35.9	37.7 /5.8	malate dehydrogenase (MDH)	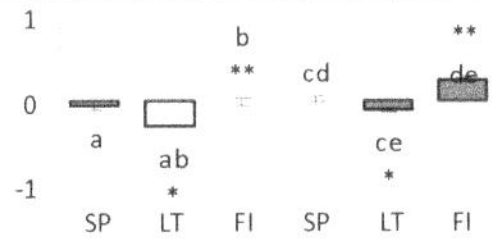
32	PITA_000037611-RA [a]	802 /11 /41.0	30.7 /6.2	proteasome subunit beta type-5 (PSMB5)	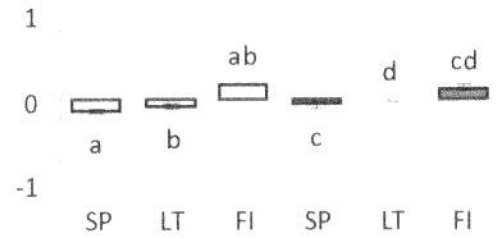

Table A1. *Cont.*

Spot No.	Accession	Mascot Score /p. no. /SC, %	Thr. M.W. /pI	Sequence Description	Protein Expression Profile
36	PITA_000004184-RA	601 /8 /17.6	57.6 /7.2	alcohol dehydrogenase partial	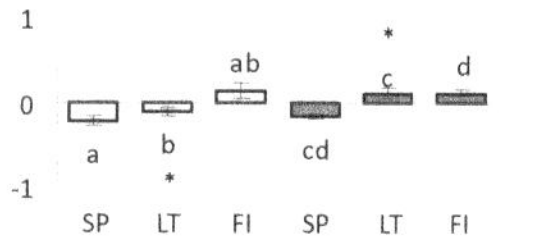
37	PITA_000055561-RA [a]	926 /15 /36.1	34.9 /5.0	40s ribosomal protein sa-like	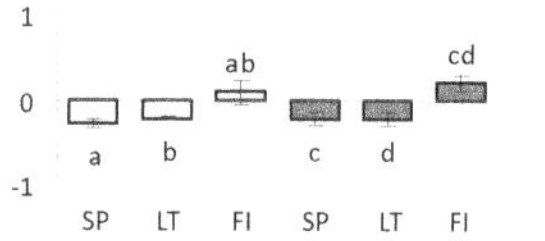
39	PITA_000052060-RA	1085 /18 /24.0	101.2 /5.7	chaperone protein 1	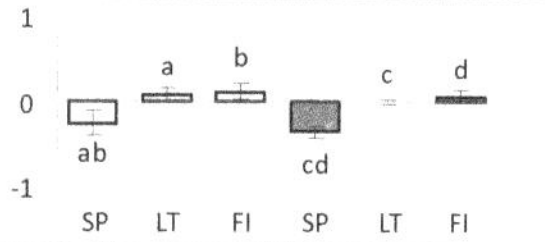
42	PITA_000009823-RA [a]	429 /8 /28.4	35.1 /4.9	Ser/Thr-protein phosphatase pp2a (PP2A) catalytic subunit	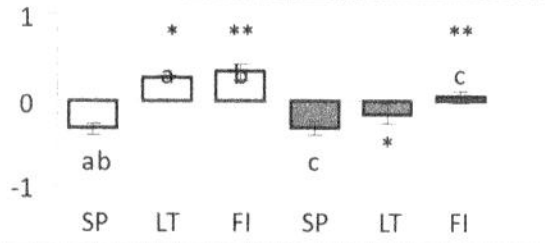
48	PITA_000019887-RA [a]	477 /10 /28.9	44.9 /5.6	guanosine-diphosphate (GDP)-mannose -epimerase 2	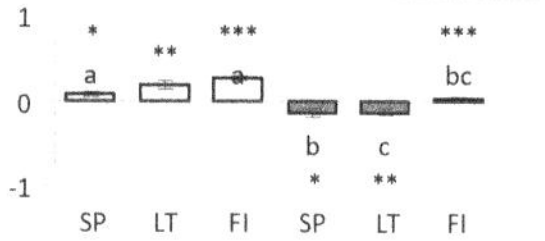
52	PITA_000011094-RA [a]	673 /10 /15.1	105 /6.0	probable linoleate 9s-lipoxygenase (LOX) 5	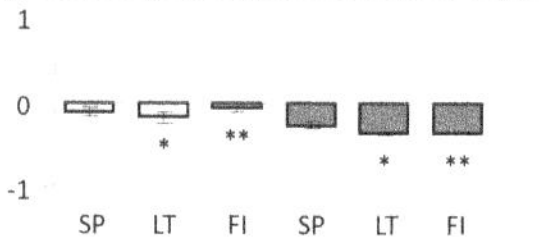
53	PITA_000020461-RA [a]	912 /19 /20.0	111.7 /6.4	aconitase cytoplasmic	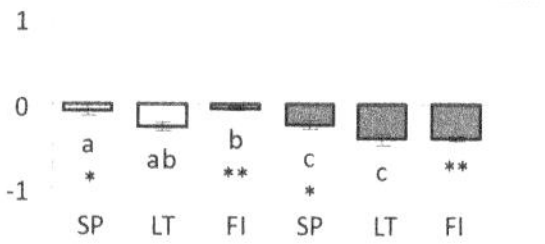
55	PITA_000016807-RA [a]	619 /11 /40.8	29.9 /4.2	tubulin beta chain-like	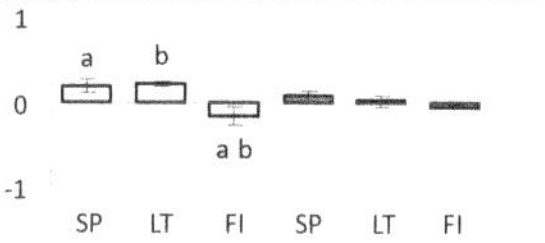
60	PITA_000006003-RA [a]	599 /12 /25.0	32.4 /6.7	tau class glutathione s-transferase (GST)	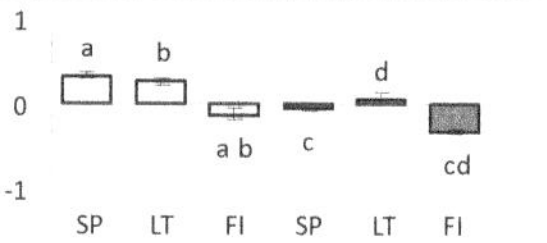
61	PITA_000059572-RA [a]	1289 /18 /41.0	56.2 /5.4	protein Tic62 chloroplastic	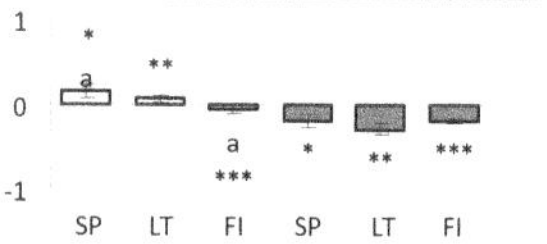

Table A1. *Cont.*

Spot No.	Accession	Mascot Score /p. no. /SC, %	Thr. M.W. /pI	Sequence Description	Protein Expression Profile
64	PITA_000047327-RA [a]	383 /7 /23.5	43.5 /6.0	oxygen-evolving enhancer (OEE) protein chloroplastic	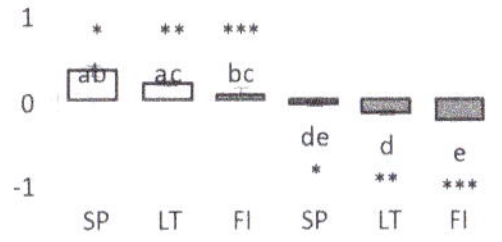
65	PITA_000055142-RA [a]	566 /9 /31.2	30.5 /8.8	nicotinamide adenine dinucleotide (NAD)-binding Rossmann-fold superfamily protein	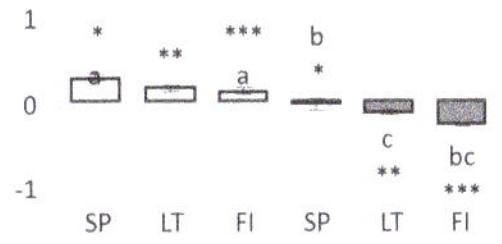
67	PITA_000008754-RA [a]	328.69 /4 /21.9	23.6 /5.1	abscisic acid water deficit stress and ripening (ASR) inducible-like partial	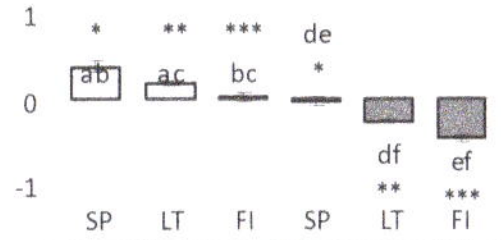
68	PITA_000022327-RA	446 /5 /50.0	17.6 /5.6	water deficit inducible lipoprotein 3 (LP3)-like partial	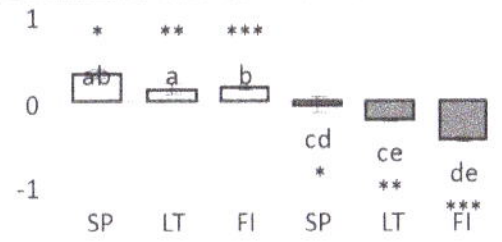
73	PITA_000023770-RA [a]	448 /9 /23.9	32.6 /5.0	lactoylglutathione lyase	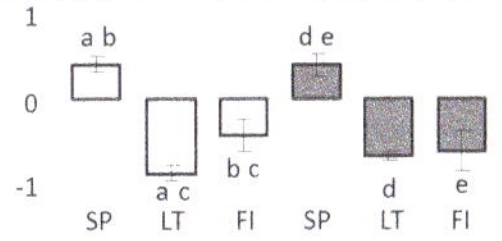
75	PITA_000008205-RA [a]	768 /14 /58.0	22.8 /8.8	PR-10 protein	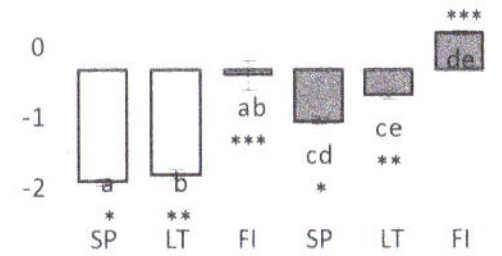
76	PITA_000077110-RA [a]	157 /3 /22.6	10.6 /5.1	cyanate hydratase	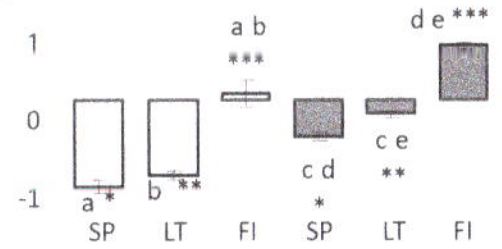

[a] Sequences that were selected as most abundant sequences from protein mixture using emPAI index. Data presented in graphs as log mean and standard error of the mean of standardized abundance, the same letters indicate statistically significant ($p < 0.01$) differences between the three populations, the same number of stars indicate significant difference between non-acclimated and acclimated experimental groups within each population. Abbreviations: p. no.—peptide number; SC—sequence coverage; Theoretical MW/pI—theoretical molecular weight and pI values; SP, LT, FI—Scots pine populations from Spain, Lithuania and Finland.

References

1. Paoletti, E.; Bytnerowicz, A.; Andersen, C.; Augustaitis, A.; Ferretti, M.; Grulke, N.; Gunthardt-Goerg, M.S.; Innes, J.; Johnson, D.; Karnosky, D.; et al. Impacts of air pollution and climate change on forest ecosystems—Emerging research needs. *Sci. World J.* **2007**, *7* (Suppl. 1), 1–8. [CrossRef]
2. Perie, C.; de Blois, S. Dominant forest tree species are potentially vulnerable to climate change over large portions of their range even at high latitudes. *PeerJ* **2016**, *4*, e2218. [CrossRef]
3. Aspinwall, M.J.; Loik, M.E.; Resco de Dios, V.; Tjoelker, M.G.; Payton, P.R.; Tissue, D.T. Utilizing intraspecific variation in phenotypic plasticity to bolster agricultural and forest productivity under climate change. *Plant Cell Environ.* **2014**, *38*, 1752–1764. [CrossRef] [PubMed]

4. Neale, D.B.; Ingvarsson, P.K. Population, quantitative and comparative genomics of adaptation in forest trees. *Curr. Opin. Plant Biol.* **2008**, *11*, 149–155. [CrossRef] [PubMed]
5. Breda, N.; Huc, R.; Granier, A.; Dreyer, E. Temperate forest trees and stands under severe drought: A review of ecophysiological responses, adaptation processes and long-term consequences. *Ann. For. Sci.* **2006**, *63*, 625–644. [CrossRef]
6. Kremer, A.; Ronce, O.; Robledo-Arnuncio, J.J.; Guillaume, F.; Bohrer, G.; Nathan, R.; Bridle, J.R.; Gomulkiewicz, R.; Klein, E.K.; Ritland, K.; et al. Long-distance gene flow and adaptation of forest trees to rapid climate change. *Ecol. Lett.* **2012**, *15*, 378–392. [CrossRef]
7. Petit, R.J.; Hampe, A. Some evolutionary consequences of being a tree. *Annu. Rev. Ecol. Evol. Syst.* **2006**, *37*, 187–214. [CrossRef]
8. Nybom, H. Comparison of different nuclear DNA markers for estimating intraspecific genetic diversity in plants. *Mol. Ecol.* **2004**, *13*, 1143–1155. [CrossRef] [PubMed]
9. Savolainen, O.; Pyhäjärvi, T.; Knürr, T. Gene flow and local adaptation in trees. *Annu. Rev. Ecol. Evol. Syst.* **2007**, *38*, 595–619. [CrossRef]
10. Eriksson, O. Evolutionary forces influencing variation among populations of *Pinus sylvestris*. *Silva Fennica* **1998**, *32*, 173–184. [CrossRef]
11. Bräutigam, K.; Vining, K.J.; Lafon-Placette, C.; Fossdal, C.G.; Mirouze, M.; Marcos, J.G.; Fluch, S.; Fraga, M.F.; Guevara, M.Á.; Abarca, D.; et al. Epigenetic regulation of adaptive responses of forest tree species to the environment. *Ecol. Evol.* **2013**, *3*, 399–415. [CrossRef] [PubMed]
12. Diz, A.P.; Martinez-Fernandez, M.; Rolan-Alvarez, E. Proteomics in evolutionary ecology: Linking the genotype with the phenotype. *Mol. Ecol.* **2012**, *21*, 1060–1080. [CrossRef] [PubMed]
13. Karr, T.L. Application of proteomics to ecology and population biology. *Heredity* **2008**, *100*, 200–206. [CrossRef] [PubMed]
14. Rockman, M.V.; Kruglyak, L. Genetics of global gene expression. *Nat. Rev. Genet.* **2006**, *7*, 862–872. [CrossRef]
15. Thiellement, H.; Zivy, M.; Plomion, C. Combining proteomic and genetic studies in plants. *J. Chromatogr. B* **2002**, *782*, 137–149. [CrossRef]
16. Chevalier, F.; Martin, O.; Rofidal, V.; Devauchelle, A.D.; Barteau, S.; Sommerer, N.; Rossignol, M. Proteomic investigation of natural variation between Arabidopsis ecotypes. *Proteomics* **2004**, *4*, 1372–1381. [CrossRef]
17. Bedon, F.; Villar, E.; Vincent, D.; Dupuy, J.W.; Lomenech, A.M.; Mabialangoma, A.; Chaumeil, P.; Barre, A.; Plomion, C.; Gion, J.M. Proteomic plasticity of two Eucalyptus genotypes under contrasted water regimes in the field. *Plant Cell Environ.* **2012**, *35*, 790–805. [CrossRef]
18. Valcu, C.M.; Lalanne, C.; Plomion, C.; Schlink, K. Heat induced changes in protein expression profiles of Norway spruce (*Picea abies*) ecotypes from different elevations. *Proteomics* **2008**, *8*, 4287–4302. [CrossRef]
19. Valcu, C.M.; Lalanne, C.; Muller-Starck, G.; Plomion, C.; Schlink, K. Protein polymorphism between 2 *Picea abies* populations revealed by 2-dimensional gel rlectrophoresis and tandem mass spectrometry. *J. Hered.* **2008**, *99*, 364–375. [CrossRef]
20. Cheddadi, R.; Vendramin, G.G.; Litt, T.; Francois, L.; Kageyama, M.; Lorentz, S.; Laurent, J.M.; De Beaulieu, J.L.; Sadori, L.; Jost, A.; et al. Imprints of glacial refugia in the modern genetic diversity of *Pinus sylvestris*. *Glob. Ecol. Biogeogr.* **2006**, *15*, 271–282. [CrossRef]
21. Sinclair, W.T.; Morman, J.D.; Ennos, R. The postglacial history of Scots pine (*Pinus sylvestris* L.) in Western Europe: Evidence from mitochondrial DNA variation. *Mol. Ecol.* **1999**, *8*, 83–88. [CrossRef]
22. Soranzo, N.; Alia, R.; Provan, J.; Powell, W. Patterns of variation at a mitochondrial sequence-tagged-site locus provides new insights into the postglacial history of European *Pinus sylvestris* populations. *Mol. Ecol.* **2000**, *9*, 1205–1211. [CrossRef]
23. Dvornyk, V.; Sirvio, A.; Mikkonen, M.; Savolainen, O. Low nucleotide diversity at the pal1 locus in the widely distributed *Pinus sylvestris*. *Mol. Biol. Evol.* **2002**, *19*, 179–188. [CrossRef]
24. Garcia-Gil, M.R.; Mikkonen, M.; Savolainen, O. Nucleotide diversity at two phytochrome loci along a latitudinal cline in *Pinus sylvestris*. *Mol. Ecol.* **2003**, *12*, 1195–1206. [CrossRef]
25. Naydenov, K.; Senneville, S.; Beaulieu, J.; Tremblay, F.; Bousquet, J. Glacial vicariance in Eurasia: Mitochondrial DNA evidence from Scots pine for a complex heritage involving genetically distinct refugia at mid-northern latitudes and in Asia Minor. *BMC Evol. Biol.* **2007**, *7*, 233. [CrossRef] [PubMed]

26. Pyhajarvi, T.; Salmela, M.J.; Savolainen, O. Colonization routes of *Pinus sylvestris* inferred from distribution of mitochondrial DNA variation. *Tree Genet. Genomes* **2008**, *4*, 247–254. [CrossRef]
27. Willis, K.J.; van Andel, T.H. Trees or no trees? The environments of central and eastern Europe during the last glaciation. *Quat. Sci. Rev.* **2004**, *23*, 2369–2387. [CrossRef]
28. Scalfi, M.; Piotti, A.; Rossi, M.; Piovani, P. Genetic variability of Italian southern Scots pine (*Pinus sylvestris* L.) populations: The rear edge of the range. *Eur. J. For. Res.* **2009**, *128*, 377. [CrossRef]
29. Hannerz, M. *Genetic and Seasonal Variation in Hardiness and Growth Rhythm in Boreal and Temperate Conifers—A Review and Annotated Bibliography*; The Forestry Research Institute of Sweden Report; SkogForsk: Uppsala, Sweden, 1998; pp. 1–140.
30. Wachowiak, W.; Perry, A.; Donnelly, K.; Cavers, S. Early phenology and growth trait variation in closely related European pine species. *Ecol. Evol.* **2018**, *8*, 655–666. [CrossRef]
31. Dietrichson, J. A summary of studies on genetic variation in forest trees grown in Scandinavia with special reference to the adaptation problem. *Norway Skogforsoksv Medd* **1971**, *29*, 25–56.
32. Eriksson, G.; Andersson, S.; Eiche, V.; Ifver, J.; Persson, A. Severity index and transfer effects on survival and volume production of *Pinus sylvestris* in northern Sweden. *Stud. For. Suecica* **1980**, *156*, 1–32.
33. Strimbeck, G.R.; Schaberg, P.G.; Fossdal, C.G.; Schroder, W.P.; Kjellsen, T.D. Extreme low temperature tolerance in woody plants. *Front. Plant Sci.* **2015**, *6*, 884. [CrossRef] [PubMed]
34. Dormling, I. Bud dormancy, frost hardiness and frost drought in seedlings of *Pinus sylvestris* and *Picea abies*. In *Advances in Cold Hardiness*; Li, P.H., Christersson, L., Eds.; CRC Press: Boca Raton, FL, USA, 1993; pp. 285–298.
35. Dormling, I. Influence of light intensity and temperature on photoperiodic response of Norway spruce provenances. In Proceedings of the IUFRO Meeting of WP Norway Spruce Provenances (S 2.03.11) and Norway Spruce Breeding (S 2.02.11), Bucharest, Romania, 24 September–1 October 1979; pp. 398–407.
36. Repo, T. Seasonal changes in frost hardiness in *Picea abies* and *Pinus sylvestris* in Finland. *Can. J. For. Res.* **1992**, *22*, 1949–1957. [CrossRef]
37. Wisniewski, M.; Nassuth, A.; Arora, R. Cold hardiness in trees: A mini-review. *Front. Plant Sci.* **2018**, *9*, 1394. [CrossRef]
38. Kontunen-Soppela, S.; Lankila, J.; Lähdesmäki, L.; Laine, K. Response of protein and carbohydrate metabolism of Scots pine seedlings to low temperature. *J. Plant Physiol.* **2002**, *159*, 175–180. [CrossRef]
39. Pukacki, P.M.; Kamińska-Rożek, E. Differential effects of spring reacclimation and deacclimation on cell membranes of Norway spruce seedlings. *Acta Soc. Bot. Pol.* **2013**, *82*, 77–84. [CrossRef]
40. Vogg, G.; Heim, R.; Hansen, J.; Schäfer, C.; Beck, E. Frost hardening and photosynthetic performance of Scots pine (*Pinus sylvestris* L.) needles. I. Seasonal changes in the photosynthetic apparatus and its function. *Planta* **1998**, *204*, 193–200. [CrossRef]
41. Vogg, G.; Heim, R.; Gotschy, B.; Beck, E.; Hansen, J. Frost hardening and photosynthetic performance of Scots pine (*Pinus sylvestris* L.) needles. II. Seasonal changes in the fluidity of thylakoid membranes. *Planta* **1998**, *204*, 201–206. [CrossRef]
42. Pukacki, P.M.; Kamińska-Rożek, E. Reactive species, antioxidants and cold tolerance during deacclimation of *Picea abies* populations. *Acta Physiol. Plant.* **2013**, *35*, 129–138. [CrossRef]
43. Bae, J.J.; Choo, Y.S.; Ono, K.; Sumida, A.; Hara, T. Photoprotective mechanisms in cold-acclimated and nonacclimated needles of *Picea glehnii*. *Photosynthetica* **2010**, *48*, 110–116. [CrossRef]
44. Zheng, Y.; Yang, Q.; Xu, M.; Chi, Y.; Shen, R.; Li, P.; Dai, H. Responses of *Pinus massoniana* and *Pinus taeda* to freezing in temperate forests in central China. *Scand. J. For. Res.* **2012**, *27*, 520–531. [CrossRef]
45. Korotaeva, N.; Romanenko, A.; Suvorova, G.; Ivanova, M.V.; Lomovatskaya, L.; Borovskii, G.; Voinikov, V. Seasonal changes in the content of dehydrins in mesophyll cells of common pine needles. *Photosynth. Res.* **2015**, *124*, 159–169. [CrossRef] [PubMed]
46. Wisniewski, M.; Basset, C.I.; Gusta, L.V. An overview of cold hardiness in woody plants: Seeing the forest through the trees. *Hortscience* **2003**, *38*, 952–959. [CrossRef]
47. Danusevicius, D.; Gabrilavicius, R. Variation in juvenile growth rhythm among *Picea abies* provenances from the Baltic states and adjacent regions. *Scand. J. For. Res.* **2001**, *16*, 305–317. [CrossRef]
48. Isaacson, T.; Damasceno, C.M.; Saravanan, R.S.; He, Y.; Catala, C.; Saladie, M.; Rose, J.K. Sample extraction techniques for enhanced proteomic analysis of plant tissues. *Nat. Protoc.* **2006**, *1*, 769–774. [CrossRef]

49. Bradford, M.M. A rapid and sensitive method for the quantitation of microgram quantities of protein utilizing the principle of protein-dye binding. *Anal. Biochem.* **1976**, *72*, 248–254. [CrossRef]
50. Tamosiune, I.; Staniene, G.; Haimi, P.; Stanys, V.; Rugienius, R.; Baniulis, D. Endophytic *Bacillus* and *Pseudomonas* spp. modulate apple shoot growth, cellular redox balance, and protein expression under in vitro conditions. *Front. Plant Sci.* **2018**, *9*, 889. [CrossRef]
51. Shevchenko, A.; Tomas, H.; Havlis, J.; Olsen, J.V.; Mann, M. In-gel digestion for mass spectrometric characterization of proteins and proteomes. *Nat. Protoc.* **2006**, *1*, 2856–2860. [CrossRef]
52. Neale, D.B.; Wegrzyn, J.L.; Stevens, K.A.; Zimin, A.V.; Puiu, D.; Crepeau, M.W.; Cardeno, C.; Koriabine, M.; Holtz-Morris, A.E.; Liechty, J.D.; et al. Decoding the massive genome of loblolly pine using haploid DNA and novel assembly strategies. *Genome Biol.* **2014**, *15*, R59. [CrossRef]
53. Ishihama, Y.; Oda, Y.; Tabata, T.; Sato, T.; Nagasu, T.; Rappsilber, J.; Mann, M. Exponentially modified protein abundance index (emPAI) for estimation of absolute protein amount in proteomics by the number of sequenced peptides per protein. *Mol. Cell Proteom.* **2005**, *4*, 1265–1272. [CrossRef]
54. Conesa, A.; Gotz, S.; Garcia-Gomez, J.M.; Terol, J.; Talon, M.; Robles, M. Blast2GO: A universal tool for annotation, visualization and analysis in functional genomics research. *Bioinformatics* **2005**, *21*, 3674–3676. [CrossRef] [PubMed]
55. Schirmer, K.; Fischer, B.B.; Madureira, D.J.; Pillai, S. Transcriptomics in ecotoxicology. *Anal. Bioanal. Chem.* **2010**, *397*, 917–923. [CrossRef] [PubMed]
56. Bigras, F.J.; Ryyppo, A.; Lindstrom, A.; Stattin, E. Cold acclimation and deacclimation of shoots and roots of conifer seedlings. In *Conifer Cold Hardiness*; Bigras, F.J., Colombo, S.J., Eds.; Springer: Dordrecht, The Netherlands, 2001; pp. 57–88.
57. Agarwal, M.; Katiyar-Agarwal, S.; Grover, A. Plant Hsp100 proteins: Structure, function and regulation. *Plant Sci.* **2002**, *163*, 397–405. [CrossRef]
58. Singla, S.L.; Pareek, A.; Grover, A. Yeast HSP 104 homologue rice HSP 110 is developmentally- and stress-regulated. *Plant Sci.* **1997**, *125*, 211–219. [CrossRef]
59. Hong, S.W.; Vierling, E. Hsp101 is necessary for heat tolerance but dispensable for development and germination in the absence of stress. *Plant J.* **2001**, *27*, 25–35. [CrossRef]
60. Thelin, L.; Mutwil, M.; Sommarin, M.; Persson, S. Diverging functions among calreticulin isoforms in higher plants. *Plant Signal Behav.* **2011**, *6*, 905–910. [CrossRef]
61. Ndimba, B.K.; Chivasa, S.; Simon, W.J.; Slabas, A.R. Identification of Arabidopsis salt and osmotic stress responsive proteins using two-dimensional difference gel electrophoresis and mass spectrometry. *Proteomics* **2005**, *5*, 4185–4196. [CrossRef]
62. Komatsu, S.; Yang, G.; Khan, M.; Onodera, H.; Toki, S.; Yamaguchi, M. Over-expression of calcium-dependent protein kinase 13 and calreticulin interacting protein 1 confers cold tolerance on rice plants. *Mol. Genet. Genom.* **2007**, *277*, 713–723. [CrossRef]
63. Stone, S.L. The role of ubiquitin and the 26S proteasome in plant abiotic stress signaling. *Front. Plant Sci.* **2014**, *5*, 135. [CrossRef]
64. Garcia-Hernandez, M.; Davies, E.; Baskin, T.I.; Staswick, P.E. Association of plant p40 protein with ribosomes is enhanced when polyribosomes form during periods of active tissue growth. *Plant Physiol.* **1996**, *111*, 559–568. [CrossRef]
65. Omidbakhshfard, M.A.; Omranian, N.; Ahmadi, F.S.; Nikoloski, Z.; Mueller-Roeber, B. Effect of salt stress on genes encoding translation-associated proteins in *Arabidopsis thaliana*. *Plant Signal Behav.* **2012**, *7*, 1095–1102. [CrossRef] [PubMed]
66. Anzlovar, S.; Dermastia, M. The comparative analysis of osmotins and osmotin-like PR-5 proteins. *Plant Biol.* **2003**, *5*, 116–124. [CrossRef]
67. Liu, J.J.; Sturrock, R.; Ekramoddoullah, A.K. The superfamily of thaumatin-like proteins: Its origin, evolution, and expression towards biological function. *Plant Cell Rep.* **2010**, *29*, 419–436. [CrossRef]
68. Zhu, B.; Chen, T.H.; Li, P.H. Expression of an ABA-responsive osmotin-like gene during the induction of freezing tolerance in *Solanum commersonii*. *Plant Mol. Biol.* **1993**, *21*, 729–735. [CrossRef]
69. Liu, J.J.; Ekramoddoullah, A.K.M. The family 10 of plant pathogenesis-related proteins: Their structure, regulation, and function in response to biotic and abiotic stresses. *Physiol. Mol. Plant Pathol.* **2006**, *68*, 3–13. [CrossRef]

70. Ukaji, N.; Kuwabara, C.; Takezawa, D.; Arakawa, K.; Fujikawa, S. Accumulation of pathogenesis-related (PR) 10/Bet v 1 protein homologues in mulberry (*Morus bombycis* Koidz.) tree during winter. *Plant Cell Environ.* **2004**, *27*, 1112–1121. [CrossRef]
71. Christie, P.J.; Hahn, M.; Walbot, V. Low-temperature accumulation of alcohol dehydrogenase-1 mRNA and protein activity in maize and rice seedlings. *Plant Physiol.* **1991**, *95*, 699–706. [CrossRef]
72. Jarillo, J.A.; Leyva, A.; Salinas, J.; Martinez-Zapater, J.M. Low temperature induces the accumulation of alcohol dehydrogenase mRNA in *Arabidopsis thaliana*, a chilling-tolerant plant. *Plant Physiol.* **1993**, *101*, 833–837. [CrossRef]
73. Tesfaye, M.; Temple, S.J.; Allan, D.L.; Vance, C.P.; Samac, D.A. Overexpression of malate dehydrogenase in transgenic alfalfa enhances organic acid synthesis and confers tolerance to aluminum. *Plant Physiol.* **2001**, *127*, 1836–1844. [CrossRef]
74. Yao, Y.X.; Dong, Q.L.; Zhai, H.; You, C.X.; Hao, Y.J. The functions of an apple cytosolic malate dehydrogenase gene in growth and tolerance to cold and salt stresses. *Plant Physiol. Biochem.* **2011**, *49*, 257–264. [CrossRef] [PubMed]
75. Nveawiah-Yoho, P.; Zhou, J.; Palmer, M.; Sauve, R.; Zhou, S. Identification of proteins for salt tolerance using a comparative proteomics analysis of tomato accessions with contrasting salt tolerance. *J. Am. Soc. Hortic. Sci.* **2013**, *138*, 382–394. [CrossRef]
76. Askari, H.; Edqvist, J.; Hajheidari, M.; Kafi, M.; Salekdeh, G.H. Effects of salinity levels on proteome of *Suaeda aegyptiaca* leaves. *Proteomics* **2006**, *6*, 2542–2554. [CrossRef]
77. Miller, J.M.; Conn, E.E. Metabolism of hydrogen cyanide by higher plants. *Plant Physiol.* **1980**, *65*, 1199–1202. [CrossRef] [PubMed]
78. Peiser, G.D.; Wang, T.T.; Hoffman, N.E.; Yang, S.F.; Liu, H.W.; Walsh, C.T. Formation of cyanide from carbon 1 of 1-aminocyclopropane-1-carboxylic acid during its conversion to ethylene. *Proc. Natl. Acad. Sci. USA* **1984**, *81*, 3059–3063. [CrossRef] [PubMed]
79. Conklin, P.L.; Norris, S.R.; Wheeler, G.L.; Williams, E.H.; Smirnoff, N.; Last, R.L. Genetic evidence for the role of GDP-mannose in plant ascorbic acid (vitamin C) biosynthesis. *Proc. Natl. Acad. Sci. USA* **1999**, *96*, 4198. [CrossRef] [PubMed]
80. Akram, N.A.; Shafiq, F.; Ashraf, M. Ascorbic acid—A potential oxidant scavenger and its role in plant development and abiotic stress tolerance. *Front. Plant Sci.* **2017**, *8*, 613. [CrossRef] [PubMed]
81. Qi, T.; Liu, Z.; Fan, M.; Chen, Y.; Tian, H.; Wu, D.; Gao, H.; Ren, C.; Song, S.; Xie, D. GDP-D-mannose epimerase regulates male gametophyte development, plant growth and leaf senescence in Arabidopsis. *Sci. Rep.* **2017**, *7*, 10309. [CrossRef]
82. Luo, J.; Shen, G.; Yan, J.; He, C.; Zhang, H. AtCHIP functions as an E3 ubiquitin ligase of protein phosphatase 2A subunits and alters plant response to abscisic acid treatment. *Plant J.* **2006**, *46*, 649–657. [CrossRef]
83. Monroy, A.; Sangwan, V.; Dhindsa, R.S. Low temperature signal transduction during cold acclimation: Protein phosphatase 2A as an early target for cold-inactivation. *Plant J.* **1998**, *13*, 653–660. [CrossRef]
84. Pais, S.M.; Gonzalez, M.A.; Tellez-Inon, M.T.; Capiati, D.A. Characterization of potato (*Solanum tuberosum*) and tomato (*Solanum lycopersicum*) protein phosphatases type 2A catalytic subunits and their involvement in stress responses. *Planta* **2009**, *230*, 13–25. [CrossRef]
85. Kalapos, M.P. The tandem of free radicals and methylglyoxal. *Chem. Biol. Interact.* **2008**, *171*, 251–271. [CrossRef] [PubMed]
86. Yadav, S.K.; Singla-Pareek, S.L.; Ray, M.; Reddy, M.K.; Sopory, S.K. Methylglyoxal levels in plants under salinity stress are dependent on glyoxalase I and glutathione. *Biochem. Biophys. Res. Commun.* **2005**, *337*, 61–67. [CrossRef] [PubMed]
87. Droog, F. Plant glutathione S-transferase, a tale of theta and tau. *J. Plant Growth Regul.* **1997**, *16*, 95–107. [CrossRef]
88. Roxas, V.P.; Smith, R.K., Jr.; Allen, E.R.; Allen, R.D. Overexpression of glutathione S-transferase/glutathione peroxidase enhances the growth of transgenic tobacco seedlings during stress. *Nat. Biotechnol.* **1997**, *15*, 988–991. [CrossRef]
89. Livanos, P.; Galatis, B.; Apostolakos, P. The interplay between ROS and tubulin cytoskeleton in plants. *Plant Signal Behav.* **2014**, *9*, e28069. [CrossRef]

90. Golan, I.; Dominguez, P.G.; Konrad, Z.; Shkolnik-Inbar, D.; Carrari, F.; Bar-Zvi, D. Tomato ABSCISIC ACID STRESS RIPENING (ASR) gene family revisited. *PLoS ONE* **2014**, *9*, e107117. [CrossRef]
91. Hsu, Y.F.; Yu, S.C.; Yang, C.Y.; Wang, C.S. Lily ASR protein-conferred cold and freezing resistance in Arabidopsis. *Plant Physiol. Biochem.* **2011**, *49*, 937–945. [CrossRef]
92. Kim, S.J.; Lee, S.C.; Hong, S.K.; An, K.; An, G.; Kim, S.R. Ectopic expression of a cold-responsive OsAsr1 cDNA gives enhanced cold tolerance in transgenic rice plants. *Mol. Cells* **2009**, *27*, 449–458. [CrossRef]
93. Padmanabhan, V.; Dias, D.M.; Newton, R.J. Expression analysis of a gene family in loblolly pine (*Pinus taeda* L.) induced by water deficit stress. *Plant Mol. Biol.* **1997**, *35*, 801–807. [CrossRef]
94. Seidler, A.; Rutherford, A.W.; Michel, H. On the role of the N-terminus of the extrinsic 33 kDa protein of Photosystem II. *Plant Mol. Biol.* **1996**, *31*, 183–188. [CrossRef]
95. Riccardi, F.; Gazeau, P.; Jacquemot, M.P.; Vincent, D.; Zivy, M. Deciphering genetic variations of proteome responses to water deficit in maize leaves. *Plant Physiol. Biochem.* **2004**, *42*, 1003–1011. [CrossRef] [PubMed]
96. Sugihara, K.; Hanagata, N.; Dubinsky, Z.; Baba, S.; Karube, I. Molecular characterization of cDNA encoding oxygen evolving enhancer protein 1 increased by salt treatment in the mangrove *Bruguiera gymnorrhiza*. *Plant Cell Physiol.* **2000**, *41*, 1279–1285. [CrossRef] [PubMed]
97. Wang, W.; Vinocur, B.; Shoseyov, O.; Altman, A. Role of plant heat-shock proteins and molecular chaperones in the abiotic stress response. *Trends Plant Sci.* **2004**, *9*, 244–252. [CrossRef]
98. Wilkinson, B.; Gilbert, H.F. Protein disulfide isomerase. *Biochim. Biophys. Acta* **2004**, *1699*, 35–44. [CrossRef]
99. Kupsch, C.; Ruwe, H.; Gusewski, S.; Tillich, M.; Small, I.; Schmitz-Linneweber, C. Arabidopsis chloroplast RNA binding proteins CP31A and CP29A associate with large transcript pools and confer cold stress tolerance by influencing multiple chloroplast RNA processing steps. *Plant Cell* **2012**, *24*, 4266–4280. [CrossRef] [PubMed]
100. Bernard, M.A.; Ray, N.B.; Olcott, M.C.; Hendricks, S.P.; Mathews, C.K. Metabolic functions of microbial nucleoside diphosphate kinases. *J. Bioenerg. Biomembr.* **2000**, *32*, 259–267. [CrossRef] [PubMed]
101. Lee, D.G.; Ahsan, N.; Lee, S.H.; Kang, K.Y.; Lee, J.J.; Lee, B.H. An approach to identify cold-induced low-abundant proteins in rice leaf. *C. R. Biol.* **2007**, *330*, 215–225. [CrossRef]
102. Krol, M.; Hurry, V.M.; Maxwell, D.P.; Malek, L.; Ivanov, A.G.; Huner, N.P.A. Low growth temperature inhibition of photosynthesis in cotyledons of jack pine seedlings (*Pinus banksianna*) is due to impaired chloroplast development. *Can. J. Bot.* **2002**, *80*, 1042–1051. [CrossRef]
103. Oquist, G.; Huner, N.P. Photosynthesis of overwintering evergreen plants. *Annu. Rev. Plant Biol.* **2003**, *54*, 329–355. [CrossRef]

Article

Bark Features for Identifying Resonance Spruce Standing Timber

Florin Dinulică [1], Cristian-Teofil Albu [2], Maria Magdalena Vasilescu [1,*] and Mariana Domnica Stanciu [3]

[1] Departament of Forest Engineering, Forest Management Planning and Terrestrial Measurements, Transilvania University of Brașov, 500123 Brașov, Romania; dinulica@unitbv.ro

[2] Gurghiu Forestry High School, 547295 Gurghiu, Romania; cristian_teofil_albu@yahoo.com

[3] Department of Mechanical Engineering, Transilvania University of Brașov, 500036 Brașov, Romania; mariana.stanciu@unitbv.ro

* Correspondence: vasilescumm@unitbv.ro; Tel.: +40-768271819

Received: 15 August 2019; Accepted: 6 September 2019; Published: 12 September 2019

Abstract: Measuring the acoustic properties of wood is not feasible for most luthiers, so identifying simple, valid criteria for diagnosis remains an exciting challenge when selecting materials for manufacturing musical instruments. This article aims to verify whether the bark qualities as a marker of resonance wood are indeed useful. The morphometric and colour traits (in CIELab space) of the bark scales were compared with the structural (width and regularity of the growth rings and of the latewood) and acoustic features (transverse sound velocity, radiation ratio, impedance, and wood basic density) of the wood from 145 standing and 10 felled spruce trees, which are considered a resource of the resonance wood in the Romanian Carpathians. It has been emphasized that the spruce trees with acoustic and structural features that match the requirements for the manufacture of violins have a bark phenotype distinguishable by colour (higher redness, lower yellowness and brightness)—as well as by scale shape (higher slenderness and width). The south-facing side of the trunk and the external side of the scale are best for identifying resonance trees by their bark. Additionally, the mature bark phenotypes denote topoclinal variations and do not depend on tree age. Moreover, the differences among bark phenotypes are noticeable to the naked eye.

Keywords: bark phenotype; bark scale; Norway spruce; resonance wood; sonic tomography

1. Introduction

There are 1.43 million ha of Norway spruce (*Picea abies* L. (Karst)) in the Romanian Carpathians [1], which host one of the most important resources of resonance wood in Europe in terms of value and volume [2,3]. Although it is continuously decreasing, the resonance wood resource still satisfies the demand for the local musical instruments industry, which are shipped across all continents [4].

Recognizing standing trees that have resonance wood has always been a challenge for luthiers. Long-time observations revealed the distinct physiognomy of resonance trees [5–7], but the topic remained somewhat unsolved until some of the physiognomic features were acoustically verified [8]. The literature provided a few morphological descriptors of the stem and crown of resonance trees [9,10] and of raw resonance wood [11–15]. However, other points of view have not placed much trust in these descriptors [16]—labelling them as folkloric [17]. Given the indicative value of some traits of trees in relation to the material's acoustic quality, we shall hereafter call them phenotypical markers or morphological descriptors of resonance spruce. Some luthiers empirically associate resonance spruce with the phenotype of smooth or thin fissured barked trees with small, soft, and rounded scales, grouped vertically [8,9,18], which is different from regular spruce trees that have thicker and deeply furrowed bark at the harvesting age of the resonance wood [4]. Furthermore, spruce with indented rings is sought after by violin makers [19]—specifically indentations imprinting the underbark side [20]. In any case, the bark descriptors of resonance wood have not yet been acoustically and statistically checked [21].

Normally, bark width returns the tree growth traits [22] and together with these, they are age- and site-related [23]. The qualitative features of bark, such as relief and colour, are predominantly hereditary [24,25] and have a certain taxonomic value so that the bark texture can allow for the digital identification of species [26,27].

Besides the ecological and functional significance [28,29], bark morphology can be a good indicator of wood properties [30]. For instance, in the case of fir trees of similar age—which have persistent smooth bark—the wood is lighter, and the cellulose amount is smaller in contrast to early rough bark trees [31]. In the case of European and Chinese pear trees, the early rough bark trees contain more lignin in the wood and have a low carbon use ratio [32]. In the case of Scots pine trees, which have deeply fissured bark in thin, square plates that are smooth and light in colour, the heartwood is red, and its amount—alongside the amount of resin—is larger [33].

Sound velocity, wood density, dynamic modulus of elasticity and their indices, as radiation ratio, specific modulus, characteristic impedance, and acoustic converting efficiency, are preferred for expressing the suitability for strings [14,33–36]. High values for specific modulus of elasticity, sound velocity and radiation ratio, lower values for impedance and internal friction, as well as the lower values for density are recommended in the choice of material for soundboards [14,37–41]. Sound propagation velocity sets the clarity of the sound emitted by the musical instrument, acoustic radiation is a measure of acoustic power—in particular of the sound loudness—and the acoustic impedance expresses the sound sprinting [42,43].

Vibrational methods have already become common in identifying damages in standing trees [44–47] and assessing tree stiffness [48,49]. Using them in examining the tree goodness for the manufacture of musical instruments is still in the early stages [8]. New advanced methods, such as X-ray light microtomography coupled with scanning electron microscopy, are involved in describing the acoustical behavior of the wood [50]. Our aim is to check the hypothesis of.the link between the bark features and the acoustic qualities of wood originating from stands that supply raw materials for violin manufacturing—specifically the possibility of an objective diagnosis of resonance wood using the bark. For this purpose, we: (1) identified the variation sources of the bark phenotype; (2) checked the connection between the bark features and the wood structure, and (3) verified the relation between the bark features and the acoustic properties of the wood.

2. Materials and Methods

2.1. Sampled Area

The materials originate from the Gurghiu Mountains, which today are home to the largest concentration of resonance wood in the Romanian Carpathians (Figure 1). The relief and the local volcanic substratum favoured the selection and settlement of the resonance spruce phenotype [4]. The resonance spruce trees are located inside the former volcanic basin, which today forms the Gurghiu Mountains. The volcanic bowl protects the trees from excess wind circulation, thus reducing the occurrence of compression wood that compromises the acoustic value of the wood [51]. The mean annual temperature ranges around 5.2 °C at 1200 to 1300 m elevation. The soil bedrock is of andesite origin, well-supplied with rain water (850–1100 mm·year^{-1}). The soils are deep, loose, and of moderate fertility.

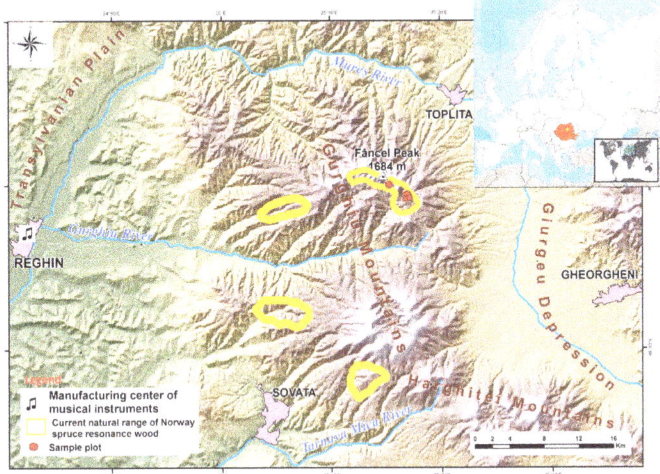

Figure 1. Location of the sampled resource of resonance spruce wood.

The local stands containing resonance wood are mixed, relatively uneven-aged or two-storied stands, and consist of spruce (*Picea abies* Karst.), beech (*Fagus sylvatica* L.), fir (*Abies alba* Mill.) and sometimes sycamore (*Acer pseudoplatanus* L.) as well. The resonance spruce trees cover the middle third of the slope—which are moderately to highly sloped—and avoid the cold air in the valleys or on the peaks.

2.2. Sampling Design

To check the link between the bark features, the wood structure, and the acoustic properties of the wood, four sample plots were established (Figure 2) in stands that had the highest occurrence of resonance spruce (identified by habitus [5,15,18]).

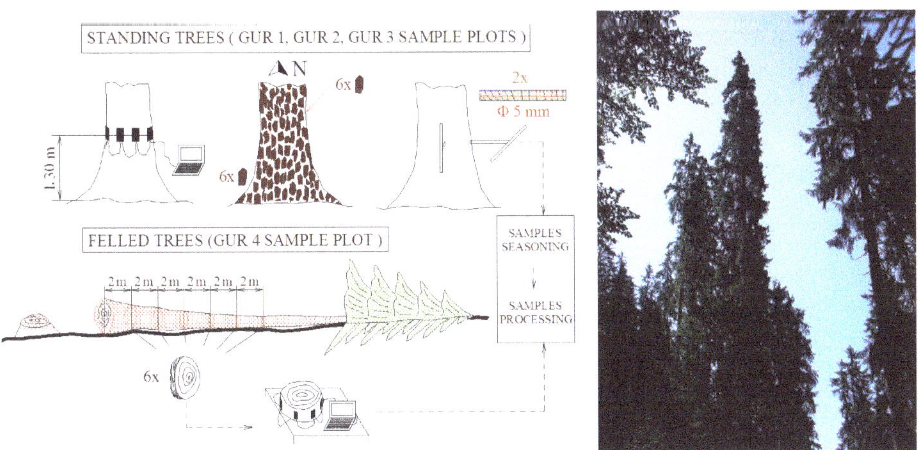

Figure 2. Scheme of the sampling design and outlook of one of the sample sites.

In the first three plots, the spruce trees with diameters larger than 20 cm were cored. The samples were collected at breast height, on two directions; one from the self-pruned sector of the bole and the other from the non-pruned sector. An increment borer with a length of 400 mm and an inner diameter of 5 mm was used (Table 1, denoted [a]). On the fourth sample plot (GUR4) which had undergone logging operations, 10 of the resonance trees used later by Gliga Company for manufacturing violins were selected. These 10 trees yielded 18 resonance logs from where discs were cut every 2 m—resulting in 65 discs that were used for the present study (Table 1, denoted [b]).

Table 1. Summary of sampled sites.

Features	Sample Plot			
	GUR1	GUR2	GUR3	GUR4
Altitude (m asl)	1380	1300	1215	1580
Facing	NE	NW	W	NE
Slope (°)	33	31	24	17
Soil type *	Cmeu	CMdy	PZrs	BOis
Species composition **	100%NS	80%NS 20%BC	80%NS 10%BC 10%SF	100%NS
Number of sampled trees	44	61	30	10
Diameter at breast height (cm) ***	25...52...81	22.5...42...81	33...58...101	42...53...62
Tree height (m) ***	23...33...39.5	20...29...40.3	21...37...49.5	28...30...35
Tree age (years at breast height) ***	144...181...298	152...182...278	122...242...418	95...168...238
Number of measured samples (cores [a]/discs [b])	66 [a]	104 [a]	58 [a]	65 [b]

* Cmeu—Eutric Cambisols; Cmdy—Distric Cambisols; PZrs—Rustic podzols—BOis—Lepti-dystric Cambisols [52]; ** BC—European beech; NS—Norway spruce; SF—Silver fir; *** Range (first quartile–third quartile) and median; [a] cores; [b] discs.

Bark samples (six bark scales) were taken at breast height from each tree and disc: three facing North and three facing South. This was done to check the influence of the various insolation conditions on the bark features [53].

2.3. Processing the Material

2.3.1. Bark Measurements

The measurements on the sampled bark were performed in the laboratory after seasoning, lasting at least three months. At the time of the measurements, the humidity of the bark scales was 11%, determined using the Ohaus MB45 halogen moisture analyzer [54] by seasoning at 104 ± 1 °C. The length LS and width WS of each scale were measured. The bark scale slenderness index (LS/WS ratio) was calculated as well. In order to quantify the colour, the CIELab chromatic system was employed [55] by using the CR-400 portable colorimeter [56]. Through this system, the colour is rendered in a space with three coordinate axes whose values are provided by the colour meter: L *—colour brightness or whiteness (%), a *—colour redness/greenness, and b *—colour yellowness/blueness. Both sides of the scales were scanned three times. A total number of 1884 scales were measured. The ability of visual perception of the colour differences between bark samples was checked using the colour difference index ΔE * resulted from calculations (relation 1). Its values were balanced to the perception scales proposed by Minemura and Umehara [57]:

$$\Delta E = \sqrt[2]{\Delta L^{*2} + \Delta a^{*2} + \Delta b^{*2}}. \quad (1)$$

2.3.2. Wood Structural Data Acquisition

After seasoning, the cores were mounted on boards. The samples from the GUR2 plot, which were used for the destructive determination of the wood density, were not glued to the board as the rings were measured after the surface of the cores was smoothed. The discs and the mounted cores were sanded to meet the requirements of 1200 dpi scanning resolution. The individual growth rings were transferred to digital format using WinDENDRO (Régent Instruments Inc., Québec City, Canada)

Density equipment [58]. The measurements were carried out on the images resulted from the scans. For each disc, the radii of the four cardinal points were measured.

The digital format of the growth rings consists of: ring width TRW, earlywood width, latewood width LWW, earlywood proportion, and latewood proportion LWP. The raw data were stacked into chronological series. Using these raw variables, two more indices were produced for identifying the resonance wood: the difference in width between the consecutive rings DBR (mm) and the ring width regularity index RI [13,59]. For the discs, the ring circumferential regularity index CI was calculated, as percentage deviation from the average of extreme values along the girth, using the relation:

$$CI_i = \frac{max\left(TR_{i_k}\right) - min\left(TR_{i_k}\right)}{average\left(TR_{i_k}\right)} \cdot 100\ [\%], \quad (2)$$

where CI_i is the circumferential irregularity of ring i, TR_{i_k} is the width of ring i on radius k of the disc.

The width of the area with resonance wood (LRE) was established according to the fluctuations TRW, LWP, and DBR along the radius [59]. The size of these features enabled the structural classification of the sampled trees (Table 2). The average structural quality class (SQC) was calculated for the trees where two cores had been extracted.

Table 2. The structural-qualitative classification of raw material intended for violin making [4].

Quality Grade (SQC)	Structural Requirements to the Resonance Wood Zone *
1. Wood of high structural quality	Ring width = 0.8 to 2.5 mm (average ≥ 1.2 mm).
	Average latewood proportion ≤ 20%.
	Average difference in width between consecutive rings ≤ 0.5 mm.
	Ring width regularity index ≤ 0.7.
	Width of resonance wood ≥ 130 mm.
2. Wood of average structural quality	Ring width = 0.8 to 2.5 mm.
	Average latewood proportion ≤ 35%.
	Average difference in width between consecutive rings ≤ 0.6 mm.
	Width of resonance wood ≥ 130 mm.
3. Wood of low structural quality	Ring width ≤ 3.0 mm (average ≥ 0.5 mm).
	Average latewood proportion ≤ 39%.
	Average difference in width between consecutive rings ≤ 1.0 mm.
	Width of resonance wood ≥ 130 mm.
4. Wood from young resonance trees	The wood meets the requirements imposed on SQC2—with the exception of the resonance zone width—which is insufficient for a violin flitch (130 mm), but exceeds 50 mm.
	Diameter at breast height ≤ 38 cm.
5. Wood from thick trees, with insufficient width of the resonance wood	Diameter at breast height > 38 cm, tree age > 150 years;
	50 ≤ width of resonance wood < 130 mm.
6. Wood with no structural-acoustic qualities	Width of resonance wood < 50 mm or;
	average ring width ≤ 0.5 mm.

* In the case of the first three classes, a maximum of seven growth rings exceeding the limits of the criteria for resonance wood by one standard deviation are admitted.

2.3.3. Acoustic Measurements

The transverse sound velocity was measured ($m \cdot s^{-1}$) in the standing trees and discs using the Arbotom sonic tomograph produced by Rinntech [60]. Although in the acoustic tests the longitudinal direction is preferred, the vibratory properties of the wood across the fibre are proportional to those along the fibre [61]. The acoustic measurements in the standing trees were done outside the vegetation season when the wood had an average moisture content gravimetrically determined, of 39% in heartwood and 98% in sapwood, and the air temperature was around 4 °C. The acoustic measurements of the discs were done in the laboratory, at the wood moisture content of 10%. The 15 sensors were put on in the breast height section of the tree, 15 to 20 cm apart (Figure 3). The position of the sensors was established according to the directions on which the coring was done in the standing trees, or the rings were measured in the discs. The sensors were hammered five times with the same force. The software associated to the Arbotom equipment generated the sound velocity matrices and the sonic tomograms.

Figure 3. Caption of a tomography in the field.

To calculate the radiation ratio and the specific acoustic resistance (impedance), the wood basic density of the cores from the GUR2 plot was measured. For this purpose, the saturation method developed by Keylwerth [62] was used. The resin was removed from the cores through classic extraction using the Soxhlet device [63] because resin biased the acoustic radiation [64] and the wood density size.

Then, the cores were divided according to the quality sections of the wood structure: wood of structural quality for manufacturing musical instruments and wood with no structural qualities, respectively (Table 2). The solvent was a mixture of benzene and absolute ethyl alcohol 2:1 [65]. The extraction lasted 12 h for each sample. Saturation was obtained by boiling in distilled water for 12 hours [66]. After boiling, the samples were weighed, resulting the mass $m_{\max}$, then dried in a drying oven at 103 °C until constant mass was achieved [67]. After cooling in an exicator, they were weighed again (mass m_0). The basic density size (ρ_c) resulted from the following calculation [68]:

$$\rho_c = \frac{1}{\frac{m_{max}}{m_0} - 0.3464} \left[\text{g}\cdot\text{cm}^{-3}\right]. \tag{3}$$

The radiation ratio R was calculated by dividing sound velocity to the basic density of the core on the direction the coring was done [38]. The impedance was calculated by multiplying sound velocity with wood basic density [38].

2.4. Data Processing

In the case of the split discs, only the values on the directions that did not cross the shake were kept in the sound velocity matrix. In the statistical processing, the sound velocity values lower than 600 $m \cdot s^{-1}$ were removed. The rough data were imported and processed using STATISTICA 8.0 [69].

Simple, multiple, and partial correlation analysis was employed in order to quantify the dependence between the variables [70]. Multiple correlation was used in order to identify and estimate the influence of the explicative variables. Only the variables with a statistically significant coefficient of partial regression (as confirmed by the t test) were considered. Some of the links were transposed into regressions. The regression model with the highest determination coefficients was adopted. To avoid multicollinearity, the relation between the predictors was verified beforehand, and only the explicative variables independent from one another were retained in the model.

The structural quality class of the basal area wood was adopted as a criterion for the stratification of the morphological tree features. Only the variables showing a link with SQC previously confirmed using a significance test were stratified. When selecting this test, the normality of the distributions was interrogated using the Shapiro-Wilks test. For the Gaussian type variables, ANOVA was adopted, and the non-parametric Kruskal-Wallis test or rank tests were adopted for the other variables [70].

3. Results

3.1. The Variability of the Bark Features

The examined bark features (scale size and colour) were moderately variable (coefficients of variation between samples from 15.7% to 32.9%)—excepting scale colour redness/greenness—which was highly variable (coefficient of variation of 82.6% and the largest amplitude). None of these variables had a Gaussian allure ($W > 0.850$, $p < 0.0001$). All were variables with continuous variations, some with high percent relative range, such as colour redness/greenness and scale length (376% and 235%, respectively, without outliers), and others with low percent relative range, such as colour brightness (91% without outliers). Even if they were continuous variables, the bark chromatics tended to separate into classes of values which distinguish the grey-barked trees from the brown-barked trees (Figure 4).

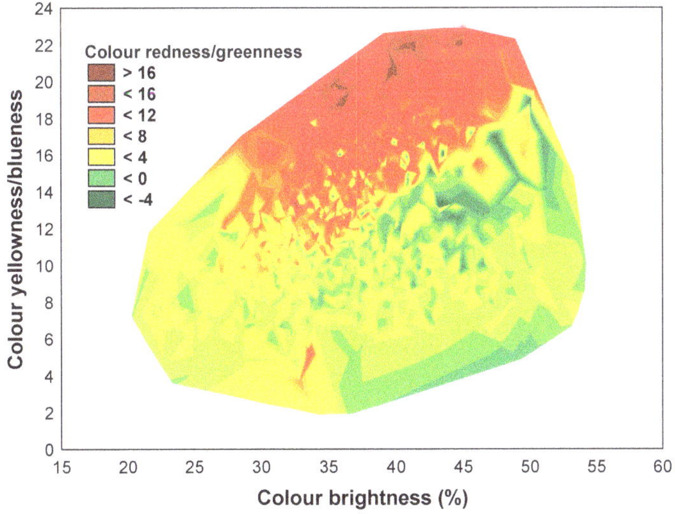

Figure 4. The spectrum of bark colours in the measured spruce trees.

Negative values for variable b * were not recorded, so blue was not detected in the composition of the bark scale colour. The negative values of a * (scale greenness) had a 7.3% frequency and were identified only on the external side of the scales—especially on the south-facing side of the trunk.

The bark features are stable among trees inside the sample plot (Table 3). The sample plot is the main source of variation for scale size, slenderness, and brightness—but not for bark scale redness/greenness and yellowness. There are differences between the northern and southern side of the bark from the trunk for redness/greenness and yellowness, and for scale width—but not for scale slenderness (Table 3). Both colour variable values are higher on the southern side. The values are more stable from plot to plot on the northern side.

Table 3. The statistical significance of the influence of some factors on the analyzed bark features, using the Kruskal-Wallis test.

Source of Variation	LS		WS		SS		L *		a *		b *	
	H	p	H	p	H	p	H	p	H	p	H	p
Sample plot	13.81	0.008	12.00	0.017	33.63	<0.001	38.86	<0.001	6.80	0.15	6.96	0.14
Tree × plot	11.00	0.25	9.88	0.36	10.82	0.29	14.75	0.10	3.12	0.96	16.49	0.05
Tree	23.08	0.006	15.52	0.08	16.04	0.007	30.03	0.0004	4.34	0.89	14.29	0.11
Cardinal point	3.02	0.08	7.22	0.007	0.29	0.59	1.93	0.17	1233.21	<0.001	938.3	<0.001
Scale	6.00	0.74	6.00	0.74	6.00	0.74	9.00	0.44	10.00	0.35	9.00	0.44
Scale × sample plot	132.23	<0.001	45.72	<0.001	18.31	<0.001	0.814	0.67	1.85	0.40	0.03	0.98

LS: bark scale length; WS: bark scale width; SS: bark scale slenderness; a * bark colour redness/greenness; b * bark colour yellowness/blueness.

The scale sizes have a total contribution of 94% to the variation of their slenderness (multiple $R \leq 0.590$, $F = 7475$, $p < 0.001$). The relation between scale shape and colour is weak: Spearman R rank order correlations between −0.156 ($p < 0.001$) and 0.048 ($p = 0.15$).

The contribution of tree age to the variation of the bark phenotype is up to 6.8% (Spearman R rank order correlations with age between −0.180 and +0.261, $p = 0.01$–0.98). The size and range of the scale slenderness index decrease with tree age ($H = 10.40$, $p = 0.03$). The influence of tree age on the bark colour was only detected on the external side of the north-facing scales ($p = 0.01$–0.05, compared to 0.06–0.98 on the southern side). On the internal side of the scales, slightly distinguishable was only the decrease of the colour brightness with age ($R = -0.176$, $p = 0.05$). The oldest trees (older than 300 years) were different from the younger trees in the low yellowness on the external side of the north-facing scales ($H = 9.41$, $p = 0.05$). The bark colour differences (ΔE *) between the northern and southern side of the stem—as well as between the internal and external side of the scales—were not influenced by tree age ($H = 1.99$, $p = 0.74$ and $H = 2.60$, $p = 0.63$, respectively).

The bark colour variation from one sample plot to another is explained as well by the altitude range. The altitude share to the scale features variation is up to 13%: Spearman R rank order correlation R with bark yellowness on the northern external side of the scale = +0.364, $p < 0.001$; R with bark redness on both the northern and southern external side of the scale = −0.288 respectively −0.264, $p < 0.001$; R with scale length = −0.159, $p = 0.04$.

All the colour differences ΔE * between the scales were, without exception, visible to the naked eye—at least at a "light" level of visual assessment (ΔE * > 0.70), according to Minemura and Umehara's classification [48]. The colour differences between the two trunk sides (northern and southern) reached "considerable" and "important" levels (3.9–9.9), while the differences between the scale sides were "very important" (10.2–13.6). The colour differences between the scale sides are considerably higher between the trunk sides and are due to the redness/greenness and yellowness ($H = 1279.45$, $p < 0.0001$ and $H = 970.15$, $p < 0.0001$, respectively). More precisely, the scale internal side is more reddish than the external side.

3.2. Linking Bark Features to the Wood Structure

The qualities of bark as a marker of wood structure were analyzed using correlation and statistical significance tests.

The length and width of the scale's variation is independent of the wood structure (Spearman rank order correlation ≤0.16, $p > 0.09$). By contrast, the scale slenderness index is strongly connected to the ring structure regularity (Table 4). The link is even better if tree age is joined (multiple R). If its influence is filtered, the contribution of the scale slenderness index to the estimation of the circumferential irregularity is at least 73% (square partial r). DBH is not determinative in the link between scale slenderness and wood structure (Table 4). The bark descriptor with the closest link to wood structure is the circumferential irregularity of latewood width (Table 4). Logistic growth fitting best describes the link between the two variables (Figure 5). Similarly, the link between scale slenderness and circumferential irregularity of ring width is better expressed by the polynomial model. All these links indicate that trees with better annual ring regularity show high slenderness scales. If the threshold value of the latewood irregularity of 70% [13,59] is applied in Figure 5, then the scale slenderness indices must be at least 1.9 for resonance spruce trees.

Table 4. Bark as predictor of spruce wood structure from the breast height (only the correlations with $R > 0.5$ are shown).

Predictor	Dependent Variable	Spearman Rank Order Simple Correlation/p	Multiple Correlation with TA or DBH/p		Partial Correlation with TA or DBH/p	
			TA/p •	DBH/p •	TA	DBH
Scale slenderness index	CI_{TRW}	−0.698/0.02	0.857/0.015	0.683/0.45	−0.855	−0.665
	CI_{LWW}	−0.914/<0.0001	0.903/0.25	0.879/0.98	−0.903	−0.879
	CI_{LWP}	−0.729/0.02	0.921/0.03	0.837/0.503	−0.875	−0.834

• p from t test for the significance of the partial regression coefficient associated with the tree age (TA) and the DBH (diameter at breast height) variable, respectively. If $p > 0.05$, then the influence of the tree age or the diameter on the predictor is not statistically proven.

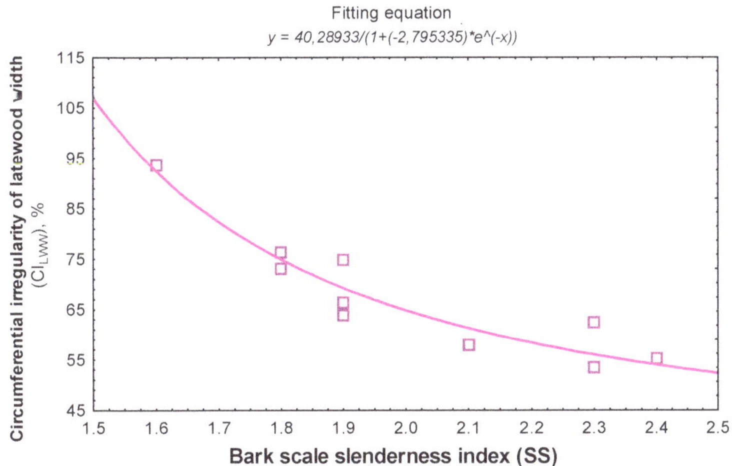

Figure 5. The regression of circumferential irregularity of latewood thickness with the bark scale slenderness index in trees sampled with discs.

The multiple regression analysis indicates that the width of the resonance zone at breast height level was rendered well by the brightness and redness of the bark scales ($F = 7.23$, $p = 0.0002$).

The trees with scales that were redder and darker on the north-facing side of the trunk have a wider resonance zone.

Some of the tree variables were grouped according to the structural quality of the wood; Figure 6 only shows the variables of interest for this research, as well as the variables of bark colour for which $p \leq 0.05$. The bark scale slenderness index and length do not appear to be connected with the wood structure quality (Figure 6). However, the bark scales of resonance structural trees are somewhat wider.

Furthermore, the values of the chromatic components were polarized according to the structural quality class (Figure 6). However, the differences between the structural classes regarding bark scale colour are not major and only refer to one of the scale sides—either the northern or the southern side. After the stratification of the colour components depending on the cardinal points and their position on the scale sides, four chromatic markers of SQC became apparent (Figure 6): bark brightness on the north side, the redness on the back of the south-facing scales, and the yellowness on both sides of the bark. The contribution of tree age was irrelevant: the correlation coefficients between the chromatic coordinates and the tree age are below 0.200, $p > 0.04$. The tree diameter has a contribution of up to 7.3% to the colour variation.

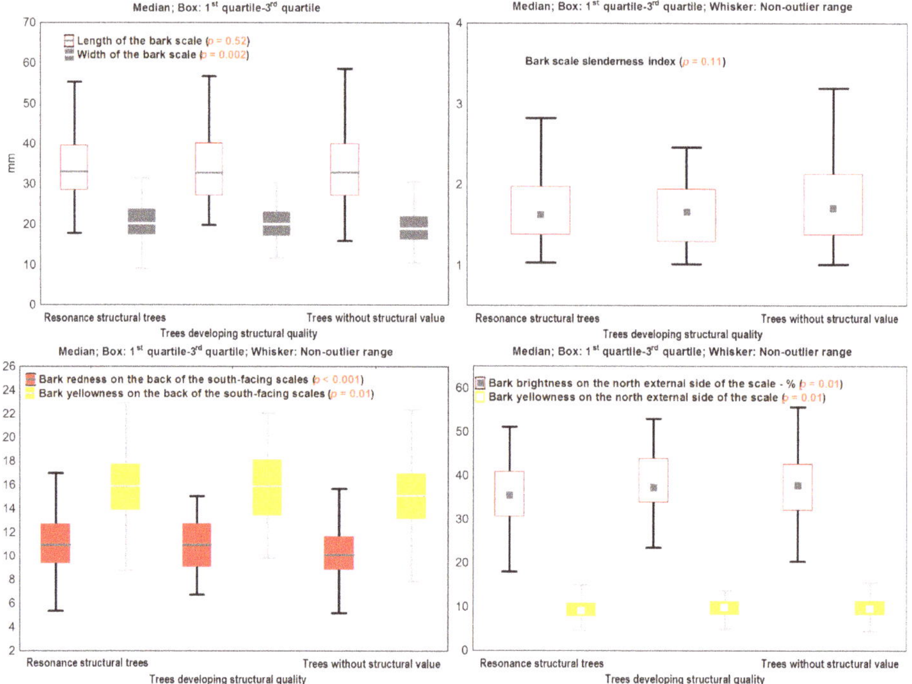

Figure 6. The stratification * of some morphological traits according to the structural quality class of the trees (p from Kruskal-Wallis test or ANOVA).

Even if the differences in bark colour between the trees are not significant—they are visible to the naked eye as proven by values ΔE^* (Table 5). The northern side of the trunk is the most suitable for this purpose. The strongest differences occur between extreme quality classes. Trees with a matured (SQC ≤ 3) or developing (SQC = 4) structural quality can be easily distinguished (at the "considerable" and "important" levels of visual assessment) from the trees with no such qualities.

Table 5. The matrix of median differences in bark colour between the structural classes of standing spruce trees (ΔE * on the northern side of the breast height section/ΔE * on the southern side).

SQC	1	2	3	4	5	6
1	-	3.19/1.48	3.17/1.86	1.15/4.32	4.48/1.00	7.52/2.51
2		-	1.13/1.56	2.14/5.79	2.63/1.42	5.59/3.78
3			-	2.03/5.56	1.63/0.94	4.73/3.05
4				-	3.40/4.79	6.49/2.75
5					-	3.13/2.46
6						-

3.3. Acoustic Screening of Bark Markers

The measured acoustic features of the wood have a low variability level (Table 6). All the acoustic variables examined are Gaussian (W from the Shapiro-Wilk test has values higher than 0.93, $p > 0.07$), which encouraged the application of ANOVA as a significance test. The values of the sound velocity can be grouped according to the sampled variation sources. Of interest is the assigning structural quality class to transverse sound velocity and radiation ratio (the highest values of these acoustics in quality class 3 and the lowest values in quality class 5, both in the standing and felled trees).

Table 6. Variation sources of acoustic features of spruce wood in stands with resonance wood from the Gurghiu Mountains (the Romanian Carpathians).

Feature	Mean/Confidence	Coefficient of Variation (%)	Source of Variation p from ANOVA F Test *		
			Stand	Tree	SQC
Transverse velocity in standing trees ** ($m \cdot s^{-1}$)	1132.99/1123.68–1142.31	14.52	<0.001	<0.001	<0.001
Transverse velocity in discs *** ($m \cdot s^{-1}$)	1334.12/1251.82–1416.41	28.25	-	0.66	0.005
Basic density ($g \cdot cm^{-3}$)	0.366/0.358–0.373	11.08	0.54	<0.001	0.69
Transverse radiation ratio in standing trees ($10^{-3} \cdot m^4/kg \cdot s$)	3047.17/2773.37–3320.98	20.78	0.99	<0.001	<0.001
Transverse acoustic impedance in standing trees ($10^3 \cdot N \cdot s/m^3$)	437.85/406.05–469.65	16.80		<0.001	0.75

* 0.05 is the threshold value for statistical significance; ** moisture content (MC) of 39% in heartwood and 98% in sapwood; *** moisture content of 10%.

Among the measured bark features, only colour offers clues about the acoustic properties of the wood. The basic density of the wood (without extractable) cannot be explained by any of the measured bark features. The extractable content in benzene-alcohol of the measured spruce cores is due to the resin (average value of 5.26% of dry weight).

The transverse sound velocity is explained by the redness of the bark: the trees which carry sound better have redder scales on the outside and less red on the inside (Table 7). Trees with a more yellow bark on the scale outside have less sound flowing wood—and therefore a lower capacity for sonic energy (Table 7).

Table 7. Bark as predictor of the acoustic properties of spruce wood (only regressions with $R > 0.5$ are shown).

Dependent Variable	Predictor (s)	Statistical Model	Adjusted R^2	F	p
Transverse sound velocity	Bark redness on the back of the north-facing scales ($a^*_{N_{in}}$), bark redness on the outside of the south-facing scales ($a^*_{S_{ext}}$)	$y = 847.54 + 30.14 \times a^*_{N_{in}} - 57.88 \times a^*_{S_{ext}}$	0.416	5.68	<0.001
Specific acoustic resistance of wood	Bark yellowness on the outside of the south-facing scales ($b^*_{S_{ext}}$), bark redness on the outside of the south-facing scales ($a^*_{S_{ext}}$)	$y = 279.85 + 25.18 \times b^*_{S_{ext}} - 36.57 \times a^*_{S_{ext}}$	0.472	5.91	0.002
Radiation ratio	Bark yellowness on the outside of the south-facing scales ($b^*_{S_{ext}}$)	$y = 3953.14 - 130.24 \times b^*_{Sext}$	0.284	5.37	0.04

The assigning of bark scale colour to wood acoustic impedance can be observed in Figure 7 and helps to conclude that trees with better acoustic emission have redder and less yellow scales.

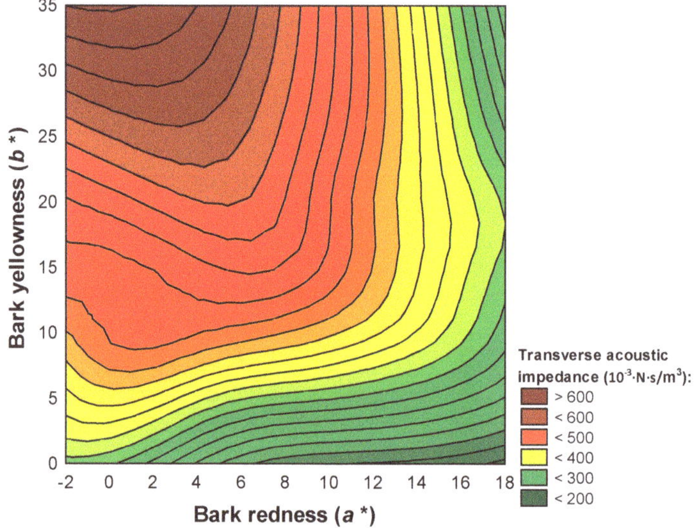

Figure 7. 3D contour plot of acoustic impedance against bark scale redness and yellowness.

4. Discussion

Joining the morphological features of the trees to the connection between wood structure-acoustic properties simplifies the identification of the trees that supply material for manufacturing musical instruments—offering luthiers expeditious criteria. These criteria should not replace the acoustic testing of the rough material, which is done by most luthiers only at the rudimentary level of hitting the trunk and examining the emitted sounds [42,71], or safer still, after the flitches are dry when the velocity of the longitudinal sound propagation is checked [4].

The relevance of the morphological bark and crown features in relation to the acoustic properties of the wood [8] lies in the contribution to the elastic-mechanical wood features of: crown metrics [72], stem knottiness [73], the repeated sway of the trees, and the factors that generate compression wood [51,74] and fibre twisting [12].

The "bark features-wood acoustics" relationship appears to be obscure. To provide an explanation, it is assumed that this relation is still manifested through the wood structure—a consequence of a common genetic control—similar to the one in spruce which guides the blooming of trees and the growth cessation simultaneously [75–77]. The comparison is not forced, since the phenology of the buds and the wood structure have common heredity [78], and resonance spruce is a late bud phenotype [5], with weak growth [59] and preference for the green colour of female strobili [6]. In spruce plantations outside the natural range, the stability of the bark features was noticeable, which explained the predominant contribution of heredity in their expression [22]. The genetic control of spruce bark features manifests through polygenes—in particular, at least two to three pairs of non-allelic genes for bark colour [79]. The weak correlation of bark colour with scale shape [79], confirmed by our analysis, suggests the control of these bark features through neighbour genes on the chromosome, with linked transmission [79].

The shoot colour in spruce has topoclinal variation [80] in which the lowland populations exhibit brownish bark, and the highland populations have grey-coloured bark [81,82]. Spruce trees from multisite comparative trials in Romania showed that the specimens with dark early rough bark originate from altitudes lower than 1000 m and are traditional populations containing resonance wood [83]. The phenotype of resonance spruce bark assigns the darker hues of scales (Figure 6)—namely more redness and less yellowness (Table 7, Figure 6). From our data, the results showed that even for a narrow altitude range (1215–1580 m) the yellowness of the bark increases with altitude while the redness decreases. Under these circumstances, the distribution of resonance trees becomes limited at medium and lower altitudes within the spruce range. In the Romanian Carpathians, resonance spruce was found at altitudes ranging between 700 and 1000 m [84], in the Metaliferi Mountains between 650 and 900 m [21], in Jura Mountains between 800 and 1000 m [53], and in the Alps between 1500 and 1900 m [10,21]. In fact, the distribution of resonance spruce is conditioned by the site ability to ensure stable soil moisture and a balanced nutrition, and to protect the trees from climatic extremes that could acoustically harm the wood structure [7,84,85]—conditions which indeed are not satisfied at high altitudes [85].

The bark redness explains the assigning of resonance wood to the *Europaea* variety (Teplouchoff) Schrotter, whose bark is brown-reddish [4]. The leaf and soil analyses conducted in natural spruce stands revealed a significantly smaller amount of nitrogen, phosphorus, and magnesium in the needles of brown-barked spruce trees as opposed to grey-barked spruce trees [86], which indicates a high efficiency in metabolizing these elements and explains the modest wood bioaccumulations in resonance wood [87].

The shape of the spruce bark scales is a feature with ecoclinal variation, conditioned by the quantity of light radiation responsible for the earliness of the rough bark [53]. Thus, differentiating the phenotypes according to the bark relief can be a result of site elevation, as well as the trees' social status and spacing. In the case of Scots pine, the association of the bark relief (longitudinally versus panel-like cracked) with the crown shape (conical and paraboidal, respectively) [88] was found. In general, resonance trees are dominant and less pressured for competition at maturity [4,9]. The resonance spruce trees' position in canopy is not the result of favouring the trees through growth, but the result of maintaining it tenaciously as the trees can reach considerable age, as shown in Table 1 [5]. In the case of poplar, it was found that MOE becomes bigger as the trees grow taller [89], so by extrapolating from spruce it can be assumed that a dominant position brings additional contribution to the elastic superiority of resonance wood. Our results show that scale shape does not appear to be relevant in relation to the acoustic properties of the wood—however, it is an indicator of its structural quality (Table 4, Figure 6). The most important clues relate to latewood regularity (Figure 5)—which is in fact a selection criterion for the rough material for violins [90]. Moreover, the latewood content has a diagnostic value superior to the growth ring width in relation to its acoustic use [91]. The regularity of the anatomical structures is a key feature of the suitability for musical instruments [13,92].

The structural features of trees are priority when choosing logs for musical instruments—even if they are occasionally viewed reluctantly [11,14,59]. Wood structure on a macroscale is a marker of acoustic features by means of wood density [90] and its direct influence on MOE [65]. In general, resonance spruce is a xylotype with a lower density than common spruce [39]. The sizes of the wood density in our sample (Table 6) are inferior to the data in the literature [14]. One explanation for this is the removal of extractables from wood and the calculation specific to the basic density (minimal weight per maximum volume). Additionally, the resonance wood in Romania is lighter than in other geographic regions from central and Eastern Europe [11], and, in the classification of spruce seed sources in Romania, the local spruce population (Gurghiu) is located in the last size class—depending on the basic density and latewood ratio [93]. In our sample, the wood density variations are independent from the variations of the bark features, as well as from those of the wood structure. In some exotics, the wood density is not relevant for the wave velocity [94].

The other acoustic properties of wood are influenced by the moisture of the material. As far as sound velocity is concerned, a ratio of 0.85 resulted between green wood and dry wood. In the case of Scots pine, this ratio was 0.92 for the tangential ultrasound velocity and 0.82 for the radial ultrasound velocity, at moistures similar to the material examined herein [95]. Thus, the depreciatory influence the moisture content has on the acoustic properties of the wood is confirmed [96]. These values refer to the transverse sound propagation, which are, in any case, relevant to the longitudinal one [49] as the wood is a high anisotropic material [97]. Additionally, there are opinions according to which the cross grain elastic properties are more important than those along the grain in the acoustic behaviour of wood [34]. By characterizing green wood (Table 6), the average found values of the transverse sound velocity and radiation ratio are similar to the data in the literature applied to European spruce selected for musical instruments, which refer to dry wood [42]. Still, the values of the transverse sound velocity we discovered in the discs—harvested from trees used later in manufacturing violins—are circa 600 $m \cdot s^{-1}$ lower than the values from the excellence class of spruce tonal wood [36]. The radiation ratio was thought to be the most important criterion for diagnosing the suitability of wood for soundboards [36].

The link of the bark features to the acoustic properties of wood changes from the northern to the southern side of the trunk, respectively (Table 7)—most likely because the acoustic properties themselves vary around stem circumference [98]. The closest relations are found on the southern side (Table 7), which explains some luthiers' preference for the sunny side of the tree [34].

The stability of the bark features inside the sample plot (Table 3) presupposes the fact that the phenotype of resonance trees, which are few [84], is in fact a mark of the population it is part of—which in turn is an elite population according to other features—such as the pruning of the trees, the branchiness and the rarity of the wood faults [99].

Out of precaution, we recommend that the diagnosis of resonance wood be given only after the examination and comparison of all the phenotypical features that individualize it, and, where possible, the acoustic testing of the wood, regardless whether the bark is a highly trusted marker of trees with acoustic properties. Consequently, recognizing and classifying resonance wood should be a multi-criteria analysis of the trees' phenotype [39].

5. Conclusions

In the case of spruce stands, which supply material for manufacturing violins, there is an important amount of phenotypical variability of the tree bark features between sites. The scale colour and length show topoclinal variations to a certain extent related to the altitude. From the variation sources analyzed, the location of the sample plot around circumference has the greatest impact on the morphometry and colour of the bark. The differences in bark among trees are to a small degree due to age, and are consistent only in relation to the sides of the trunk, the south-facing side being the most relevant for diagnosing resonance wood. On the north-facing side of the trunk, the differences in scale colour from plot to plot are insignificant. On the south-facing side of the trunk, the influence of the tree age is not manifested, which increases its value when characterizing the bark phenotype for spruce. The most significant differences in colour are recorded between the external and internal sides of the bark scales—which are stable with age. The differences between phenotypes regarding bark colour are significant enough to be visible to the naked eye.

The sizes of the bark are not enlightening with regard to the annual ring structure specific to resonance wood, but their ratio is a valuable marker of the regularity of this structure—especially of the latewood width regularity. The relation shows that the trees with elongated bark scales have better ring regularity around circumference. The bark colour provides valuable clues for the identification of structures with acoustic value. The resonance structural trees have less brightness and yellowness, but more redness in the bark. The differences between the wood structure qualities are recorded on the external side of the north-facing scales and on the internal side of the south-facing scales.

The size of the bark scale is not relevant with regard to the acoustic properties of the wood, but, on the other hand, the scale colour is an important marker of the acoustic quality of the material. The bark redness and yellowness on the external side of the south-facing scales have a high diagnostic value in relation to the acoustic performance of the material expressed by the sound velocity, acoustic impedance, and radiation ratio—all in transversal direction. The bark does not offer any clues regarding wood density. The phenotype of trees with acoustic properties is signaled by brown bark, high redness, and low yellowness on the external side of the scales—especially on the south-facing side of the trunk.

At least by analyzing scale colour and slenderness, the bark can be used as a criterion for diagnosing standing resonance wood.

Author Contributions: Conceptualization, F.D. and C.T.A.; methodology, F.D.; software, F.D.; validation, F.D. and M.D.S.; investigation, C.T.A.; data curation, F.D. and M.M.V.; writing—original draft preparation, F.D.; writing—review and editing, M.M.V.; visualization, M.D.S.; supervision, F.D.

Funding: The authors received no specific funding for this work.

Acknowledgments: The authors thank Petru Maran who assisted us in the map analysis, and our colleague Aureliu Florin Hălălișan, who helped us with the bark scale analysis. We are grateful as well to Alexandra Stan, for her contribution in English editing.

Conflicts of Interest: The authors declare no conflict of interest.

References

1. Şofletea, N.; Curtu, L. *Dendrology*; Pentru Viaţă Publishing House: Braşov, Romania, 2007; pp. 40–53.
2. Surmiński, J. Wood properties and uses. In *Biology and Ecology of Norway Spruce*; Tjoelker, M.G., Boratyński, A., Bugała, W., Eds.; Springer: Dordrecht, The Netherlands, 2007; pp. 333–342.
3. Ege, K. The Piano Soundboard—Modal Studies in the Low- and Mid-Frequency Ranges. Ph.D. Thesis, École Polytechique, Palaiseau, France, 2009.
4. Albu, C.T. Research on the Characteristics of Resonance Spruce Wood from Gurghiu River Basin in Conjunction with the Requirements of the Musical Instruments Industry. Ph.D. Thesis, Transilvania University of Brașov, Brașov, Romania, 2010.
5. Ștefănescu, P. A Norway spruce forest site in Gurghiu Mountains containing resonance wood, located in the range of Sovata Forest District. *Rev. Pădurilor* **1961**, *76*, 85–92.
6. Ștefănescu, P. Contributions to the knowledge of resonance Norway spruce from Gurghiu Mountains. *Rev. Pădurilor* **1964**, *79*, 511–517.
7. Domont, P. Spruce Resonance Wood: A Jewel of Mountain Forests. Available online: https://www.waldwissen.net/waldwirtschaft/holz/verarbeitung/wsl_klangholz/index_FR (accessed on 10 July 2019).
8. Fedyukov, V.I.; Saldaeva, E.Y.; Chernova, M.S.; Chernov, V.Y. Biomorphology of Spruce Trees as a Diagnostic Attribute for Non-Destructive Selection of Resonant Wood in a Forest. *South-East Eur. For.* **2018**, *9*, 147–153. [CrossRef]
9. Grapini, V.; Constantinescu, N. *Resonance Spruce*; Forest Research Institute: Bucharest, Romania, 1968; p. 19.

10. Zugliani, G.; Dotta, L. Resonance wood. Technical features and ecological conditions. *Sherwood* **2009**, *154*, 7–13.
11. Krzysik, F. Problems of resonance wood sorting and use. *Holzindustrie* **1968**, *1*, 3–7.
12. Hutchins, C.M. Wood for violins. *Catgut Acoust. Soc. Newsl.* **1978**, *29*, 14–18.
13. Rocaboy, F.; Bucur, V. About the physical properties of wood of twentieth century violins. *J. Acoust. Soc. Am.* **1990**, *1*, 21–28.
14. Bucur, V. *Acoustics of Wood*, 2nd ed.; Springer: Berlin, Germany, 2006; pp. 173–196.
15. Zugliani, G.; Dotta, L. Resonance wood. Management, selection and manufacture in the state forest of Paneveggio. *Sherwood* **2009**, *155*, 14–18.
16. Buksnowitz, C.; Teischinger, A.; Müller, U.; Pahler, A.; Evans, R. Resonance wood [*Picea abies* (L.) Karst.]: Evaluation and prediction of violin makers' quality-grading. *J. Acoust. Soc. Am.* **2007**, *121*, 2384–2395. [CrossRef]
17. Darnton, M. Violin Making. Available online: http://darntonviolins.com (accessed on 12 May 2009).
18. Pașcovici, N. Spruce as resonance and keyboard wood. I Resonance spruce in forests. *Rev. Pădurilor* **1930**, *46*, 85–99.
19. Nocetti, M.; Romagnoli, M. Seasonal cambial activity of spruce (*Picea abies* Karst.) with indented rings in the Paneveggio Forest (Trento, Italy). *Acta Biol. Crac. Ser. Bot.* **2008**, *50*, 27–34.
20. Bonamini, G.; Uziellli, L. A simple non-destructive method to recognize resonance spruce. *Monti e Boschi* **1998**, *6*, 50–53.
21. Domont, P. Development of Resonance Wood in Switzerland, Rep. no. 5.162. 2000. Available online: http://www.waldwissen.net/waldwirtschaft/holz/verarbeitung/wsl_klangholz/wsl_klangholz_rapport_final.pdf (accessed on 29 February 2016).
22. Popescu, G.A. *Conifers in the Hills of the Eastern and Curvature Subcarpathians*; Ceres Publishing House: Bucharest, Romania, 1984; pp. 122–126, 192–200.
23. Arco Molina, J.G.; Hadad, M.A.; Domínguez, D.P. Tree age and bark thickness as traits linked to frost ring probability on *Araucaria araucana* trees in northern Patagonia. *Dendrochronologia* **2016**, *37*, 116–125. [CrossRef]
24. Otegbeye, G.O.; Kellison, R.C. Genetics of wood and bark characteristics of *Eucalyptus viminalis*. *Silvae Genet.* **1980**, *29*, 27–31.
25. Bdeir, R.; Muchero, W.; Yordanov, Y.; Tuskan, G.A.; Busov, V.; Gailing, O. Quantitative trait locus mapping of Populus bark features and stem diameter. *BMC Plant Biol.* **2017**, *17*, 224. [CrossRef] [PubMed]
26. Kalbande, R.B. Digital visual bark: An image based tool for plant diversity research, diversity in bark supports in tree identification. *Int. J. Life Sci. Pharma Res.* **2014**, *4*, 17–24.
27. Boudra, S.; Yahiaoui, I.; Behloul, A. A Comparison of Multi-scale Local Binary Pattern Variants for Bark Image Retrieval. In *Advanced Concepts for Intelligent Vision Systems*; Battiato, S., Blanc-Talon, J., Gallo, G., Philips, W., Popescu, D., Scheunders, P., Eds.; Springer: Cham, Switzerland, 2015; Volume 9386, pp. 764–775.
28. Nicolai, V. The bark of trees: Thermal properties, microclima and fauna. *Oecologia* **1986**, *69*, 148–160. [CrossRef]
29. Rossel, J.A.; Castorena, M.; Laws, C.A.; Westoby, M. Bark ecology of twigs vs. main stems: Functional traits across eighty-five species angiosperms. *Oecologia* **2015**, *178*, 1033–1043. [CrossRef]
30. Wang, X.H.; Ma, X.H.; Jin, G.Q.; Chen, L.Y.; Zhou, Z.C. Variation pattern of individual types and wood characters in natural stands of *Schima superba*. *Sci. Silvae Sin.* **2011**, *47*, 133–139.
31. Mihai, D. Research on Fir Variability in Romania with a View to Expand Valuable Specimens to Plantations. Ph.D. Thesis, University of Brașov, Brașov, Romania, 1978.
32. Ogata, A.; Itai, A.; Nishiyama, M.; Ikeda, H.; Kanahama, K. Analyses of early rough bark phenotype found in seedlings of a cross between the European pear 'Bartlett' and the Chinese pear 'Yali'. *Sci. Hortic.* **2012**, *148*, 1–8. [CrossRef]
33. Carlisle, A. A guide to the named variants of Scots pine (*Pinus silvestris*, Linnaeus). *Forestry* **1958**, *31*, 203–224. [CrossRef]
34. Schelleng, J.C. Wood for violins. *Catgut Acoust. Soc. Newsl.* **1982**, *37*, 8–19.
35. Ono, T.; Norimoto, M. On physical criteria for the selection of wood for soundboards of musical instruments. *Rheol. Acta* **1984**, *23*, 652–656. [CrossRef]
36. Meyer, H.G. A practical approach to the choice of tone wood for the instruments of the violin family. *Catgut Acoust. Soc. J.* **1995**, *2*, 9–13.

37. Brémaud, I. *What Do We Know on "Resonance Wood" Properties? Selective Review and Ongoing Research*; Societé Française d'Acoustique: Nantes, France, 2012; pp. 2760–2764.
38. Wegst, U.G.K. Wood for sound. *Am. J. Bot.* **2006**, *93*, 1439–1448. [CrossRef] [PubMed]
39. Brémaud, I. Acoustical properties of wood in string instruments soundboards and tuned idiophones: Biological and cultural diversity. *J. Acoust. Soc. Am.* **2012**, *131*, 807–818. [CrossRef]
40. Woodhouse, J. The acoustics of the violin: A review. *Rep. Prog. Phys.* **2014**, *77*, 115901. [CrossRef]
41. Mania, P.; Fabisiak, E.; Skrodzka, E. Investigation of modal behaviour of resonance spruce wood samples (*Picea abies* L.). *Arch. Acoust.* **2017**, *42*, 23–28. [CrossRef]
42. Haines, D.W. On musical instruments wood. *Catgut Acoust. Soc. Newsl.* **1979**, *31*, 23–32.
43. Bucur, V. Towards an objective assessment of violin wood properties. *Rev. For. Fr.* **1983**, *32*, 130–137. [CrossRef]
44. Ouis, D. Vibrational and acoustical experiments on logs of spruce. *Wood Sci. Technol.* **1999**, *33*, 151–184. [CrossRef]
45. Gilbert, E.A.; Smiley, E.T. Picus Sonic tomography for the quantification of decay in white oak (Quercus Alba) and Hickory (*Carya* spp.). *J. Arboric.* **2004**, *30*, 277–281.
46. Rinn, F. Central Defects in Sonic Tomography. 2015. Available online: http://download.rinntech.com/RINN_CentralDefectsInSonicTreeTomography_WesternArborist_Spring_2015.pdf (accessed on 20 June 2019).
47. Gilbert, G.S.; Ballesteros, J.O.; Barrios-Rodriguez, C.A.; Bonadies, E.F.; Cedeno-Sanchez, M.L.; Fossatti-Caballero, N.J.; Trejos-Rodriguez, M.M.; Perez-Suniga, J.M.; Holub-Young, K.S.; Henn, L.A.W.; et al. Use of Sonic Tomography to Detect and Quantify Wood Decay in Living Trees. *Appl. Plant Sci.* **2016**, *4*, 1600060. [CrossRef] [PubMed]
48. Lindström, H.; Harris, P.; Nakada, R. Methods for measuring stiffness of young trees. *Holz Als Roh- Werkstoff* **2002**, *60*, 165–174. [CrossRef]
49. Wang, X.; Carter, P.; Ross, R.J.; Brashaw, B.K. Acoustic assessment of wood quality of raw forest materials—A path to increased profitability. *For. Prod. J.* **2007**, *57*, 6–14.
50. Caniato, M.; Favretto, S.; Bettarello, F.; Schmid, C. Acoustic characterization of resonance wood. *Acta Acust. United Acust.* **2018**, *104*, 1030–1040. [CrossRef]
51. Brémaud, I.; Ruelle, J.; Thibaut, A.; Thibaut, B. Changes in viscoelastic vibrational properties between compression and normal wood: Role of microfibril angle and of lignin. *Holzforschung* **2013**, *67*, 75–85. [CrossRef]
52. Food and Agriculture Organization of the United Nations. *World Reference Base for Soils Resources*; World Soil Research Report No. 84; FAO: Rome, Italy, 1998.
53. Bouvarel, P. Variability of Norway spruce (*Picea excelsa* Link.) in the French Jura Mountains. Distribution and characters of different types. *Rev. For. Fr.* **1954**, *6*, 85–98. [CrossRef]
54. Ohaus Corporation. *MB45 Moisture Analyzer. Instruction Manual*; Ohaus Co.: Parsippany, NJ, USA, 2001; p. 71.
55. Hunt, R.W.G. *Measuring Colour*, 3rd ed.; Fountain Press: Kingston-upon-Thames, UK, 1998; p. 153.
56. Konica-Minolta. *Chroma Meter CR-400/410. Instruction Manual*; Konica-Minolta Sensing Inc.: Osaka, Japan, 2007; p. 156.
57. Dirckx, O.; Triboulot-Trouy, M.C.; Merlin, A.; Deglise, X. Changes of wood colour of *Abies grandis* exposed to solar radiation. *Ann. For. Sci.* **1992**, *49*, 425–447. [CrossRef]
58. WinDENDRO. *WinDENDRO^{TM}2006 for Tree-Ring Analysis. Manual of Exploitation*; Régent Instruments Inc.: Québec City, QC, Canada, 2007; p. 133.
59. Dinulică, F.; Albu, C.T.; Borz, S.A.; Vasilescu, M.M.; Petritan, I.C. Specific structural indexes for resonance Norway spruce wood used for violin manufacturing. *Bioresources* **2015**, *10*, 7525–7543. [CrossRef]
60. Rinn, F. *ARBOTOM®: 3-D Tree Impulse Tomograph. User Manual*; Rinntech: Heidelberg, Germany, 2012; p. 57.
61. Shen, J. Relationships between longitudinal and radial Picea genera sound vibration parameters. *Front. For. Chin.* **2006**, *4*, 431–437. [CrossRef]
62. Keylwerth, R. A contribution to qualitative increment analysis. *Eur. J. Wood Wood Prod.* **1954**, *12*, 77–83. [CrossRef]
63. Keith, C.T. Resin content of red pine wood and its effect on specific gravity determinations. *For. Chron.* **1969**, *45*, 338–343. [CrossRef]

64. Zaițev, E.V. The acoustic properties of cedar wood. *Деревообрабатывающая* промышленность **1969**, *7*, 18–19.
65. Brémaud, I. Diversity of Woods Used or Usable in Musical Instruments Making. Ph.D. Thesis, Montpellier II University, Montpellier, France, 2006.
66. Polge, H. The quality of wood of the principal exotic conifers used in French plantations. *Ann. Eaux For.* **1963**, *20*, 403–469.
67. Williamson, G.B.; Wiemann, M.C. Measuring wood specific gravity ... Correctly. *Am. J. Bot.* **2010**, *97*, 519–524. [CrossRef]
68. Smith, D.M. *Maximum Moisture Content Method for Determining Specific Gravity of Small Wood Samples*; Rep. no. 2014. USDA; Forest Service: Madison, WI, USA, 1954; p. 9. Available online: http://hdl.handle.net/1957/2445 (accessed on 10 February 2019).
69. StatSoft Inc. *STATISTICA (Data Analysis Software System)*, version 8.0; StatSoft Inc.: Tulsa, OK, USA, 2007.
70. Zar, J.H. *Biostatistical Analysis*, 5th ed.; Pearson Prentice-Hall: Upper Saddle River, NJ, USA, 2010.
71. Beldeanu, E.C. *Forest Products and Wood Science*; Transilvania University Publishing House: Brașov, Romania, 1999.
72. Kuprevicius, A.; Auty, D.; Achim, A.; Caspersen, J.P. Quantifying the influence of live crown ratio on the mechanical properties of clear wood. *Forestry* **2013**, *86*, 361–369. [CrossRef]
73. Lundström, T.; Heiz, U.; Stoffel, M.; Stöckli, V. Fresh-wood bending: Linking the mechanical and growth properties of a Norway spruce stem. *Tree Physiol.* **2007**, *27*, 1229–1241. [CrossRef]
74. Norimoto, M.; Ono, T.; Watanabe, Y. Selection of wood used for piano soundboards. *J. Soc. Rheol. Jpn.* **1984**, *9*, 115–119. [CrossRef]
75. Gyllenstrand, N.; Clapham, D.; Källman, T.; Lagercrantz, U. A Norway spruce FLOWERING LOCUS T Homolog is implicated in control of growth rhytm in conifers. *Plant Physiol.* **2007**, *144*, 248–257. [CrossRef]
76. Lagercrantz, U. At the end of the day: A common molecular mechanism for photoperiod responses in plants? *J. Exp. Bot.* **2009**, *60*, 2501–2515. [CrossRef] [PubMed]
77. Karlgren, A.; Gyllenstrand, N.; Clapham, D.; Lagercrantz, U. FLOWERING LOCUS T/TERMINAL FLOWER1-LIKE genes affect growth rhytm and bud set in Norway spruce. *Plant Physiol.* **2013**, *163*, 792–803. [CrossRef] [PubMed]
78. Vaganov, E.A.; Hughes, M.K.; Shashkin, A.V. *Growth Dynamic of Conifer Tree Rings. Images of Past and Future Environments*; Springer: Berlin/Heidelberg, Germany, 2006; p. 354.
79. Stănescu, V.; Șofletea, N.; Popescu, O. Phetotypical circumstances concerning the Norway spruce genome. *Rev. Silvic. Cineg.* **1997**, *2*, 3–5.
80. Stănescu, V.; Şofletea, N. Ecological genetic study in mountain spruce forests. *Rev. Pădurilor* **1990**, *105*, 114–119.
81. Przybylski, T. Morphology. In *Biology and Ecology of Norway Spruce*; Tjoelker, M.G., Boratyński, A., Bugała, W., Eds.; Springer: Dordrecht, The Netherlands, 2007; pp. 9–14.
82. Godet, J.D. *Guide of Barks of European Trees. Identify and Compare the Species*; Delachaux et Niestlé: Paris, France, 2012; pp. 36–37.
83. Niţu, C.; Răiescu, V.; Creangă, I. *Research Concerning the Behaviour of Norway Spruce Provenances Tested in Different Site Conditions*; Forest Research and Management Institute: Bucharest, Romania, 1984; p. 36.
84. Geambașu, N. *Research on the Management of Norway Spruce Resonance and Keyboard Stands*; Tehnică Silvică Publishing House: Bucharest, Romania, 1995; pp. 64–77.
85. Pașcovici, N. The spruce as resonance and keyboard wood. II The forest site conditions for the resonance spruce. *Rev. Pădurilor* **1930**, *46*, 279–305.
86. Bolea, V.V.; Popescu, N.E.; Surdu, A.; Mandai, M. Ecological significances of nutrition biodiversity in Norway spruce. 1 Chemical composition of needles in relation to soil, branching, bark colour and health condition of the tree. *Rev. Silvic. Cineg.* **1996**, *1*, 9–17.
87. Hejnowicz, A. Anatomy, embryology, and karyology. Bud structure and shoot development. In *Biology and Ecology of Norway Spruce*; Tjoelker, M.G., Boratyński, A., Bugała, W., Eds.; Springer: Dordrecht, The Netherlands, 2007; pp. 49–342.
88. Józefaciukowa, W.; Ubysz-Borucka, L. Variation in habit forms of Scots pine (*Pinus silvestris* L.) on the area of Poland. *Silvae Genet.* **1972**, *21*, 9–17.

89. Yin, Y.; Jiang, X.; Wang, L.; Bian, M. Predicting wood quality of green logs by resonance vibration and stress wave in plantation-grown *Populus × euramericana*. *For. Prod. J.* **2011**, *61*, 136–142. [CrossRef]
90. Ghelmeziu, N.; Beldie, I.P. On the characteristics of resonance spruce wood. *Bull. Transilvania Univ. Brașov* **1970**, *12*, 315–326.
91. Spycher, M.; Schwarze, F.W.M.R.; Steiger, R. Assesment of resonance wood quality by comparing its physical and histological properties. *Wood Sci. Technol.* **2008**, *42*, 325–342. [CrossRef]
92. Brancheriau, L.; Baillères, H.; Détienne, P.; Gril, J.; Kronland, R.; Metzger, B. Classifying xylophone bar materials by perceptual, signal processing and wood anatomy analysis. *Ann. For. Sci.* **2006**, *63*, 73–81. [CrossRef]
93. Budeanu, M. Testing the Genetic Value of Some Norway Spruce Seed Sources in Multisite Comparative Trials. Ph.D. Thesis, Transilvania University of Braşov, Brașov, Romania, 2012.
94. Baar, J.; Tippner, J.; Gryc, V. The influence of wood density on longitudinal wave velocity determined by the ultrasound method in comparison to the resonance longitudinal method. *Eur. J. Wood Prod.* **2012**, *70*, 767–769. [CrossRef]
95. Vorobicov, A.M.; Evdokimova, N.I.; Okolykhina, L.V. The influence of density and moisture on the velocity of ultrasonic oscillations transmission through pine wood. *Lesnoy Zhurnal* **1968**, *135*, 106–109.
96. Macareva, T.A. Factors affecting the constant acoustic of resonance wood. Деревообрабатывающая промышленность **1968**, *10*, 14–15.
97. Bucur, V.; Lanceleur, P.; Roge, B. Acoustic properties of wood in tridimensional representation of slowness surfaces. *Ultrasonics* **2002**, *40*, 537–541. [CrossRef]
98. Lachenbruch, B.; Jonson, G.R.; Downes, G.M.; Evans, R. Relationship of density, microfibril angle and sound velocity with stiffness and strength in mature wood of Douglas-fir. *Can. J. For. Res.* **2010**, *40*, 55–64. [CrossRef]
99. Dinulică, F.; Albu, C.T.; Zdrob, G.S. What and how much do we know about the determinism of resonance Norway spruce? *Rev. Pădurilor* **2015**, *130*, 23–40.

Article

Assessment of Genetic Diversity of Tea Germplasm for Its Management and Sustainable Use in Korea Genebank

Kyung Jun Lee, Jung-Ro Lee, Raveendar Sebastin, Myoung-Jae Shin, Seong-Hoon Kim, Gyu-Taek Cho and Do Yoon Hyun *

National Agrobiodiversity Center, National Institute of Agricultural Sciences (NAS), RDA, Jeonju 54874, Korea
* Correspondence: dyhyun@korea.kr; Tel.: +82-63-238-4912

Received: 24 July 2019; Accepted: 4 September 2019; Published: 8 September 2019

Abstract: Tea (*Camellia sinensis* (L.) O. Kuntze) is cultivated in many developing Asian, African, and South American countries, and is the most widely consumed beverage in the world. It is of critical importance to understand the genetic diversity and population structure of tea germplasm for effective collection, conservation, and utilization. In this study, 410 tea accessions collected from South Korea were analyzed using 21 simple sequence repeat (SSR) markers. Among 410 tea accessions, 85.4% (350 accessions) were collected from Jeollanam-do. A total of 286 alleles were observed, and the genetic diversity and evenness were estimated to be on average 0.79 and 0.61, respectively, across all the tested samples. Using discriminant analysis of principal components, four clusters were detected in 410 tea accessions. Among them, cluster 1 showed a higher frequency of rare alleles (less than 1%). Using the calculation of the index of association and rbaD value, each cluster showed a clonal mode of reproduction. The result of analysis of molecular variance (AMOVA) showed that most of the variation observed was within populations (99%) rather than among populations (1%). The present study revealed the presence of lower diversity and simpler population structure in Korean tea germplasms. Consequently, more attention should be focused on collecting and conserving the new tea individuals to broaden genetic variation of new cultivars in future breeding of the tea plant.

Keywords: *Camellia sinensis*; genetic diversity; population structure; SSR

1. Introduction

Tea (*Camellia sinensis* (L.) O. Kuntze, $2n = 2x = 30$) is one of the most popular non-alcoholic beverages worldwide, and is consumed by approximately 70% of the world's population for its refreshing taste, attractive aroma, therapeutic uses, and mildly stimulating properties [1]. It is an economically important tree crop, grown in over 52 countries in Asia, Africa, and South America [2,3]. The tea is a woody ever-green perennial plant and recorded to be native to Yunnan and Sichuan provinces in China and the northern part of Myanmar [4]. In Korea, although tea was introduced from China as early as the seventh century, the development of the tea industry was slow, and production was small [5].

The importance of using genetic resources in breeding programs to enhance crop genetic potential has been well recognized [6]. Many germplasm appraisal methods, such as morphology, biochemistry, molecular markers, and sensory evaluation, have been used to evaluate the resources of tea germplasm [7–9]. The phenotype can be referred to as a good standard for the evaluation of tea germplasm, because this method is simply based on the morphological traits to analyze the genetic diversity assessment [10]. Recently, the technology of using molecular markers has been proven to be one of the most effective methods for identifying different tea varieties [2,7,11–14].

Tea is an out-crossing species, and selected elite genotypes are propagated vegetatively and released as clonal varieties [13,15,16]. Clonal identification is traditionally based on morphological descriptors such as plant shape, stem width, leaf shape, young leaf type, and fruit shape [15,17]. However, as in many out-crossing crops, tea is highly heterozygous with most of its morphological, physiological, and biochemical descriptors showing continuous variation and high plasticity [18,19]. Korir et al. reported that morphological traits are associated with drawbacks such as the influences of environment on trait expressions, epistatic interactions, and pleiotropic effects among others despite the value of their advantages [17]. On the contrary, molecular markers are used as they are least affected by environmental factors and indefinite presence. In addition, they offer a possibility to observe the genome directly and thus eliminate the shortcomings inherent in a phenotype observation [15]. In previous studies, genetic diversity, discrimination and differentiation of tea germplasms have been assessed using different DNA markers such as restriction fragment length polymorphism (RFLP), randomly amplified polymorphic DNA (RAPD), amplified fragment length polymorphism (AFLP), inter-simple sequence repeat (ISSR), and simple sequence repeat (SSR) [3,5,7,11–15,20–22].

In Korea, the national research institutes collected and conserved excellent tea individuals and investigated their morphological characteristics [23]. Also, some studies analyzed the genetic diversity of Korean tea germplasm using RFLP and RAPD [5,14,16]. However, the analysis of genetic diversity in Korean tea germplasm is not sufficient as it included a very small population (approximately 20–50 individuals). In the present study, 21 SSR primer pairs were used to analyze 410 tea accessions from Korea, and the aim was: (1) to evaluate the genetic diversity and population structure of Korean tea accessions and (2) to estimate the genetic differentiation and variation source among inferred populations. It is hypothesized that the results of the present study would be helpful to gain a deeper understanding on the genetic diversity, population structure, and differentiation of tea germplasm to guide effective collection, conservation, and application of tea genetic resources in Korea.

2. Materials and Methods

2.1. Plant Materials

A total of 410 tea accessions were obtained from the National Agrobiodiversity Center (NAC) at the Rural Development Administration in South Korea. Among 410 tea accessions, 400 accessions were collected from 31 cities at three provinces in South Korea, while ten accessions lacked the data of collecting area (Table S1).

2.2. DNA Extraction

Genomic DNA was extracted from the tea leaves using a Qiagen DNA extraction kit (Qiagen, Hilden, Germany). DNA quality and quantity were measured using 1% (w/v) agarose gel and spectrophotometrically (Epoch, BioTek, Winooski, VT, USA). Extracted DNA was diluted to 30 ng/μL and stored at −20 °C until further PCR amplification.

2.3. SSR Genotyping

For SSR analysis, a total of 21 SSRs were fluorescently labeled (6-FAM, HEX and NED) and used for the detection of amplification products (Table 1 and Table S2). PCR reactions were carried out using 25 μL reaction mixture, containing 30 ng template DNA, 1.5 mM $MgCl_2$, 0.2 mM of each dNTPs, 0.5 μm of each primer, and 1 U Taq polymerase (Inclone, Korea). The amplification was performed with the cycling conditions of: initial denaturation at 94 °C for 5 min, followed by 35 cycles of denaturation at 95 °C for 30 s, annealing at 55 °C for 30 s, extension at 72 °C for 1 min, and a final extension step at 72 °C for 10 min. Each amplicon was resolved on ABI prism 3500 DNA sequence (ABI3500, Thermo Fisher Scientific Inc., Wilmington, DE, USA) and scored using Gene Mapper Software (Version 4.0, Thermo Fisher Scientific Inc.).

Table 1. List of 21 simple sequence repeat (SSR) primers used in this study.

Locus	Sequence (5′→ 3′)	Dye
MSG0258	F: ACTCATCACCATGCCTTCTCCATC R: GTTTAGCTCAACTGGTGGAACCTCAACT	FAM
MSE0173	F: GTGTTCACCAACAACTCACCAAGG R: TGTCGAACAAAGATACACCCCAAA	FAM
MSE0313	F: TGCTATGCCGCCTAACAAAAACTT R: ACCACCAACAACAATTCCCACTCT	FAM
MSG0403	F: ATGATCGCCGGTTTAGAGATGAAT R: GTTTAAGCTGGCTAACCTACACGGAGC	FAM
MSE0083	F: GAGGAAAGAGATTATGCGGTGTGG R: GTGAGCCTTCAAAAGACAGCAACG	FAM
TM604	F: TCTCAATCAAGGTTCCGAGG R: TGACATATTCGCCCAACAAA	FAM
TM530	F: CCGTGTTTACCACCACCCT R: CCCTGGGAACAAGAAAGTGA	FAM
MSG0361	F: AGATGGAGGTAGAGAGAGAGGCAG R: GTTTGTCCCTCTCATTTTCAACGC	HEX
MSE0029	F: ATAGCCAATCAAGCTCCTCCTCCT R: AGTCTGTTCCTCCCTTGATGATCG	HEX
MSG0610	F: ACAGAGGAGGAAGATGATCGGTAA R: GTTTGAAGAAGAAGAAAACTCCCGCCAT	HEX
MSE0291	F: AATCAAATAACACTTGCACCCGC R: AAAAAGAGAAAGTCACGTCCACGG	HEX
MSG0681	F: AGGGTTTGCGTCTTCAAAGAGAGA R: GTTTGTAACACTTGCCACGTTTCG	HEX
MSG0470	F: ATAGGGTTCGAAAATGGCAGG R: GTTTGAGGTGGCAAGTTTGTGACTGT	HEX
MSG0699	F: ATGCGACAGTGTTGCTGAGATTTT R: GTTTCAAAAATGGGGTGTCTACAGAGGG	HEX
MSG0429	F: AGGACCGTTCTTCCCTACCTGTAA R: GTTTGAGATTGAGGATGTGGTCGTTGT	NED
MSG0380	F: ACAGACCTTCACCCTCTCCATTTC R: GTTTACCTCTGCCTTCGTTCTTCAGC	NED
MSG0423	F: ACTCCATGTGCTGCTCTGTAGTTC R: GTTTGCAGGAAGTTGAGCCAGAC	NED
MSE0237	F: CTCTCCTTCTTCACACCCTCCAAA R: TTGTTCTCAAAGAACCTCCTTCGC	NED
MSE0113	F: TACCTTCTGCAACTCCAGCAATCC R: TGAGATTGACCATCTTTCATCGGA	NED
MSE0107	F: TCTCTCTACTCCTGCGCAATCTCA R: TCAAAGATGTTGCTCTCGTCAACC	NED
TM382	F: TCTCAAAACCAAATAGGCTCAA R: TTGCGTTATGATTTCTGGGA	NED

2.4. Population Structure and Genetic Diversity

The Number of alleles (Na), Shannon index (I), Nei's unbiased gene diversity (GD), and Eveness were calculated using *poppr* package for *R* software [24]. The analysis of molecular variance (AMOVA) and calculation of the coefficient of genetic differentiation among populations (PhiPT) were done using GenAlEx software (6.5 version) with 999 permutations [25].

The population structure was analyzed by a discriminant analysis of principal components (DAPC) using the *adegenet* package for *R* software [26,27]. The *find.clusters* function was used to detect the number of clusters in the population. It uses K-means clustering which decomposes the total variance of a variable into between-group and within-group components. The best number of subpopulations has the lowest associated Bayesian Information Criterion (BIC). A cross-validation function (*Xval. dapc*) was used to confirm the correct number of principal component (PC) to be retained. In this analysis, the data is divided into two sets: a training set (90% of the data) and a validation set (10% of the data) The member of each group is selected by stratified random sampling, which ensures that at least one member of each group or population in the original data is represented in both training and validation sets. DAPC is carried out on the training set by retaining variable numbers of PCs, and the degree to which the analysis is able to accurately predict the group membership of excluded individuals (those in the validation set) is used to identify the optimal number of PCs to be retained. At each level of PC retention, the sampling and DAPC procedures are repeated many times [28]. The best number of PCs that should be retained is associated with the lowest root mean square error. The resultant clusters were plotted in a scatter plot of the first and second linear discriminants of DAPC.

2.5. Estimation of Reproduction Mode among 410 Korean Tea Accessions

Linkage disequilibrium was calculated to test for the evidence of sexual reproduction. Linkage among loci can be caused by clonal reproduction and selection events; and as linkage increases populations fall into linkage disequilibrium, while recombination from sexual reproduction breaks up linkage among loci and generates linkage equilibrium. To quantify linkage, *poppr* package calculates the indices Ia (The index of association) and rbarD (The standardized index of association). High values of Ia, i.e., values that differ strongly from 0, can be interpreted as evidence of strong linkage and linkage disequilibrium [29]. The rbarD value has been shown to be a more reliable estimator of linkage equilibrium than Ia since it is not influenced by sample size [30], but to be thorough both metrics were calculated. Significance was tested by creating a null dataset (999 random permutations) and if the observed rbarD value lies outside the null dataset then the null hypothesis that no linkage exists would be rejected [24,29].

3. Results

3.1. Regional Distribution of 410 Tea Accessions in South Korea

The collection areas of 410 tea accessions were seven cities in Gyeongsangnam-do (GN), 19 in Jeollanam-do (JN), and five in Jeollabuk-do (JB) (Table 2). Among 410 tea accessions, 350 tea accessions (85.4%) were collected from JN, 7.1% (29 accessions) from GN, and 5.1% (21 accessions) from JB. Of 31 cities, the largest tea accessions were collected from Boseong (18.8%), followed by Suncheon (13.9%), and less than 10% from other cities.

Table 2. Distribution of collected locations and clusters from discriminant analysis of principal components (DAPC) analysis in 410 tea accessions.

Location	Cluster [1]				Total (%)
	1	2	3	4	
GN [2]					
Gimhae	1	1	3	1	6 (1.5)
Namhae	1			2	3 (0.7)
Sacheon			2	1	3 (0.7)
Sancheong	2		1		3 (0.7)
Yangsan	2	2	1	1	6 (1.5)
Jinju		2	1	1	4 (1.0)
Hadong		2	1	1	4 (1.0)
Subtotal	6	7	9	7	29 (7.1)
JN					
Gangjin	6	2	5	11	24 (5.9)
Goheung	2	1			3 (0.7)
Gokseong	1		6	2	9 (2.2)
Gwangyang	2	1		2	5 (1.2)
Gwangju	11	3			14 (3.4)
Gurye	6	7	2	3	18(4.4)
Naju	10	3	10	6	29 (7.1)
Damyang			2	3	5 (1.2)
Mu'an	7	1	1	2	11 (2.7)
Boseong	28	17	16	16	77 (18.8)
Suncheon	17	7	19	14	57 (13.9)
Yeosu	1	3	3	3	10 (2.4)
Yeong'am	4	5	1	2	12 (2.9)
Jangseong	3	3	9	5	20 (4.9)
Jangheung	3		2	3	8 (2.0)
Jindo		1	1		2 (0.5)
Hampyeong	3	1	2	2	8 (2.0)
Haenam	11	1	3	2	17 (4.1)
Hwasun	11	4	4	2	21 (5.1)
Subtotal	126	60	86	78	350 (85.4)
JB					
Gochang		2	1	1	4 (1.0)
Gimje	1		1	1	3 (0.7)
Sunchang	2		3		5 (1.2)
Iksan	1				1 (0.2)
Jeongeup	1	3	3	1	8 (2.0)
Subtotal	5	5	8	3	21 (5.1)
Unknown	1	3	5	1	10 (2.4)
Total	138	75	108	89	410

[1] Clusters were from the result of DAPC analysis, [2] GN—Gyeongsangnam-do; JN—Jeollanam-do; JB—Jeollabuk-do.

3.2. Population Structure and Mode of Reproduction of 410 Tea Accessions

In order to understand the genetic relationship among 410 tea accessions, DAPC analysis was performed (Figure 1). Four clusters were detected in coincidence with the lowest BIC value using *find.clusters* function. DAPC analysis was carried out using the detected number of clusters. Typically, 50 first PCs (68.4% of variance conserved) of PCA and three discriminant eigenvalues were retained. These values were confirmed by a cross-validation analysis. The number of accessions in each cluster was 138, 75, 108, and 88 corresponding to clusters 1 to 4, respectively (Table 2).

Cluster 1 was the highest in JN with 36%, while JB and GN had a similar level of 21% and 24%, respectively. Unlike Cluster 1, cluster 2 accounted for the lowest percentage with 17% in JN and JB and

24% in GN. Cluster 3 had the highest ratio in JB (38%), followed by GN (31%) and JN (25%). The ratio of cluster 4 in GN and JN was almost similar accounting for 24% and 22%, respectively, while in JB it was only 14%. Among the ten unknown accessions, five accessions were in cluster 3 (50%), followed by cluster 2 (30%).

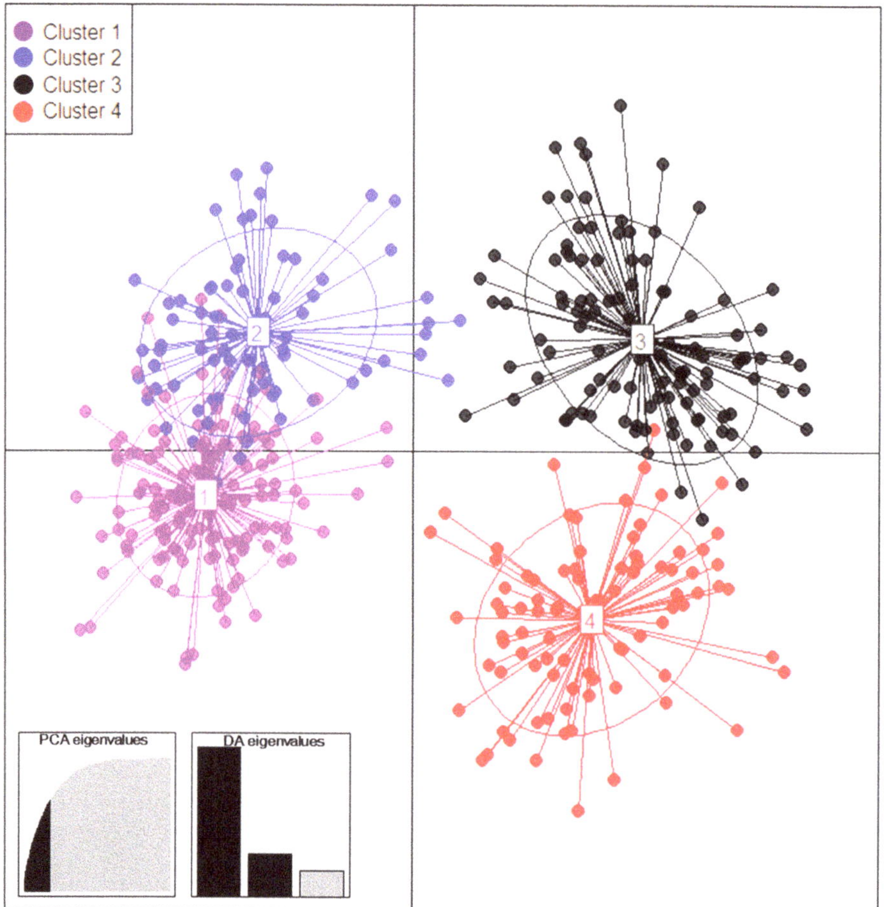

Figure 1. Discriminant analysis of principal components (DAPC) for 410 tea accessions. The axes represent the first two Linear Discriminants (LD). Each circle represents a cluster and each dot represents an individual. Numbers represent the different subpopulations identified by DAPC analysis.

Among 410 tea accessions, 409 multilocus SSR genotypes were found (Table 3). The Simpson's Dominance (λ), a Simpson's diversity (1-λ), and Nei's Gene diversity (GD) were calculated as 0.998, 0.002, and 0.792 in 410 tea accessions, respectively. In four clusters, 1-λ and GD ranged from 0.007 (C1) to 0.013 (C2) and 0.695 (C4) to 0.805 (C2), respectively. Each cluster was tested for linkage disequilibrium and evidence of sexual reproduction (Table 3 and Figure S3). All accessions showed the index of association (Ia) = 0.792 and rbarD = 0.0583. The ranges of Ia and rbarD in four clusters were from 0.123 (C1) to 0.999 (C4) and from 0.0062 (C1) and 0.0508 (C4), respectively. The hypothesis of no linkage among markers was rejected for all the accessions ($p = 0.001$) and each cluster (C1, $p = 0.018$; C2, C3, and C4, $p = 0.001$), thus supporting a clonal mode of reproduction.

Table 3. Genetic diversity and linkage of 410 tea accessions.

Cluster	N [a]	MLG	λ	GD	Ia	rbarD	p-Value (rbarD)
C1	138	138	0.993	0.778	0.123	0.0062	0.018
C2	75	75	0.987	0.805	0.554	0.0282	0.001
C3	108	108	0.991	0.724	0.462	0.0235	0.001
C4	89	88	0.989	0.695	0.999	0.0508	0.001
Total	410	409	0.998	0.792	1.154	0.0583	0.001

[a] N, Number of accessions; MLG, Number of multilocus genotypes; λ, Simpson's index; GD, Nei's unbiased gene diversity; Ia, The index of association; rbadD, The standardized index of association.

3.3. Genetic Diversity

A total of 286 alleles were detected among 410 tea accessions by polymorphic 21 SSR markers (Table S3). On average, 13.6 alleles varying from five (TM604) to 20 (MSE0083 and MSG0699) were amplified by each marker. The Shannon index (I) varied from 0.98 (MSE0237) to 2.31 (MSE0083) among 21 SSR markers. The average gene diversity (GD) was estimated to be 0.79 and varied from 0.52 (MSE0237) to 0.87 (MSE0083 and MSG0361). The average evenness (E) was 0.75 varied from 0.61 (MSG0429) to 0.87 (TM604), respectively.

Among four clusters, the lowest number of alleles was recorded in the cluster 3 (190, mean = 9.1), while the highest was registered in the cluster 1 (247, mean = 11.8) (Table 4). The average I, GD, and E varied from 1.50 (C4) to 1.89 (C2), 0.69 (C4) to 0.81 (C2), and 0.68 (C4) to 0.76 (C2), respectively.

Seventy-six rare alleles, defined with a frequency less than 1%, were observed in 21 SSR markers (Table 4). All the rare alleles were observed in 119 accessions from four clusters. Typically, 53.9% of rare alleles were observed in 41 accessions from cluster 1, followed by 39.5% in 30 accessions in cluster 2. Unique private alleles were found in 35 accessions and 40% of them belonged to cluster 1.

Table 4. Distribution of rare alleles observed in four clusters.

Locus	C1	C2	C3	C4	Total
MSG0258	2 (2)	1 (1)	1 (1)	1 (1)	3 (5)
MSE0173	0 (0)	2 (2)	1 (1)	0 (0)	2 (3)
MSE0313	5 (5)	1 (1)	0 (0)	1 (1)	5 (7)
MSE0403	3 (3)	2(3)	1 (2)	1 (1)	6 (9)
MSE0083	6 (9)	3 (4)	1 (2)	0 (0)	6 (15)
TM604	1 (1)	0 (0)	0 (0)	0 (0)	1 (1)
TM530	0 (0)	0 (0)	0 (0)	1 (1)	1 (1)
MSG0361	0 (0)	0 (0)	0 (0)	0(0)	6 (0)
MSE0029	1 (3)	1 (1)	0 (0)	0 (0)	1 (4)
MSE0291	1 (4)	1 (2)	1 (1)	1 (1)	4 (8)
MSG0681	2 (2)	1 (1)	2 (2)	0 (0)	5 (5)
MSG0470	5 (7)	2 (2)	0 (0)	0 (0)	6 (9)
MSG0699	1 (1)	1 (2)	4 (4)	1 (1)	6 (2)
MSG0429	1 (3)	0 (0)	0 (0)	1 (1)	1 (4)
MSG0380	2 (2)	2 (2)	1 (1)	0 (0)	2 (5)
MSG0423	2 (3)	1 (2)	1 (1)	1 (1)	4 (7)
MSE0237	0 (0)	1 (2)	2 (4)	0 (0)	3 (6)
MSE0113	3 (5)	4 (6)	2 (2)	2 (2)	6 (15)
MSE0107	3 (3)	4 (5)	0 (0)	0 (0)	5 (8)
TM382	2 (2)	1 (1)	0 (0)	2 (2)	3 (5)
Total	41 (56)	30 (39)	20 (24)	16 (16)	76 (119)

3.4. Gene Flow

The sources of genetic differentiation were revealed among different inferred clusters by the AMOVA method. Results indicated that 1% of variations could be attributed to differentiation among clusters and 99% of variations could be attributed to differentiation within inferred clusters (Table 5). PhiPT and gene flow (Nm) for 410 tea accessions was 0.014 ($p < 0.001$) and 36.156, respectively. Pairwise population PhiPT values for four clusters ranged from 0.01 (C2–C4) to 0.021 (C2–C3) (Table 6). Pairwise population estimates of gene flow (Nm) for four clusters ranged from 23.717 to 47.872 migrants per clusters.

Table 5. Analysis of molecular variance (AMOVA) of 410 tea accessions.

Source	df	SS	MS	Est. Var.	%		Value	*p* Value
Among Pops	3	201.375	67.125	0.388	1%	PhiPT	0.014	0.001
Within Pops	406	11391.250	28.057	28.057	99%	Nm	36.156	
Total	409	11592.624		28.445	100%			

Table 6. Pairwise population PhiPT values (Below diagonal) and Nm values based on 999 permutations (above diagonal) from AMOVA. All PhiPT values were significantly greater than 0 ($p < 0.0001$).

	C1	C2	C3	C4
C1	-	45.120	33.525	40.760
C2	0.011	-	23.717	47.872
C3	0.015	0.021	-	37.797
C4	0.012	0.010	0.013	-

4. Discussion

Erosion of plant genetic diversity is a very serious problem caused by modernization and replacement of wild plants or landraces with a few elite varieties [31,32]. Therefore, collection and preservation of plant genetic resources are of immense importance for crop breeding to support the demands of a growing human population. Effective management and utilization of plant genetic resources require information about the origin of strains, phenotypic traits, and genetic diversity (identified by molecular techniques) [33]. In this study, analysis of genetic diversity of 410 tea accessions collected and conserved in the Korea genebank was performed. Genetic diversity provides an assurance of future genetic progress and insurance against unforeseen threats to agricultural production such as disease epidemics or climate changes. Thus, the fate of genetic diversity in these gene pools is of utmost importance if plant breeding will continue to address the pressing needs of society such as increased yield, genetic resistance to diseases and pests, improved nutritional and processing quality of crop products, and reduction in environmental effects [34].

In the present study, about 85% of tea accessions were collected from Boseong and Suncheon in Jeollanam-do (JN) (Table 2). According to Eom and Kim, the tea seeds obtained from China were firstly cultivated in Mount Jiri in JN and so a majority of tea plants are included in the Honam region (Jeollanam-do and Jeollabuk-do) [35]. In addition, tea experiment stations in Boseong experiment station (BES) and Mokpo experiment station (MES, a city close to Suncheon) have collected the tea accessions since the late 1990s [18]. The two experiment stations have probably collected tea accessions around the area where the institute is located, and thus the largest number of tea accessions was collected in the region. The tea accessions of the two research institutes have been managed as registered tea germplasms of the NAC and appear to cause a regional collectivity imbalance of tea accessions in Korea.

In this study, the mean Nei's gene diversity (GD, 0.792) across 21 SSR markers was higher than other studies; 0.652 in 280 tea accessions using 23 SSR markers [7], 0.640 in 450 tea accessions using 96 EST-SSR marker [22], 0.543 in 185 Chinese tea cultivars using 48 SSR markers [13], and 0.680 in 64 Sri Lankan tea cultivars using 33 EST- or genomic-SSR markers [3]. The gene diversity of a locus, also known expected heterozygosity, is a fundamental measure of genetic variation in a population, and describes the proportion of heterozygosis expected under Hardy–Weinberg equilibrium [36]. As tea is an open pollinated plant, the tea plant shows highly heterogeneous and consequently broad genetic variation [13]. The obtained results also showed high gene diversity in a manner similar to the previously reported data. However, Yao et al., mentioned that the comparison of the degree of genetic diversity between different studies is difficult as the analysis may be affected by various factors like sampling schemes, number of SSR markers, sizes of SSR repeats, and location of SSR in the genome [22].

Contrary to the result of higher gene diversity, 410 tea accessions in this study were characterized by an extreme dearth of genetic diversity as revealed by an overall Simpson's Dominance (λ) of 0.998. Furthermore, the AMOVA revealed there was no significant difference among populations, suggesting low genetic diversity across the entire collected region. In addition, the standardized index of association (rbarD, 0.0583, ($p < 0.001$)) supported the hypothesis of clonal population structure based on the linkage disequilibrium tests, where the null hypothesis of random mating was rejected for all populations. Under clonal propagation, heterozygosis and allelic diversity at each locus are expected to increase [37,38]. While high levels of clonality tend to increase genetic variation within the population, an opposite effect is expected on genetic differentiation among populations and on genotypic diversity, both decreasing with the rate of clonal reproduction [37,39]. Indeed, the 410 tea accessions in this study were landraces and are likely to have been collected from private farms. As breeding a reliable cultivar for a private farm is nearly impossible, almost all the tea gardens consist of seedling tea plants from the local and wild origin with great morphological variations [23]. BES and MES also collected and investigated the morphological characteristics of tea germplasm and many variations were observed in the number of stems, stem length, leaf area, and leaf color which were within the limits of the investigation of the morphological characteristics [40]. Due to the lack of sufficient studies on the genetic diversity of Korean tea germplasm, a few researchers argued over the importance of genetic collection and preservation of tea accessions [14,21].

Previous studies performed the analysis of genetic diversity of different tea accessions using molecular markers like RFLP, RAPD, and SSR [3,5,21,22]. In addition, the STRUCTURE software was used to analyze the population structure of tea germplasm [2,3,13,22]. To analyze the genetic diversity and population structure of Korean tea accessions, 21 SSR markers and DAPC analysis were used in this study. The DAPC method provides an interesting alternative to STRUCTURE software as it does not require that populations should be in Hard-Weinberg equilibrium and can handle large sets of data without using parallel processing software [41]. DAPC analysis divided the population into well-defined clusters associated with provenance, ploidy, taxonomy and breeding program of the genotypes and related to their genetic structure [42]. According to Rosyara et al., STRUCTURE, EIGENSTRAT, and DAPC exhibit the ability to control population structure in association with mapping studies [43]. EIGENSTRAT and DAPC were slightly better than STRUCTURE but DAPC led to a better separation among populations. In this study, DAPC (four clusters) analysis provided a more detailed clustering within tea accessions than STRUCTURE (two populations) (Figure S1). Campoy et al., reported that their results of population structure in sweet cherry using STRUCTURE and DAPC showed good consistency between the two methods and DAPC analysis provided a more detailed clustering among the populations compared to STRUCTURE analysis [41].

As per the result of DAPC, 410 tea accessions were divided into four clusters (Figure 1). Among them, cluster 1 and 2 showed a higher frequency of rare alleles and genetic diversity, and there was high gene flow (Nm = 45.120) between two clusters. Yao et al., reported that a majority of rare SSR alleles and higher diversity were observed in the tea accessions from Yunnan and its neighboring provinces, considered as an original center of the tea plant in China [22]. They also reported that the allele number, genetic diversity, and PIC value of tea germplasm significantly decreased with the distance away from the origin center of the tea plant. Although a particular collection area cannot be designated as an origin, tea accessions contained in cluster 1 and 2 are thought to be the origin of the Korean tea germplasm due to their ratio of rare alleles and higher genetic diversity.

Kaundun et al., reported that tea accessions collected from Korea showed higher genetic diversity than those from Taiwan and Japan [21]. On the other hand, Jeong and Park mentioned that the genetic variation in Korean tea population is smaller compared to Chinese or Japanese wild tea populations [18]. The results of the present study confirmed that 410 tea accessions collected and conserved in Korea genebank exhibit the narrow genetic variations. Park et al., suggested that the low genetic diversity of Korean tea was established from a limited gene stock from China [14]. The short history and relatedly homogeneous environment in which they were introduced in the southwestern part of the country did not favor population differentiation. In addition, loss of diversity was exacerbated by the mass destruction of tea plantations in the fourteenth century due to political and religious reasons [44]. Consequently, tea being a highly outcrossing species, variability is mostly expected within rather than between the populations as predicted by Hamrick [45].

5. Conclusions

In this study, genetic diversity and population structure of 410 Korean tea accessions collected and conserved in Korea genebank were analyzed using 21 SSR markers. The results provided molecular evidence for the narrow genetic base of the Korean tea accessions. According to the database in NAC, initially, 4223 tea accessions were collected and conserved (http://genebank.rda.go.kr). However, only 510 tea accessions were conserved in NAC as many tea accessions were destroyed by the extreme cold condition in the winter season. Among them, 427 tea accessions were collected from Korea, 56 from China, 22 from Japan, and five from Indonesia. In conclusion, there exists an urgent need for broadening the genetic base of tea accessions in Korea genebank and the necessitation can be achieved by not only collecting tea plants in Korea but also introducing the tea germplasm from other countries. Additionally, it is necessary to analyze the biochemical components of tea accessions in order to gain an understanding of their effects on the quality characteristics of tea varieties and promote utilization of tea germplasm for tea breeding.

Supplementary Materials: The following are available online at http://www.mdpi.com/1999-4907/10/9/780/s1, Table S1: List of 410 tea accessions used in this study; Table S2: Repeat motif and product size of 21 SSR primers used in this study; Table S3: Genetic diversity of 21 SSR markers in each cluster; Figure S1: (A) Relationship between delta K and K as revealed by STRUCTURE harvester. (B) Population structure analysis of 410 tea accessions inferred using STRUCTURE software based on 21 SSR markers for delta K = 2.

Author Contributions: Conceptualization, K.J.L.; Data curation, K.J.L., R.S., and M.-J.S.; Formal analysis, K.J.L., R.S., and M.-J.S.; Investigation, J.-R.L., S.-H.K., and G.-T.C.; Project administration, D.Y.H.; Resources, J.-R.L., S.-H.K., and G.-T.C.; Writing—original draft, K.J.L.; Writing—review and editing, D.Y.H.

Funding: This research was funded by [Research Program for Agricultural Science & Technology Development, National Institute of Agricultural Sciences, Rural Development Administration, Republic of Korea] grant number [PJ01355702].

Conflicts of Interest: The authors declare no conflict of interest.

References

1. Karak, T.; Bhagat, R.M. Trace Elements in Tea Leaves, Made Tea and Tea Infusion: A Review. *Food Res. Int.* **2010**, *43*, 2234–2252. [CrossRef]
2. Wambulwa, M.C.; Meegahakumbura, M.K.; Kamunya, S.; Muchugi, A.; Möller, M.; Liu, J.; Xu, J.-C.; Li, D.-Z.; Gao, L.-M. Multiple origins and a narrow genepool characterise the African tea germplasm: Concordant patterns revealed by nuclear and plastid DNA markers. *Sci. Rep.* **2017**, *7*, 4053. [CrossRef] [PubMed]
3. Karunarathna, K.H.T.; Mewan, K.M.; Weerasena, O.V.D.S.J.; Perera, S.A.C.N.; Edirisinghe, E.N.U.; Jayasoma, A.A. Understanding the genetic relationships and breeding patterns of Sri Lankan tea cultivars with genomic and EST-SSR markers. *Sci. Hortic.* **2018**, *240*, 72–80. [CrossRef]
4. Wight, W. Nomenclature and Classification of the Tea Plant. *Nature* **1959**, *183*, 1726–1728. [CrossRef]
5. Matsumoto, S.; Kiriiwa, Y.; Yamaguchi, S. The Korean tea plant (*Camellia sinensis*): RFLP analysis of genetic diversity and relationship to Japanese tea. *Breed. Sci.* **2004**, *54*, 231–237. [CrossRef]
6. Thomas, T.A.; Mathur, P.N. Germplasm Evaluation and Utilzation. In *Plant Genetic Resources: Conservation and Management. Concepts and Approaches*; Paroda, R.S., Arora, R.K., Eds.; IBPGR Regional Office: New Delhi, India, 1991.

7. Wambulwa, M.C.; Meegahakumbura, M.K.; Samson, K.; Alice, M.; Michael, M.; Liu, J.; Xu, J.C.; Sailesh, R.; Li, D.Z.; Gao, L.M.; et al. Insights into the genetic relationships and breeding patterns of the african tea germplasm based on nSSR markers and cpDNA sequences. *Front. Plant Sci.* **2016**, *7*, 1244. [CrossRef] [PubMed]
8. Li, Y.; Chen, C.; Li, Y.; Ding, Z.; Shen, J.; Wang, Y.; Zhao, L.; Xu, M.; Li, Y.; Chen, C.; et al. The identification and evaluation of two different color variations of tea. *J. Sci. Food Agric.* **2016**, *96*, 4951–4961. [CrossRef]
9. Feng, L.; Gao, M.J.; Hou, R.Y.; Hu, X.Y.; Zhang, L.; Wan, X.C.; Wei, S.; Feng, L.; Gao, M.J.; Hou, R.Y.; et al. Determination of quality constituents in the young leaves of albino tea cultivars. *Food Chem.* **2014**, *155*, 98–104. [CrossRef]
10. Gunasekare, M.T.K. Applications of molecular markers to the genetic improvement of *Camellia sinensis* L. (tea)—A review. *J. Pomol. Hort. Sci.* **2007**, *82*, 161–169.
11. Chen, L.; Zhou, Z.X.; Chen, L.; Zhou, Z.X. Variations of main quality components of tea genetic resources [*Camellia sinensis* (L.) O. Kuntze] preserved in the China National Germplasm Tea Repository. *Plant Foods Hum. Nutr.* **2005**, *60*, 31–35. [CrossRef]
12. Chen, L.; Yamaguchi, S.; Chen, L.; Yamaguchi, S. RAPD markers for discriminating tea germplasms at the inter-specific level in China. *Plant Breed.* **2010**, *124*, 404–409. [CrossRef]
13. Fang, W.; Cheng, H.; Duan, Y.; Jiang, X.; Li, X. Genetic diversity and relationship of clonal tea (*Camellia sinensis*) cultivars in China as revealed by SSR markers. *Plant Syst. Evol.* **2012**, *298*, 469–483. [CrossRef]
14. Park, Y.-G.; Kaundun, S.S.; Zhyvoloup, A. Use of the bulked genomic DNA-based RAPD methodology to assess the genetic diversity among abandoned Korean tea plantations. *Genet. Resour. Crop Evol.* **2002**, *49*, 159–165.
15. Lai, J.A.; Yang, W.C.; Hsiao, J.Y. An assessment of genetic relationships in cultivated tea clones and native wild tea in Taiwan using RAPD and ISSR markers. *Bot. Bull. Acad. Sin.* **2001**, *42*, 93–100.
16. Oh, C.J.; Lee, S.; You, H.C.; Chae, J.G.; Han, S.S. Genetic Diversity of Wild Tea (*Camellia sinensis* L.) in Korea. *Korean J. Plant Resour.* **2008**, *21*, 41–46.
17. Korir, N.K.; Han, J.; Shangguan, L.; Wang, C.; Kayesh, E.; Zhang, Y.; Fang, J. Plant variety and cultivar identification: Advances and prospects. *Crit. Rev. Biotechnol.* **2013**, *33*, 111–125. [CrossRef]
18. Jeong, B.-C.; Park, Y.-G. Tea Plant (*Camellia sinensis*) Breeding in Korea. In *Global Tea Breeding: Achievements, Challenges and Perspectives*; Springer: Berlin/Heidelberg, Germany, 2012; pp. 263–288.
19. Gai, Z.; Wang, Y.; Jiang, J.; Xie, H.; Ding, Z.; Ding, S.; Wang, H. The quality evaluation of tea (*Camellia sinensis*) varieties based on the metabolomics. *Horticulture* **2019**, *54*, 409. [CrossRef]
20. Chen, L.; Yao, M.; Zhao, L.P.; Wang, X. Recent Research Progresses on Molecular Biology of Tea Plant (*Camellia sinensis*). In *Floriculture, Ornamental and Plant Biotechnology: Advances and Topical Issues*, 1st ed.; da Silva, J.A.T., Ed.; Global Science Books: London, UK, 2006; Volume 4, pp. 425–436.
21. Kaundun, S.S.; Zhyvoloup, A.; Park, Y.-G. Evaluation of the genetic diversity among elite tea (*Camellia sinensis* var. sinensis) accessions using RAPD markers. *Euphytica* **2000**, *115*, 7–16. [CrossRef]
22. Yao, M.-Z.; Ma, C.-L.; Qiao, T.-T.; Jin, J.-Q.; Chen, L. Diversity distribution and population structure of tea germplasms in China revealed by EST-SSR markers. *Tree Genet. Genomes* **2012**, *8*, 205–220. [CrossRef]
23. Jeong, B.-C.; Song, Y.S.; Moon, Y.H.; Han, S.K.; Bang, J.K. *Collection and Multiplication of Superior Germplasm for Development of the New Tea Tree Cultivar*; Joint Research Report for Tea germplasm in Korea; RDA: Mokpo, Korea, 2007; pp. 1–46.
24. Kamvar, Z.N.; Tabima, J.F.; Grünwald, N.J. Poppr: An R package for genetic analysis of populations with clonal, partially clonal, and/or sexual reproduction. *PeerJ.* **2014**, *2*, e281. [CrossRef]
25. Peakall, R.; Smouse, P.E. GenAlEx 6.5: Genetic analysis in Excel. Population genetic software for teaching and research±an update. *Bioinformatics* **2012**, *28*, 2537–2539. [CrossRef] [PubMed]
26. Ivandic, V.; Hackett, C.A.; Nevo, E.; Keith, R.; Thomas, W.T.B.; Forster, B.P. Analysis of simple sequence repeats (SSRs) in wild barley from the Fertile Crescent: Associations with ecology, geography and flowering time. *Plant Mol. Biol.* **2002**, *48*, 511–527. [CrossRef] [PubMed]
27. Jombart, T. adegenet: A R package for the multivariate analysis of genetic markers. *Bioinformatics* **2008**, *24*, 1403–1405. [CrossRef] [PubMed]
28. Jombart, T.; Collins, C. A tutorial for Discriminant Analysis of Principal Components (DAPC) using adegenet 2.0.0. Available online: http://adegenet.r-forge.r-project.org/files/tutorial-dapc.pdf (accessed on 6 September 2019).

29. Grunwald, N.J.; Kamvar, Z.N.; Everhart, S.E.; Tabima, J.F.; Knaus, B.J. Population Genetic and Genomics in R. Available online: http://grunwaldlab.github.io/Population_Genetics_in_R/ (accessed on 3 April 2019).
30. Agapow, P.-M.; Burt, A. Indices of multilocus linkage disequilibrium. *Mol. Ecol. Notes* **2001**, *1*, 101–102. [CrossRef]
31. van de Wouw, M.; Kik, C.; van Hintum, T.; van Treuren, R.; Visser, B. Genetic erosion in crops: Concept, research results and challenges. *Plant Genet. Resour.* **2010**, *8*, 1–15. [CrossRef]
32. van Heerwaarden, J.; Hellin, J.; Visser, R.F.; van Eeuwijk, F.A. Estimating maize genetic erosion in modernized smallholder agriculture. *Theor. Appl. Genet.* **2009**, *119*, 875–888. [CrossRef] [PubMed]
33. Kim, O.G.; Sa, K.J.; Lee, J.R.; Lee, J.K. Genetic analysis of maize germplasm in the Korean Genebank and association with agronomic traits and simple sequence repeat markers. *Genes Genom.* **2017**, *39*, 843–853. [CrossRef]
34. Gepts, P. Plant genetic resources conservation and utilization: The accomplishments and future of a societal insurance policy. *Crop Sci.* **2006**, *46*, 2278–2292. [CrossRef]
35. Eom, B.-C.; Kim, J.-W. Phytocoenosen and Distribution of a Wild Tea (*Camellia sinensis* (L.) Kuntze) Population in South Korea. *Korean J. Plant Resour.* **2017**, *30*, 176–190. [CrossRef]
36. Harris, A.M.; DeGiorgio, M. An Unbiased Estimator of Gene Diversity with Improved Variance for Samples Containing Related and Inbred Individuals of any Ploidy. *G3 Genes Genom. Genet.* **2017**, *7*, 671–691. [CrossRef] [PubMed]
37. Balloux, F.; Lehmann, L.; de Meeûs, T. The Population Genetics of Clonal and Partially Clonal Diploids. *Genetics* **2003**, *164*, 1635–1644. [PubMed]
38. Birky, C.W., Jr. Heterozygosity, heteromorphy, and phylogenetic trees in asexual eukaryotes. *Genetics* **1996**, *144*, 427–437. [PubMed]
39. Halkett, F.; Simon, J.C.; Balloux, F. Tackling the population genetics of clonal and partially clonal organisms. *Trends Ecol. Evol.* **2005**, *20*, 194–201. [CrossRef] [PubMed]
40. Song, Y.S.; Moon, Y.H.; Han, S.K.; Jeong, B.C.; Bang, J.K. Morphological characteristics of progeny population in collected wild tea (*Camellia sinensis*). *J. Korean Tea Soc.* **2005**, *11*, 93–105.
41. Campoy, J.A.; Lerigoleur-Balsemin, E.; Christmann, H.; Beauvieux, R.; Girollet, N.; Quero-García, J.; Dirlewanger, E.; Barreneche, T. Genetic diversity, linkage disequilibrium, population structure and construction of a core collection of *Prunus avium* L. landraces and bred cultivars. *BMC Plant Biol.* **2016**, *16*, 49. [CrossRef] [PubMed]
42. Deperi, S.I.; Tagliotti, M.E.; Bedogni, M.C.; Manrique-Carpintero, N.C.; Coombs, J.; Zhang, R.; Douches, D.; Huarte, M.A. Discriminant analysis of principal components and pedigree assessment of genetic diversity and population structure in a tetraploid potato panel using SNPs. *PLoS ONE* **2018**, *13*, e0194398. [CrossRef]
43. Rosyara, U.R.; De Jong, W.S.; Douches, D.S.; Endelman, J.B. Software for Genome-Wide Association Studies in Autopolyploids and Its Application to Potato. *Plant Genome* **2016**, *9*. [CrossRef] [PubMed]
44. Choi, S. Studies of the aroma of Korean green tea. In Proceedings of the International Symposium of Tea Science; ISTS: Shizuoka, Japan, 1991; pp. 130–134.
45. Hamrick, J. Isozymes and the analysis of genetic structure in plant populations. In *Isozymes in Plant Biology*; Soltis, D.E., Soltis, P.S., Dudley, T.R., Eds.; Springer: Dordrecht, The Netherlands, 1989; pp. 87–105.

Review

Mutation Mechanism of Leaf Color in Plants: A Review

Ming-Hui Zhao, Xiang Li, Xin-Xin Zhang, Heng Zhang and Xi-Yang Zhao *

State Key Laboratory of Tree Genetics and Breeding, School of Forestry, Northeast Forestry University, Harbin 150040, China; zhaominghui66@163.com (M.-H.Z.); lx2016bjfu@163.com (X.L.); zhangxinxin@nefu.edu.cn (X.-X.Z.); zhangheng815@nefu.edu.cn (H.Z.)

* Correspondence: zhaoxyphd@163.com; Tel.: +86-0451-8219-2225

Received: 3 July 2020; Accepted: 3 August 2020; Published: 6 August 2020

Abstract: Color mutation is a common, easily identifiable phenomenon in higher plants. Color mutations usually affect the photosynthetic efficiency of plants, resulting in poor growth and economic losses. Therefore, leaf color mutants have been unwittingly eliminated in recent years. Recently, however, with the development of society, the application of leaf color mutants has become increasingly widespread. Leaf color mutants are ideal materials for studying pigment metabolism, chloroplast development and differentiation, photosynthesis and other pathways that could also provide important information for improving varietal selection. In this review, we summarize the research on leaf color mutants, such as the functions and mechanisms of leaf color mutant-related genes, which affect chlorophyll synthesis, chlorophyll degradation, chloroplast development and anthocyanin metabolism. We also summarize two common methods for mapping and cloning related leaf color mutation genes using Map-based cloning and RNA-seq, and we discuss the existing problems and propose future research directions for leaf color mutants, which provide a reference for the study and application of leaf color mutants in the future.

Keywords: color mutation; pigment metabolism; chlorophyll; anthocyanin; mutation mechanism; RNA-seq

1. Introduction

As part of photosynthesis, leaves play an important role in the growth and development of plants. Leaf color mutation is a high frequency character variation that is easy to recognize and ubiquitous in various higher plants [1]. Leaf color mutations usually affect the plant's photosynthetic efficiency, resulting in stunted growth and even death. Therefore, leaf color mutations have been considered harmful mutations with no practical value by researchers in the past [2,3]. Since Granick used the Chlorella vulgaris green mutant *W5* to validate that protoporphyrin III was a precursor to chlorophyll synthesis in 1948, research related to leaf color mutants has gradually gained attention, especially research related to chlorophyll synthesis [4,5]. Physical and chemical mutagenesis and tissue culture are usually used to induce leaf color mutations and obtain leaf color mutants [6]. In general, the mutant genes of leaf color mutants can directly or indirectly affect the pigment (such as chlorophyll and anthocyanin) synthesis, degradation, content and proportion, which can block photosynthesis and lead to abnormal leaf color. Leaf color mutations are generally expressed at the seedling stage and can be divided into eight types: albino, greenish-white, white emerald, light green, greenish-yellow, etiolation, yellow-green, and striped [1]. In addition, leaf color mutants can also be divided into four types: total chlorophyll increased type, total chlorophyll deficient type, chlorophyll a deficient type, and chlorophyll b deficient type [7]. As an especially ideal material, leaf color mutants play an important role in the research of photosynthetic mechanisms, the chlorophyll biosynthesis pathway, chloroplast development and genetic control mechanisms [8,9]. Leaf color mutants have been obtained in *Zea mays* [10,11], *Populus* L. [12,13], *Nicotiana tabacum* [14–16], *Arabidopsis thaliana* [17–20], *Oryza sativa* [21–24], *Rosa multiflora* [25,26], and *Cucumis melo* [27], resulting in many studies being conducted in these species.

Genetic changes in plant cells are usually the cause of leaf color mutations, although the underlying mechanism is more complicated. There are more than 700 sites involved in leaf color mutations in higher plants, all of which are involved in the development, metabolism or signal transduction of leaf color formation. We can therefore analyze and identify gene functions and understand gene interactions by using these mutants [28]. This paper mainly reviews studies related to leaf color mutants, describing some genes that control or affect pigment metabolism, chloroplast development and differentiation, and photosynthesis. We summarize two methods for mapping and identifying leaf color mutation genes, and preliminarily elucidate the formation mechanism of leaf color mutations. We also present the achievements and challenges inherent to the study of leaf color mutants and future research directions to provide a reference for the future study and application of leaf color mutants.

2. Genetic Model of Plant Leaf Color Mutants

Many studies have proved that the genetic modes of plant leaf color mutants are mainly divided into three types: nuclear heredity, cytoplasmic heredity and nuclear-plasmid gene interaction heredity, among which nuclear heredity is the most important genetic mode of leaf color. Studies of leaf color mutants mainly focus on recessive mutants controlled by nuclear genes. Such mutants follow Mendelian inheritance laws, including single-gene inheritance and multigene inheritance, among which the single gene recessive inheritance mode results in the majority of leaf color mutants and leaf color inheritance. A series of white leaf color mutants *virescen1*, *virescen3*, and yellow-green leaf *YSA* mutants were found in rice, all of which were proved to be controlled by a pair of recessive nuclear genes [28–30]. In addition, leaf color mutations controlled by a single recessive gene were also found in such species as *Capsicum annuum*, *Cucumis sativus*, and *C. melo* [27,31,32]. Moreover, it was found that the Sesamum indicum yellow-green mutation was controlled by an incompletely dominant nuclear gene, *Siyl-1* [33]. Conversely, few studies and reports exist related to leaf color mutations caused by mutations of the cytoplasmic genes and nuclear cytoplasmic gene interactions in such species as *A. thaliana*, *N. tabacum*, and *Lycopersicon esculentum* [34–36]. These mutations may be related to the fact that plant cells contain multiple organelles (e.g., chloroplasts and mitochondria) with their own DNA molecules.

3. Molecular Mechanisms of Plant Leaf Color Mutations

The molecular mechanisms of leaf color mutations are complex. The mutated genes can directly or indirectly interfere with pigment synthesis and stability, resulting in various sources of leaf color mutations. In this paper, we summarized the molecular mechanisms underlying the formation of leaf color mutants through the following aspects.

3.1. *Abnormal Chlorophyll Metabolism Pathway*

3.1.1. Mutations of Genes Related to the Chlorophyll Synthesis Pathway

Chlorophyll in higher plants includes chlorophyll a and chlorophyll b. The biosynthesis of chlorophyll begins when L-Glutamyl-tRNA produces chlorophyll a, which is then oxidized by chlorophyllide, an oxygenase, to form the chlorophyll b synthesis cycle. This process involves a total of fifteen steps, and the whole synthesis process involves the participation of fifteen enzymes and the expression of twenty-seven genes encoding these enzymes (Figure 1 and Table 1) [37,38]. Any obstacle in this synthetic process will affect chlorophyll biosynthesis, resulting in leaf color mutants.

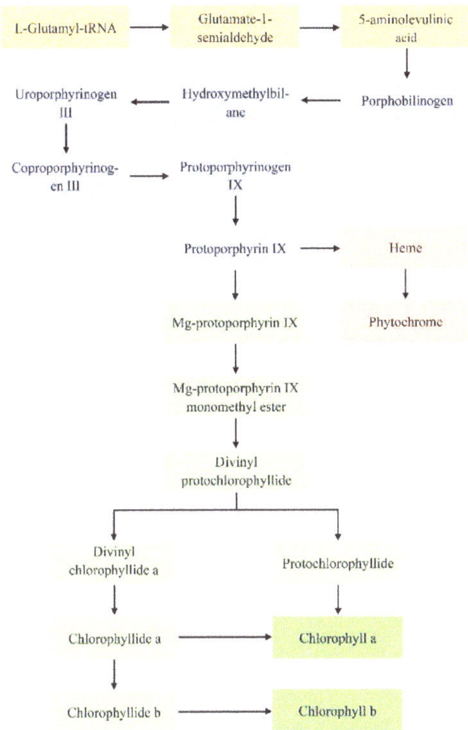

Figure 1. Biosynthetic pathway of angiosperm chlorophyll.

The entire process of chlorophyll synthesis is divided into two main parts: the first part includes the process of L-Glutamyl-tRNA synthesizing protoporphyrin IX (proto IX), and the second part encompasses from protoporphyrin IX to chlorophyll biosynthesis. The two synthetic parts can be divided into three stages: the first stage is from L-Glutamyl-tRNA to 5-aminolevulinic acid (ALA), the second stage is from ALA to proto IX biosynthesis, and the third stage is from proto IX to chlorophyll biosynthesis. Furthermore, the entire process of chlorophyll synthesis is localized at three tissues: the synthesis of ALA to proto IX is completed in the middle of the chloroplast stroma; the synthesis of proto IX to chlorophyllide occurs on the chloroplast membrane; and the process of chlorophyll a and chlorophyll b synthesis is completed on the thylakoid membrane [39]. Problems with any of genes in Table 1 may lead to changes in enzyme activity and function within chlorophyll synthesis process, resulting in the accumulation of excessive intermediates, affecting the normal metabolism of chlorophyll, and changing the proportion of pigments in chloroplasts, which can lead to oxidative damage, causing different plants to produce leaf color mutations or possibly even causing plant death [40]. For example, after glutamate-tRNA synthetase is silenced by the virus, the mutant phenotype is extremely yellow [41]. When the function of chlorophyll a oxidase is abnormal, chlorophyll b synthesis is reduced, or it is not synthesized [42]. In the fifteen steps of the chlorophyll metabolism pathway, the earlier the occurrence of the mutation, the more pronounced the leaf color mutations, which generally present as a yellow or white phenomenon. If the mutation occurs in the later stages of chlorophyll synthesis, it usually only results in patches, stripes, etc. [43].

In the chlorophyll biosynthesis pathway, the synthesis of ALA and the insertion of the Mg ion into proto IX are the two main control points that directly affect chlorophyll synthesis [44]. ALA is catalyzed by Glutamyl-tRNA reductase (GluTR) and Glutamate-1-semialdehyde 2,1-aminomutase (GSA-AM), which controls the rate of chlorophyll and heme synthesis. This is the rate-limiting step of the tetrapyrrole synthesis pathway, and it plays a key role in chlorophyll synthesis [45]. For example, the overexpression of the *HEMA1* gene in yellowing plants leads to an increase in protochlorophyllide content [46]. The synthesis of ALA is influenced by GluTR reductase, and GluTR is encoded by the *HEMA* gene, of which plants contain at least two (*HEMA 1, HEMA 2*) [47,48]. The antisense *HEMA1* gene was transferred into *A. thaliana*, and it was found that the antisense *HEMA1* gene inhibited the formation of ALA, which resulted in decreases of protochlorophyllide synthesis and chlorophyll content [49]. The activity of GluTR is regulated by heme, which as a terminal product can also provide feedback regulation on the activity of GluTR. For example, gene mapping of the *A. thaliana 'ulf'* mutant was located at *Hy1* site, which encodes for the synthesis of heme oxidase, and the reduced heme oxidase activity in the mutant led to the accumulation of heme, which inhibited the activity of GluTR [50].

Table 1. Genes encoding and enzymes involved in the chlorophyll synthesis pathway in *A. thaliana*.

Step	Enzyme Name	Abbreviated Name of Enzyme	Gene Name	Locus Name in *Arabidopsis*
1	Glutamyl-tRNA reductase	GluTR	*HEMA1*	AT1G58290
			HEMA2	AT1G09940
			HEMA3	AT2G31250
2	Glutamate-1-semialdehyde 2, 1-aminomutase	GSA-AM	*GSA1 (HEML1)*	AT5G63570
			GSA2 (HEML2)	AT3G48730
3	5-Aminolevulinate dehydratase	PBGS (ALAD)	*HEMB1*	AT1G69740
			HEMB2	AT1G44318
4	Porphobilinogen deaminase	PBGD	*HEMC*	AT5G08280
5	Uroporphyrinogen III synthase	UROS	*HEMD*	AT2G26540
6	Uroporphyrinogen III decarboxylase	UROD	*HEME1*	AT3G14930
			HEME2	AT2G40490
7	Coproporphyrinogen III oxidase	CPOX	*HEMF1*	AT1G03475
			HEMF2	AT4G03205
8	Protoporphyrinogen oxidase	PPOX	*HEMG1*	AT4G01690
			HEMG2	AT5G14220
9	Magnesium chelatase H subunit	MgCh	*CHLH*	AT5G13630
	Magnesium chelatase I subunit		*CHL11*	AT4G18480
			CHL12	AT5G45930
	Magnesium chelatase D subunit		*CHLD*	AT1G08520
10	Mg-protoporphyrin IX methyltransferase	MgPMT	*CHLM*	AT4G25080
11	Mg-protoporphyrin IX monomethyl ester	MgPME	*CRD1 (ACSF)*	AT3G56940
12	3,8-Divinyl protochlorophyllide a 8-vinyl reductase	DVR	*DVR*	AT5G18660
13	Protochlorophyllide oxidoreductase	POR	*PORA*	AT5G54190
			PORB	AT4G27440
			PORC	AT1G03630
14	Chlorophyll synthase	CHLG	*CHLG*	AT3G51820
15	Chlorophyllide a oxygenase	CAO	*CAO (CHL)*	AT1G44446

Magnesium chelatase is the key factor in guaranteeing the biosynthesis of chlorophyll, which can catalyze the formation of Mg protoporphyrin by combining a magnesian ion with protoporphyrin [51]. The reaction step of metal ion insertion into protoporphyrin IX is the branching point for the synthesis of chlorophyll, heme and plant pigments. Magnesium chelatase catalyzes magnesian ion insertion into protoporphyrin IX to form the chlorophyll branch, while ferric chelatase catalyzes ferrous ion to be inserted into protoporphyrin IX to form the heme and plant pigment branches. Magnesium chelatase and ferric chelatase compete for protoporphyrin IX at the branch point [45,52]. Studies have shown that the main reason for the formation of chlorophyll mutants is problems with magnesium chelatase, which prevents the synthesis of chlorophyll from proceeding normally [53]. The decrease in magnesium chelatase activity leads to a decrease in chlorophyll content in the mutant, and the phenotype of the mutant will also be affected by the enzyme activity [54]. The function of the magnesium chelatase is complex, which consists of the *D*, *I*, and *H* subunits, depends on the synergy of the three functional subunits. For example, in *Arabidopsis*, a single base mutation in the third exon of the *GUN5* gene encoding the magnesium chelatase *H* subunit, will cause the formation of albinos [55]. If the *CHL1* gene encoding the magnesium chelatase *D* subunit mutates, the mutant will present as a light yellow-green at the seedling stage, and the expression of the nuclear gene *LHCP II* will also be affected through a feedback regulation mechanism [56]. In addition, chlorophyll synthesis may be regulated by other factors besides the expression of the twenty-seven gene encoding enzymes. For example, in a study of *Pak-choi* yellow leaf mutants, the expression of *CHLG* encoding chlorophyll synthase was found to be significantly decreased, but there was no difference between wild type and mutant *CHLG* sequences, indicating that the chlorophyll synthetase activity was also regulated by other regulatory factors [57].

3.1.2. Mutations of Genes Related to the Chlorophyll Degradation Pathway

In leaves with no color mutation, chlorophyll degradation is coupled with simultaneous chlorophyll synthesis in a dynamic balance. Chlorophyll degrades during the transformation of chloroplasts into chromoplast, which indicates that leaves will begin to age [58]. The reaction process of chlorophyll degradation is mainly divided into two stages. The first stage is the degradation of chlorophyll into the primary fluorescent chlorophyll catabolite (pFCC). In the second stage, the pFCC in the vacuole forms nonfluorescent chlorophyll catabolite (NCC), which is finally transformed into the oxidation degradation product, pyrrole (Figure 2) [59]. This degradation pathway involves a series of complex reactions, and an interruption in any step may change the chlorophyll content and produce leaf color mutations.

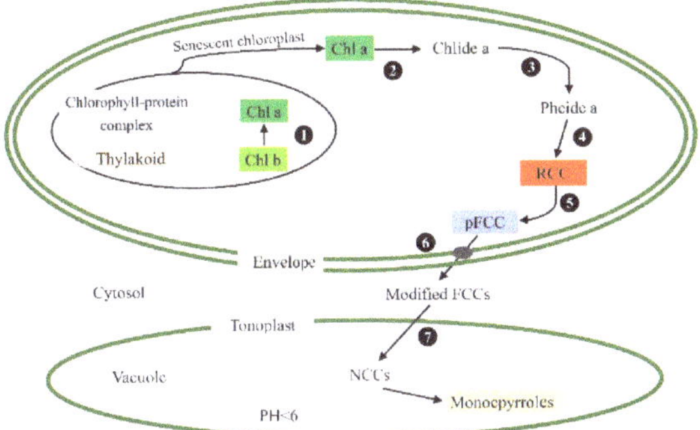

Figure 2. Biodegradation pathway of chlorophyll in higher plant. (1) Chl b reductase; (2) Chlorophyllase; (3) Metal chelating substance; (4) Pheide a oxygenase; (5) RCC reductase; (6) Catabolite transporter; (7) ABC transporter. Chl a, Chlorophyll a; Chl b, Chlorophyll b; Chlide a, chlorophyllide a; Pheide a, Pheophorbide a; RCC, red Chl catabolite; pFCC, primer fluorescent Chl catabolite; FCCs, fluorescent Chl catabolites; NCCs, nonfluorescent Chl catabolites.

Mutation of the chlorophyll degradation pathway will lead to a lack of chlorophyll degradation in plants, resulting in the stay-green phenomenon in plant leaves; this kind of mutation is called a stay-green mutation. In the *Z. mays 'fs854'* mutant, the inhibition of the chlorophyll degradation pathway led to the occurrence of the stay-green phenomenon in maize, which promoted the increase of maize yield due to its high photosynthesis [60]. If the process of chlorophyll degradation is accelerated, leaf color mutants may also be produced. The expression of *CHL2* and *RCCR* encoding key chlorophyll degradation enzymes in *Cymbidium sinense* mutants was higher than wild-type, resulting in a decrease in chlorophyll content, which may be the reason for the yellow color mutation of Moran leaf [61]. Stay-green mutations have a distinct phenotype that can delay the leaf senescence of crops and increase crop yield. It has a much higher potential application value compared with yellow-green seedlings, and it therefore attracts significant research attention.

3.1.3. Mutations of Genes Related to the Heme Metabolism Pathway

Tetrapyrrole is the skeleton structure and common precursor of chlorophyll and heme biosynthesis in plants, the biosynthetic pathway of which is divided into two branches, the chlorophyll biosynthetic pathway and the heme biosynthetic pathway, at protoporphyrin IX. The first branch chelates protoporphyrin IX with a magnesian ion to produce Mg-protoporphyrin IX, which finally forms chlorophyll; the second branch chelates proto IX with a ferrous ion to form heme [40]. Heme is a type of iron-containing cyclic tetrapyrrole that is the intermediate product of the synthetic pathway of phytochrome and phycobilin [62]. After a series of complex reactions, heme finally forms a photosensitive pigment chromophore (Figure 3) [40]. As heme and chlorophyll in plants have the same precursor and share part of the same pathway, they are the branch products of chlorophyll biosynthesis. Therefore, the synthesis of chlorophyll is regulated by the feedback inhibition of intracellular heme content. When the heme metabolism branch of plants is disturbed, it may lead to an increase in heme content in the cells. High concentrations of heme inhibit the synthesis of ALA through feedback regulation, which is a common precursor of chlorophyll and heme, resulting in the inhibition of chlorophyll synthesis and variations in leaf color [63].

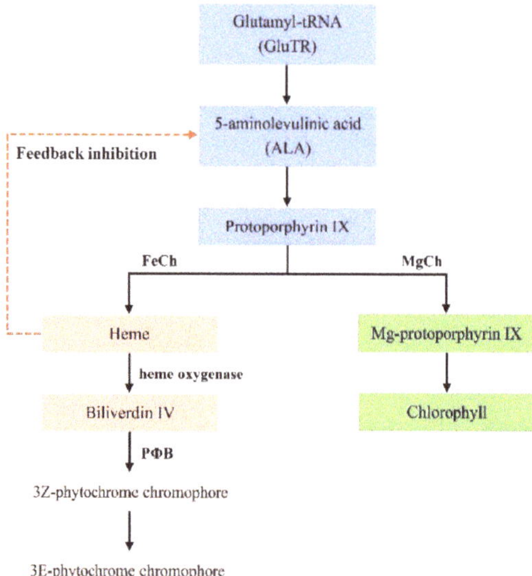

Figure 3. Heme metabolism pathway in high plants. The 3Z-phytochrome chromophore could spontaneously form a 3E-phytochrome chromophore or be an enzyme catalyst.

Heme oxygenase (HO) catalyzes the degradation of heme to synthesize phytochrome precursors in higher plants, which controls the rate of heme degradation. Both pea and *Arabidopsis* leaf color mutants are caused by *HO1* deficiency [64]. The *Arabidopsis 'flu'* mutant gene is located at the *Hy1* site, which encodes a heme oxidase. A decrease in heme oxygenase activity in the mutant results in the accumulation of heme, while the accumulation of heme inhibits the activity of glutamyl tRNA reductase, which indicates that heme can provide feedback regulation for the activity of glutamyl tRNA reductase [50]. In the study of the rice *YELLOW-GREEN LEAF2* mutant, a 7-kb insertion was found in the first exon of *YGL2/HO1*, resulting in a significant reduction in the expression of *ygl2* in the *ygl2* mutant, which hinders chlorophyll synthesis and leads to the yellow leaf phenotype [65]. A 45 bp fragment was found to be inserted into the gene encoding heme oxygenase through map-based cloning, which inhibited the expression of the heme oxygenase encoding gene and resulted in leaf color mutation [66]. In addition, leaf color mutants lacking phytochrome chromophores can be used to study the regulation of heme on chlorophyll synthesis, as well as may be used as an ideal model with which to study the photomorphogenesis of higher plants [67].

3.2. Abnormal Chloroplast Development and Differentiation

Chloroplasts originate from protoplasts and are autonomous organelles in plant cells. Their differentiation and development are controlled by the interaction and co-regulation of nuclear genes that encode related proteins and plastid genes involving gene transcription, RNA processing, and protein translation, folding and transportation [68]. This process can be divided into seven steps: nuclear gene transcription, chloroplast protein input and processing, chloroplast gene transcription and translation, thylakoid formation, pigment synthesis, plasmid–nuclear signal transduction, and chloroplast division [69]. Additionally, chloroplast development is closely related to chlorophyll content, and any obstructed pathway in the process of chloroplast development leads to chloroplast hypoplasia, thus affecting chlorophyll content in plants and causing leaf color mutations.

The regulatory pathway of the sesame yellow leaf character mutation was first analyzed in the *'Siyl-1'* sesame mutant with yellow-green leaf color. The results showed that the number of chloroplasts and the morphological structure of the mutants changed significantly, and the chlorophyll content also decreased significantly [33]. The generation of leaf color variation is also closely related to the obstruction of the plastid-nuclear signal transduction pathway. Nuclear genes can encode chloroplast proteins, regulate the metabolic state of chloroplasts, and regulate the transcription and translation of chloroplast genes. Plastids can also regulate nuclear gene expression via retrograde signaling pathways [70]. In the study of *Arabidopsis 'cue'* mutants, the leaves of *'cue'* mutants with the visible phenotypes (virescent, yellow-green, pale), defective chloroplast development, delayed differentiation of chloroplasts, and defects in mesophyll structure. The results show that there were complex gene interactions in the cells, and nuclear-cytoplasmic genes coordinated the growth and development of plants through signal molecules [71]. The dynamic balance of chloroplast development, protein synthesis, and degradation is also an important factor affecting leaf color mutation [72]. The results of rice pale green mutants indicated that the decrease of chlorophyll content in leaves was mainly caused by the mutation of the protein *CSP41b* encoding chloroplast development [73].

In addition to being an important part of chloroplast structure, the thylakoid also plays an important role in the function of chloroplasts. One study showed that variation in thylakoid and vesicle structure can lead to the formation of *'vipp1'* in *Arabidopsis* mutants [74]. While researching the barley yellow-green leaf color mutant *'ygl9'*, it was found that grana and stroma thylakoids were severely linear, grana could not be stacked normally, and the chloroplast ultrastructure was damaged in leaves of *'ygl9'* at seeding stage. [75]. In the study of a watermelon *yellow leaf (YL)* mutant, compared with ZK, the chloroplasts of the YL plant underwent incomplete development, resulting in a small number of grana thylakoids in the chloroplasts in which the arrangement was disordered and the cell metabolism was weak; these changes ultimately affected the light harvesting ability of the plants [76].

3.3. Abnormal Carotenoid Metabolism Pathway

Carotenoids are mainly divided into yellow carotene and orange lutein. Carotenoids have the function of absorbing and transmitting light energy in plants, which can play an important role in the protection of chlorophyll [77]. Regulation of carotenoid biosynthesis is mainly achieved through regulation of carotenoid content via the level of transcription of the enzymes and genes involved in carotenoid synthesis, and through regulating the type and quantity of carotenoids produced [78].

The main genes involved in plant carotenoid synthesis are phytoene synthase (PSY), phytoene desaturase (PDS), ζ-carotene desaturase (ZDS), lycopeneβ-cyclase (LCYB) and lycopene epsilon-cyclase (LYCE), among others. Phytoene synthase is considered the most important regulatory enzyme in the carotenoid synthesis pathway [79]. In *Arabidopsis*, the overexpression of *PSY* can lead to a significant increase in β-carotene content in leaves, and a corresponding increase in total carotenoids, which facilitates greening when etiolated seedlings emerge from the soil [80]. Ectopic expression of *PSY1* in tobacco leads to leaves showing abnormal pigmentation, and very young leaves are sometimes colored bright orange, before rapidly turning green [81]. The *immutans (im)* variegation mutant of *Arabidopsis* has light-dependent variegated phenotypes (green and white leaf sectors), the green leaf sectors contain normal chloroplasts, while the white leaf sectors contain abnormal chloroplasts that lack carotenoids due to a defect in phytoene desaturase activity [82,83]. In the study of *PHS* mutants in rice, the main enzymes involved in the anabolic pathway of carotene and lutein were mutated, resulting in photooxidative damage of leaf photosystem II (PSII). The core proteins CP43, CP47, and D1 of PSII were all reduced, accompanied by the accumulation of reactive oxygen species (ROS) in the plants. This result showed that damage to carotenoid biosynthesis would result in photooxidation damage and abscisic acid (ABA) deficiency, which would lead to the sprouting of spikes and the appearance of albinism in leaves [84]. Park et al. [85] showed that carotenoids have a feedback regulating effect on plastids, such that changes in carotenoid synthesis can regulate the development of plastids.

3.4. *Abnormal Anthocyanin Metabolism Pathway*

Anthocyanins are a class of water-soluble pigments widely present in plants that belongs to a group of flavonoids produced by the secondary metabolism of plants. It mainly accumulates in the form of glycosides in plant vacuoles and can be transformed from chlorophyll. The generalized term anthocyanin refers to all kinds of anthocyanin glycosides [86]. The anthocyanin biosynthesis pathway is an important branch of the flavonoid metabolism pathway, which is closely related to the presentation of plant color [87]. Its biosynthetic pathway has been clearly demonstrated through the research of model plants such as *Arabidopsis*, corn, petunia, etc.

3.4.1. Mutations of Structural Genes Related to the Anthocyanin Synthesis Pathway

Anthocyanin biosynthesis is a direct precursor of phenylalanine in the cytoplasm, which is catalyzed by a series of enzymes encoded by structural genes. After various modifications, it is transported to the vacuole and other places for storage. The synthetic pathway from phenylalanine to anthocyanins can be divided into three stages: (1) phenylalanine produces 4-coumaroyl CoA, which is catalyzed by phenylalanine ammonia-lyase; (2) 4-coumaroyl CoA produces colorless anthocyanins, which are catalyzed by biological enzymes such as chalcone synthase; and (3) colorless anthocyanins catalyze the production of anthocyanins through a series of enzymes in the end-stage metabolic pathway. In this process, a series of structural, modification and transport-related genes are directly and positively regulated by a complex (MBW complex for short) composed of MYB (v-myb avian myeloblastosis viral oncogene homolog), bHLH (basic helIX-loop-helIX) and WDR (WD repeat) regulatory factors [88,89]. Therefore, the genes that affect anthocyanin metabolism can be divided into two categories: structural genes and regulatory genes. Anthocyanin synthesis in plants is strongly influenced by the structural genes in the metabolic pathway, because the structural genes directly encode the enzymes needed in the biosynthesis pathway of anthocyanin metabolism, including phenylalanine ammonia-lyase (PAL), chalcone synthase (CHS), chalcone isomerase (CHI), flavanone 3′–5′ hydroxylase (F3′5′H) and anthocyanin synthase (ANS), etc. (Figure 4) [90,91]. Variations in these structural or regulatory genes can lead to the formation of various leaf color mutations.

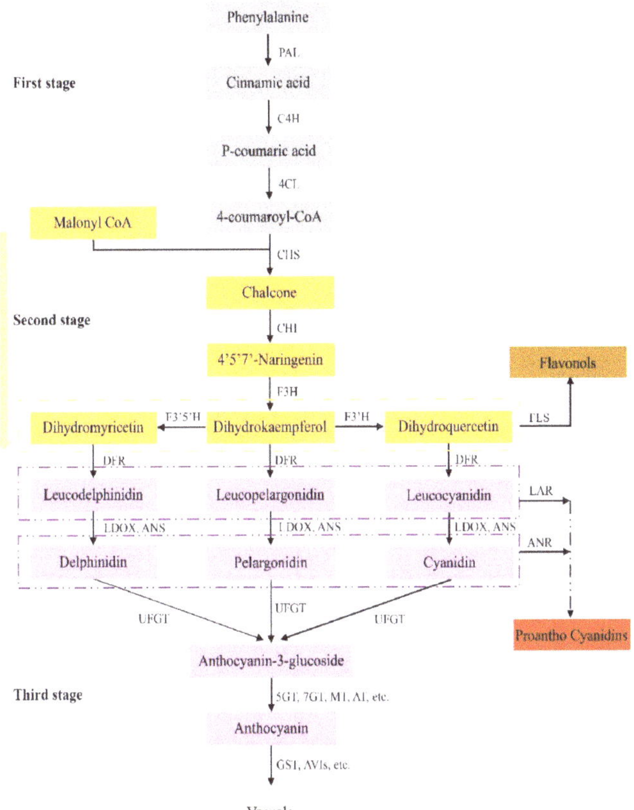

Figure 4. Biosynthesis pathway of anthocyanins in plants. PAL, Phenylalanine ammonia lyase; C4H, Cinnamate 4-hydroxylase, 4CL, 4-coumarate CoA ligase; CHS, Chalcone synthase; CHI, Chalcone isomerase; F3H, Flavanone 3-hydroxylase; F3′H, Flavonoid 3′-hydroxylase; F3′5′H, Flavonoid 3′,5′-hydroxylase; DFR, Dihydroflavonol 4-reductase; ANS, Anthocyanidin synthase; LDOX, leucoanthocyanidin dioxygenase; UFGT, Flavonoid 3-O-glucosyltransferase; 5GT, Anthocyanin 5-O-glucosyltransferase; 7GT, Flavonoid 7-O-glucosyltransferase; MT, Methyl transferase; AT, Anthocyanin acyltransferase; GST, Glutathione S-transferase; AVIs, Anthocyanic vacuolar inclusions.

In the anthocyanin metabolism pathway, the PAL is the initial enzyme of anthocyanin synthesis and an important regulatory site. Its expression is regulated by its own development and environmental factors. The PAL activity and anthocyanin content of purple-foliage plum leaves were found to increase under light induction, and the leaves gradually turned purplish red [92]. CHS is another key enzyme in the anthocyanin biosynthetic pathway. By using RNA interference to inhibit the expression of *Torenia hybrida CHS*, the blue *T. hybrida* rich in mallow pigment and peony pigment can be transformed into white *T. hybrida* with anthocyanin deficiency [93]. When *Freesia hybrida CHS1* was imported into *Petunia hybrida*, its color changed from white to pink [94]. CHI is also related to the performance of plant leaf color. Inhibiting its expression or reducing its enzyme activity causes a large accumulation of 4′-hydroxychalcone in the anthocyanin synthesis pathway, which affects the rate of anthocyanin synthesis [95]. For example, *CHI* silencing can turn the color of carnation and tobacco to yellow [96]. Moreover, the ANS regulates the oxidation of downstream colorless proanthocyanin into colored anthocyanin. In a study of the onion mutant *ANSPS*, the gene encoding the ANS was found to have a base insertion mutation, which caused the mutant material to have a different yellow color from the wild type [97]. Overexpression of the *F3′5′H* in roses and chrysanthemums can lead to the synthesis of delphinium in flowers, thus generating blue-hued flowers [98,99]. Therefore, disorders of the anthocyanin metabolic pathway are an important mechanism through which to produce leaf color mutations.

3.4.2. Abnormal Regulatory Factors Related to the Anthocyanin Synthesis Pathway

Anthocyanin synthesis regulatory genes play an important role in plant metabolism. Although they are not structural genes, they are very important for the regulation of structural genes. These transcription factors regulate the metabolic pathways of anthocyanin biosynthesis by binding to the cis-acting elements in the promoters of structural genes to regulate the expression of one or more genes in the biosynthesis pathway. There are three types of transcription factors: MYB (v-myb avian myeloblastosis viral oncogene homolog), bHLH (basic helIX-loop-helIX), and WDR (WD40 repeat proteins). The expression of structural genes is directly controlled by the MBW (MYB-bHLH-WD40) protein complex, which is formed by three types of transcription factors: MYB, bHLH and WDR. The specific combinations of MYB, bHLH and WDR in the MBW protein complex determine the target and intensity of the complex regulation [100,101].

In the process of anthocyanin biosynthesis, most *MYB* transcription factors have positive regulatory effects and usually exist in plant genomes in the form of gene families, including two motifs, such as *R2* and *R3*, or one motif in *R3*. The helIX-helIX-turn-helIX structure at the C-terminal of the conserved domain can specifically recognize and bind to DNA, while its highly conserved C-terminal is of great significance for anthocyanin regulation [102,103]. *PtrMYB119* function as transcriptional activators of anthocyanin accumulation in both Arabidopsis and poplar. Overexpression of *PtrMYB119* in hybrid poplar resulted in elevated accumulation of anthocyanins in whole plants, which had strong red-color pigmentation, especially in leaves relative to non-transformed control plants [12]. Additionally, *PAP1* (production of anthocyanin segment 1) is a key *MYB* class transcription factor that regulates anthocyanin synthesis and affects the expression of structural genes such as *CHS*, *CHI* and *ANS*. For example, activating the overexpression of *Arabidopsis PAP1* can induce the accumulation of anthocyanins in large quantities, which leads to the dark purple color of the leaves of *Arabidopsis PAP1* [104]. In blood-fleshed peach, the coloring BL upstream of the MYB can activate the anthocyanin MYB transcription, resulting in anthocyanin accumulation in the flesh, which then becomes blood-colored [105].

The bHLH transcription factor is involved in the regulation of various physiological pathways, among which the regulation of flavonoid and anthocyanin synthesis is one of its most important functions [106]. In *Setaria italica*, *PPLS1* is a bHLH transcription factor that regulates the purple color of pulvinus and leaf sheath in foxtail millet. Co-expression of both *PPLS1* and *SiMYB85* in tobacco leaves increased anthocyanin accumulation and the expression of genes involved in anthocyanin biosynthesis (*NtF3H*, *NtDFR* and *Nt3GT*), resulting in purple leaves [107]. Moreover, the bHLH transcription factor can also coregulate the target gene with MYB. For example, expression of *Medicago truncatula MtTT8* in *A. thaliana tt8* mutant can produce anthocyanins and restore the phenotype of anthocyanin deficiency of the mutant [108]. The regulatory R2R3-MYB transcription factor *AmRosea1* and the bHLH transcription factor *AmDelila* in *Antirrhinum majus* have been extensively studied. The simultaneous expression of the *AmRosea1* and *AmDelila* activates anthocyanin production in the orange carrot, which young leaves were dark purple and matured into dark green leaves during the subsequent growth relative to non-transformed plants [109]. In addition, the chrysanthemum *CmbHLH* transcription factor can activate *DFR* expression and promote anthocyanin synthesis when co-expressed with the *CmMYB6* transcription factor [110].

In all the WDR proteins studied, their functions were more similar to those of an interaction platform between other proteins, and their role may that of an intermediate medium connecting MYB and bHLH to form a complex [111]. For example, the *bHLH* transcription factors *TT8*, *EGl3* and WD40 repeat proteins *TTG1* of *Arabidopsis* form MBW complexes, which regulate the expression of

DFR, *ANS* and *tt19*, thus affecting anthocyanin biosynthesis [112]. In addition, WD40 can stabilize the MBW protein complex and directly regulate the transcriptional regulation of the anthocyanin synthesis pathway [113]. For example, in the study of *M. truncatula*, the silencing of *MYB5* and *MYB1* in *Arabidopsis* was found to directly change the expression pattern of the structural genes, resulting in a decrease in anthocyanin content [114].

4. Mapping and Cloning of Leaf Color Mutational Genes

With the rapid development of next-generation genome sequencing technologies, the application of genome, pangenome and transcriptome is increasingly extensive in revealing the genetic variation traits and dissecting the genetic regulation mechanism of key genes in plants. Current research is focused on revealing the mechanisms of growth, fruit quality, stress response, and mutation traits, as well as gender determination and flowering, in plants [115–119]. Currently, map-based cloning and RNA-seq are mainly used to mine genes related to leaf color mutations. By using next-generation genome sequencing technologies for the analysis of genomic and transcriptomic data, it is possible to greatly improve the efficiency and precision of the identification of leaf color mutation related genes in plant. The identification, cloning and functional study of these mutant genes plays an important role in elucidating the molecular mechanisms of leaf color mutation, plant photosynthesis, pigment synthesis, etc.

4.1. Using Map-Based Cloning to Discover Mutant Genes

Map-based cloning, also known as positional cloning, is a new technology for identifying and isolating genes through forward genetics. This technology can gradually locate and clone the gene of interest based on the traits of the organism and the position of closely linked molecular markers on the chromosome without knowing the sequence of the gene of interest in advance. Map-based cloning technology includes the following three steps: the first is to build positioning groups through inbreeding, hybridization, or back-crossing, etc.; the second is to find the markers linked with the gene of interest through molecular marker technology to build a genetic map; and the third is to predict and screen candidate genes and determine the function of the gene of interest by means of molecular biology [120].

The genes controlling leaf colors in plants are usually closely related to genes related to chlorophyll metabolism or chloroplast development. For instance, in the study of rice mutants, gene mapping of yellow-green leaf color genes in rice were performed using map cloning. The results showed that the gene encoding the chloroplast signal recognition particle had a mutation from *A* to *T*, which led to the change in leaf color [121]. In the study of tomato mutants, a single base mutation in the first intron was found using map-based cloning to result in downregulation of *WV* expression, which inhibited the expression of chloroplast-encoded genes and blocked chloroplast formation and chlorophyll synthesis, thereby producing leaf color mutants [122]. At present, map-based cloning technology is increasingly being used to study mutants, and many genes related to chlorophyll synthesis and chloroplast development have been cloned in rice, barley, pear, *Brassica oleracea*, and cucumber [23,123–126].

4.2. Using RNA-Seq to Discover Mutant Genes

RNA-seq, also known as mRNA-seq, is used to study changes in gene expression at the mRNA level and can directly perform transcriptome analysis on a species without reference to genome information. At present, it is considered to be an effective method for discovering new genes and for annotating coding and noncoding genes in the study of plant leaf color mutants, based on the different expression levels of transcripts among different samples to obtain the differentially expressed genes (DEGs) [127]. Through the functional annotation and enrichment analysis of differentially expressed genes, we can discover some key candidate genes, which is helpful in clarifying the molecular mechanisms of leaf color mutation [128].

RNA-seq is widely used in leaf color mutants of ornamental plants. For example, in the mutants (etiolated, rubescent, and albino) of *Anthurium andraeanum* 'Sonate', the ultrastructure of chloroplasts in mutant leaves was disrupted, and very few intact chloroplasts were observed. The ratio of carotenoid/total Chl in all three mutants was higher than that of the wild type and the content of anthocyanin was at a similar level in the leaf of etiolated mutant and wild type plants, while the albino leaves had the lowest anthocyanin content. After RNA sequencing, most genes related to chlorophyll synthesis, anthocyanin transport and pigment biosynthesis were down-regulated compared with the wild type. These results indicate that the abnormal leaf color of mutants was caused by changes in the expression pattern of genes responsible for pigment biosynthesis [129]. Therefore, it can be concluded that the chloroplast development, division and the mutation of chlorophyll and anthocyanin synthesis pathway in anthocyanin mutants affected the synthesis of chlorophyll and anthocyanin, and finally led to abnormal leaf color [130]. In addition, we also found the main differentially expressed genes related to chlorophyll metabolism, chloroplast development and the photosynthetic system after analyzing the transcriptome data of cucumber [32], tomato [131], barley [132], wheat [133], Birch [134], etc.

5. Conclusions

A growing number of genes that control or influence pigment metabolism, chloroplast development and differentiation, and photosynthesis have been identified from a variety of leaf color mutants. Through the study of their function and interaction relationships, the formation mechanism of leaf color mutants has been clarified, which is of great significance to the study of leaf color mutants. At present, research on the formation of leaf color mutants has focused on the functions of related genes and transcription factors, such as chlorophyll metabolism, chloroplast development and anthocyanin metabolism, while few studies have examined nuclear-cytoplasmic gene interactions, transcription factors, regulatory elements, and other pigment metabolism. Moreover, gene transcription and translation are influenced not only by transcription factors but also by non-coding RNA (such as miRNA and lncRNA), epigenetics (such as DNA methylation and histone acetylation), and protein spatial structure, although few studies have reviewed these influences. Therefore, in future studies, we should strengthen the research on other pigment molecules, nuclear-plasmid signal transduction, and transcription regulation, as well as examining noncoding RNA, DNA methylation, etc. This will help us to obtain more relevant information on the regulation of pigment anabolic metabolism, and to analyze the molecular regulation mechanisms of plant leaf color mutations from a system network level.

Author Contributions: Conceptualization, X.-Y.Z. and M.-H.Z.; methodology, X.L., X.-X.Z. and H.Z.; validation, M.-H.Z., X.L. and X.-X.Z.; resources, M.-H.Z. and H.Z.; writing—original draft preparation, M.-H.Z.; writing—review and editing, R.R.-S. and X.-Y.Z.; supervision, X.-Y.Z.; project administration, X.-Y.Z.; funding acquisition, X.-Y.Z. All authors have read and agreed to the published version of the manuscript.

Funding: This research was funded by the Fundamental Research Funds for the Central Universities, grant number 2572017DA02.

Conflicts of Interest: The authors declare no conflict of interest.

References

1. Awan, M.A.; Konzak, C.F.; Rutger, J.N.; Nilan, R.A. Mutagenic effects of sodium azide in rice. *Crop Sci.* **1980**, *20*, 663–668. [CrossRef]
2. Wang, L.; Yue, C.; Cao, H.; Zhou, Y.; Zeng, J.; Yang, Y.; Wang, X. Biochemical and transcriptome analyses of a novel chlorophyll-deficient chlorina tea plant cultivar. *BMC Plant Biol.* **2014**, 14. [CrossRef] [PubMed]
3. Beale, S.I.; Appleman, D. Chlorophyll synthesis in chlorella: Regulation by degree of light limitation of growth. *Plant Physiol.* **1971**, *47*, 230–235. [CrossRef] [PubMed]
4. Brestic, M.; Zivcak, M.; Kunderlikova, K.; Allakhverdiev, S.I. High temperature specifically affects the photoprotective responses of chlorophyll b-deficient wheat mutant lines. *Photosynth. Res.* **2016**, *130*, 251–266. [CrossRef] [PubMed]

5. Wu, Z.M.; Zhang, X.; Wang, J.L.; Wan, J.M. Leaf chloroplast ultrastructure and photosynthetic properties of a chlorophyll-deficient mutant of rice. *Photosynthetica* **2014**, *52*, 217–222. [CrossRef]
6. Seetharami Reddi, T.V.; Reddi, V.R. Frequency and spectrum of chlorophyll mutants induced in rice by chemical mutagens. *Theor. Appl. Genet.* **1984**, *67*, 231–233. [CrossRef]
7. Falbel, T.G.; Staehelin, L.A. Partial blocks in the early steps of the chlorophyll synthesis pathway: A common feature of chlorophyll b-deficient mutants. *Physiol. Plant.* **1996**, *97*, 311–320. [CrossRef]
8. Tanaka, A.; Tanaka, R. Chlorophyll metabolism. *Curr. Opin. Plant Biol.* **2006**, *9*, 248–255. [CrossRef]
9. Chen, G.; Bi, Y.R.; Li, N. EGY1 encodes a membrane-associated and ATP-independent metalloprotease that is required for chloroplast development. *Plant J.* **2005**, *41*, 364–375. [CrossRef]
10. Lonosky, P.M.; Zhang, X.S.; Honavar, V.G.; Dobbs, D.L.; Fu, A.; Rodermel, S.R. A proteomic analysis of maize chloroplast biogenesis. *Plant Physiol.* **2004**, *134*, 560–574. [CrossRef]
11. Hu, Y.; Chen, B. Arbuscular mycorrhiza induced putrescine degradation into γ-aminobutyric acid, malic acid accumulation, and improvement of nitrogen assimilation in roots of water-stressed maize plants. *Mycorrhiza* **2020**, *30*, 329–339. [CrossRef] [PubMed]
12. Cho, J.-S.; Van Phap, N.; Jeon, H.-W.; Kim, M.-H.; Eom, S.H.; Lim, Y.J.; Kim, W.-C.; Park, E.-J.; Choi, Y.-I.; Ko, J.-H. Overexpression of *PtrMYB119*, a R2R3-MYB transcription factor from *Populus trichocarpa*, promotes anthocyanin production in hybrid poplar. *Tree Physiol.* **2016**, *36*, 1162–1176. [CrossRef] [PubMed]
13. Hu, Y.; Peuke, A.D.; Zhao, X.; Yan, J.; Li, C. Effects of simulated atmospheric nitrogen deposition on foliar chemistry and physiology of hybrid poplar seedlings. *Plant Physiol. Biochem.* **2019**, *143*, 94–108. [CrossRef] [PubMed]
14. Wu, S.; Guo, Y.; Adil, M.F.; Sehar, S.; Cai, B.; Xiang, Z.; Tu, Y.; Zhao, D.; Shamsi, I.H. Comparative proteomic analysis by iTRAQ reveals that plastid pigment metabolism contributes to leaf color changes in tobacco (Nicotiana tabacum) during curing. *Int. J. Mol. Sci.* **2020**, *21*, 2394. [CrossRef] [PubMed]
15. Zhou, h.; Wang, K.; Wang, F.; Espley, R.V.; Ren, F.; Zhao, J.; Ogutu, C.; He, H.; Jiang, Q.; Allan, A.C.; et al. Activator-type R2R3-MYB genes induce a repressor-type R2R3-MYB gene to balance anthocyanin and proanthocyanidin accumulation. *New Phytol.* **2019**, *221*, 1919–1934. [CrossRef] [PubMed]
16. Bae, C.H.; Abe, T.; Matsuyama, T.; Fukunishi, N.; Nagata, N.; Nakano, T.; Kaneko, Y.; Miyoshi, K.; Matsushima, H.; Yoshida, S. Regulation of chloroplast gene expression is affected in ali, a novel tobacco albino mutant. *Ann. Bot.* **2001**, *88*, 545–553. [CrossRef]
17. Tominaga-Wada, R.; Nukumizu, Y.; Wada, T. Flowering is delayed by mutations in homologous genes CAPRICE and TRYPTICHON in the early flowering Arabidopsis cpl3 mutant. *J. Plant Physiol.* **2013**, *170*, 1466–1468. [CrossRef] [PubMed]
18. Moon, J.; Zhu, L.; Shen, H.; Huq, E. *PIF1* directly and indirectly regulates chlorophyll biosynthesis to optimize the greening process in Arabidopsis. *Proc. Natl. Acad. Sci. USA* **2008**, *105*, 9433–9438. [CrossRef]
19. Aluru, M.R.; Yu, F.; Fu, A.; Rodermel, S. Arabidopsis variegation mutants: New insights into chloroplast biogenesis. *J. Exp. Bot.* **2006**, *57*, 1871–1881. [CrossRef]
20. Jarvis, P.; Chen, L.J.; Li, H.; Peto, C.A.; Fankhauser, C.; Chory, J. An *Arabidopsis* mutant defective in the plastid general protein import apparatus. *Science* **1998**, *282*, 100–103. [CrossRef]
21. Zheng, J.; Wu, H.; Zhu, H.; Huang, C.; Liu, C.; Chang, Y.; Kong, Z.; Zhou, Z.; Wang, G.; Lin, Y.; et al. Determining factors, regulation system, and domestication of anthocyanin biosynthesis in rice leaves. *New Phytol.* **2019**, *223*, 705–721. [CrossRef]
22. Deng, L.; Qin, P.; Liu, Z.; Wang, G.; Chen, W.; Tong, J.; Xiao, L.; Tu, B.; Sun, Y.; Yan, W.; et al. Characterization and fine-mapping of a novel premature leaf senescence mutant yellow leaf and dwarf 1 in rice. *Plant Physiol. Biochem.* **2017**, *111*, 50–58. [CrossRef] [PubMed]
23. Zhu, X.; Guo, S.; Wang, Z.; Du, Q.; Xing, Y.; Zhang, T.; Shen, W.; Sang, X.; Ling, Y.; He, G. Map-based cloning and functional analysis of YGL8, which controls leaf colour in rice (Oryza sativa). *BMC Plant Biol.* **2016**, 16. [CrossRef] [PubMed]
24. Jung, K.H.; Hur, J.; Ryu, C.H.; Choi, Y.; Chung, Y.Y.; Miyao, A.; Hirochika, H.; An, G.H. Characterization of a rice chlorophyll-deficient mutant using the T-DNA gene-trap system. *Plant Cell Physiol.* **2003**, *44*, 463–472. [CrossRef]
25. Wang, S.; Wang, P.; Gao, L.; Yang, R.; Li, L.; Zhang, E.; Wang, Q.; Li, Y.; Yin, Z. Characterization and complementation of a chlorophyll-less dominant mutant GL1 in Lagerstroemia indica. *DNA Cell Biol.* **2017**, *36*, 354–366. [CrossRef] [PubMed]

26. Li, Y.; Zhang, Z.; Wang, P.; Wang, S.; Ma, L.; Li, L.; Yang, R.; Ma, Y.; Wang, Q. Comprehensive transcriptome analysis discovers novel candidate genes related to leaf color in a Lagerstroemia indica yellow leaf mutant. *Genes Genom.* **2015**, *37*, 851–863. [CrossRef]
27. Liu, L.; Sun, T.; Liu, X.; Guo, Y.; Huang, X.; Gao, P.; Wang, X. Genetic analysis and mapping of a striped rind gene (st3) in melon (Cucumis melo L.). *Euphytica* **2019**, *215*. [CrossRef]
28. Kusumi, K.; Sakata, C.; Nakamura, T.; Kawasaki, S.; Yoshimura, A.; Iba, K. A plastid protein NUS1 is essential for build-up of the genetic system for early chloroplast development under cold stress conditions. *Plant J.* **2011**, *68*, 1039–1050. [CrossRef]
29. Su, N.; Hu, M.-L.; Wu, D.-X.; Wu, F.-Q.; Fei, G.-L.; Lan, Y.; Chen, X.-L.; Shu, X.-L.; Zhang, X.; Guo, X.-P.; et al. Disruption of a rice pentatricopeptide repeat protein causes a seedling-specific albino phenotype and its utilization to enhance seed purity in hybrid rice production. *Plant Physiol.* **2012**, *159*, 227–238. [CrossRef]
30. Yoo, S.-C.; Cho, S.-H.; Sugimoto, H.; Li, J.; Kusumi, K.; Koh, H.-J.; Iba, K.; Paek, N.-C. Rice virescent3 and stripe1 encoding the large and small subunits of ribonucleotide reductase are required for chloroplast biogenesis during early leaf development. *Plant Physiol.* **2009**, *150*, 388–401. [CrossRef]
31. Cheng, G.-X.; Zhang, R.-X.; Liu, S.; He, Y.-M.; Feng, X.-H.; Ul Haq, S.; Luo, D.-X.; Gong, Z.-H. Leaf-color mutation induced by ethyl methane sulfonate and genetic and physio-biochemical characterization of leaf-color mutants in pepper (Capsicum annuum L.). *Sci. Hortic.* **2019**, *257*. [CrossRef]
32. Cao, W.; Du, Y.; Wang, C.; Xu, L.; Wu, T. Cscs encoding chorismate synthase is a candidate gene for leaf variegation mutation in cucumber. *Breed. Sci.* **2018**, *68*, 571–581. [CrossRef] [PubMed]
33. Gao, T.-M.; Wei, S.-L.; Chen, J.; Wu, Y.; Li, F.; Wei, L.-B.; Li, C.; Zeng, Y.-J.; Tian, Y.; Wang, D.-Y.; et al. Cytological, genetic, and proteomic analysis of a sesame (Sesamum indicum L.) mutant Siyl-1 with yellow-green leaf color. *Genes Genom.* **2020**, *42*, 25–39. [CrossRef] [PubMed]
34. La Rocca, N.; Rascio, N.; Oster, U.; Rudiger, W. Amitrole treatment of etiolated barley seedlings leads to deregulation of tetrapyrrole synthesis and to reduced expression of Lhc and RbcS genes. *Planta* **2001**, *213*, 101–108. [CrossRef] [PubMed]
35. Monde, R.A.; Zito, F.; Olive, J.; Wollman, F.A.; Stern, D.B. Post-transcriptional defects in tobacco chloroplast mutants lacking the cytochrome b6/f complex. *Plant J.* **2000**, *21*, 61–72. [CrossRef]
36. Hosticka, L.P.; Hanson, M.R. Induction of plastid mutations in tomatoes by nitrosomethylurea. *J. Hered.* **1984**, *75*, 242–246. [CrossRef]
37. Li, J.; Yu, X.; Cai, Z.; Wu, F.; Luo, J.; Zheng, L.; Chu, W. An overview of chlorophyll biosynthesis in higher plants. *Mol. Plant Breed.* **2019**, *17*, 6013–6019. [CrossRef]
38. Beale, S.I. Green genes gleaned. *Trends Plant Sci.* **2005**, *10*, 309–312. [CrossRef]
39. Matringe, M.; Camadro, J.M.; Block, M.A.; Joyard, J.; Scalla, R.; Labbe, P.; Douce, R. Localization within chloroplasts of protoporphyrinogen oxidase, the target enzyme for diphenylether-like herbicides. *J. Biol. Chem.* **1992**, *267*, 4646–4651.
40. Tanaka, R.; Kobayashi, K.; Masuda, T. Tetrapyrrole Metabolism in Arabidopsis thaliana. *Arab. Book* **2011**, *9*, e0145. [CrossRef]
41. Sampath, P.; Mazumder, B.; Seshadri, V.; Gerber, C.A.; Chavatte, L.; Kinter, M.; Ting, S.M.; Dignam, J.D.; Kim, S.; Driscoll, D.M.; et al. Noncanonical function of glutamyl-prolyi-tRNA synthetase: Gene-specific silencing of translation. *Cell* **2004**, *119*, 195–208. [CrossRef] [PubMed]
42. Lee, S.; Kim, J.H.; Yoo, E.; Lee, C.H.; Hirochika, H.; An, G.H. Differential regulation of chlorophyll a oxygenase genes in rice. *Plant Mol. Biol.* **2005**, *57*, 805–818. [CrossRef] [PubMed]
43. Sakuraba, Y.; Schelbert, S.; Park, S.-Y.; Han, S.-H.; Lee, B.-D.; Andres, C.B.; Kessler, F.; Hoertensteiner, S.; Paek, N.-C. Stay-green and chlorophyll catabolic enzymes interact at light-harvesting complex ii for chlorophyll detoxification during leaf senescence in arabidopsis. *Plant Cell* **2012**, *24*, 507–518. [CrossRef] [PubMed]
44. Wang, P.-R.; Zhang, F.-T.; Gao, J.-X.; Sun, X.-Q.; Deng, X.-J. An overview of chlorophyll biosynthesis in higher plants. *Acta Bot. Boreali Occident. Sin.* **2009**, *29*, 629–636. [CrossRef]
45. Cornah, J.E.; Terry, M.J.; Smith, A.G. Green or red: What stops the traffic in the tetrapyrrole pathway? *Trends Plant Sci.* **2003**, *8*, 224–230. [CrossRef]
46. Schmied, J.; Hedtke, B.; Grimm, B. Overexpression of *HEMA1* encoding glutamyl-tRNA reductase. *J. Plant Physiol.* **2011**, *168*, 1372–1379. [CrossRef]

47. Kumar, A.M.; Csankovszki, G.; Soll, D. A second and differentially expressed glutamyl-tRNA reductase gene from Arabidopsis thaliana. *Plant Mol. Biol.* **1996**, *30*, 419–426. [CrossRef]
48. Ilag, L.L.; Kumar, A.M.; Soll, D. Light regulation of chlorophyll biosynthesis at the level of 5-aminolevulinate formation in Arabidopsis. *Plant Cell* **1994**, *6*, 265–275. [CrossRef]
49. Kumar, A.M.; Soll, D. Antisense HEMA1 RNA expression inhibits heme and chlorophyll biosynthesis in Arabidopsis. *Plant Physiol.* **2000**, *122*, 49–56. [CrossRef]
50. Goslings, D.; Meskauskiene, R.; Kim, C.H.; Lee, K.P.; Nater, M.; Apel, K. Concurrent interactions of heme and FLU with Glu tRNA reductase (HEMA1), the target of metabolic feedback inhibition of tetrapyrrole biosynthesis, in dark- and light-grown Arabidopsis plants. *Plant J.* **2004**, *40*, 957–967. [CrossRef]
51. Walker, C.J.; Willows, R.D. Mechanism and regulation of Mg-chelatase. *Biochem. J.* **1997**, *327 Pt 2*, 321–333. [CrossRef]
52. Strand, A.; Asami, T.; Alonso, J.; Ecker, J.R.; Chory, J. Chloroplast to nucleus communication triggered by accumulation of Mg-protoporphyrinIX. *Nature* **2003**, *421*, 79–83. [CrossRef] [PubMed]
53. Rissler, H.M.; Collakova, E.; DellaPenna, D.; Whelan, J.; Pogson, B.J. Chlorophyll biosynthesis. Expression of a second chl I gene of magnesium chelatase in Arabidopsis supports only limited chlorophyll synthesis. *Plant Physiol.* **2002**, *128*, 770–779. [CrossRef] [PubMed]
54. Papenbrock, J.; Mock, H.P.; Tanaka, R.; Kruse, E.; Grimm, B. Role of magnesium chelatase activity in the early steps of the tetrapyrrole biosynthetic pathway. *Plant Physiol.* **2000**, *122*, 1161–1169. [CrossRef] [PubMed]
55. Mochizuki, N.; Brusslan, J.A.; Larkin, R.; Nagatani, A.; Chory, J. Arabidopsis genomes uncoupled 5 (GUN5) mutant reveals the involvement of Mg-chelatase H subunit in plastid-to-nucleus signal transduction. *Proc. Natl. Acad. Sci. USA* **2001**, *98*, 2053–2058. [CrossRef]
56. Shi, Y.; He, Y.; Guo, D.; Lv, X.; Huang, Q.; Wu, J. Genetic analysis and gene mapping of a pale green leaf mutant HM133 in rice. *Chin. J. Rice Sci.* **2016**, *30*, 603–610. [CrossRef]
57. Zhang, K.; Liu, Z.; Shan, X.; Li, C.; Tang, X.; Chi, M.; Feng, H. Physiological properties and chlorophyll biosynthesis in a Pak-choi (Brassica rapa L. ssp chinensis) yellow leaf mutant, pylm. *Acta Physiol. Plant.* **2017**, *39*. [CrossRef]
58. Wenkai, D.; Yuan, S.; Hu, F. Research progress on molecular mechanisms of the leaf color mutation. *Mol. Plant Breed.* **2019**, *17*, 1888–1897. [CrossRef]
59. Hoertensteiner, S. Chlorophyll degradation during senescence. *Annu. Rev. Plant Biol.* **2006**, *57*, 55–77. [CrossRef]
60. You, S.-C.; Cho, S.-H.; Zhang, H.; Paik, H.-C.; Lee, C.-H.; Li, J.; Yoo, J.-H.; Lee, B.-W.; Koh, H.-J.; Seo, H.S.; et al. Quantitative trait loci associated with functional stay-green SNU-SG1 in rice. *Mol. Cells* **2007**, *24*, 83–94.
61. Zhu, C.; Yang, F.; Shi, S.; Li, D.; Wang, Z.; Liu, H.; Huang, D.; Wang, C. Transcriptome characterization of cymbidium sinense 'dharma' using 454 pyrosequencing and its application in the identification of genes associated with leaf color variation. *PLoS ONE* **2015**, *10*, e0128592. [CrossRef] [PubMed]
62. Wilks, A. Heme oxygenase: Evolution, structure, and mechanism. *Antioxid. Redox Signal.* **2002**, *4*, 603–614. [CrossRef] [PubMed]
63. Terry, K. Feedback inhibition of chlorophyll synthesis in the phytochrome chromophore-deficient aurea and yellow-green-2 mutants of tomato. *Plant Physiol.* **1999**, *119*, 143–152. [CrossRef] [PubMed]
64. Linley, P.J.; Landsberger, M.; Kohchi, T.; Cooper, J.B.; Terry, M.J. The molecular basis of heme oxygenase deficiency in the pcd1 mutant of pea. *FEBS J.* **2006**, *273*, 2594–2606. [CrossRef] [PubMed]
65. Chen, H.; Cheng, Z.; Ma, X.; Wu, H.; Liu, Y.; Zhou, K.; Chen, Y.; Ma, W.; Bi, J.; Zhang, X.; et al. A knockdown mutation of YELLOW-GREEN LEAF2 blocks chlorophyll biosynthesis in rice. *Plant Cell Rep.* **2013**, *32*, 1855–1867. [CrossRef]
66. Li, J.; Wang, Y.; Chai, J.; Wang, L.; Wang, C.; Long, W.; Wang, D.; Wang, Y.; Zheng, M.; Peng, C.; et al. Green-revertible Chlorina 1 (grc1) is required for the biosynthesis of chlorophyll and the early development of chloroplasts in rice. *J. Plant Biol.* **2013**, *56*, 326–335. [CrossRef]
67. Fankhauser, C.; Chory, J. Light control of plant development. *Annu. Rev. Cell Dev. Biol.* **1997**, *13*, 203–229. [CrossRef]
68. Chan, K.X.; Crisp, P.A.; Estavillo, G.M.; Pogson, B.J. Chloroplast-to-nucleus communication: Current knowledge, experimental strategies and relationship to drought stress signaling. *Plant Signal. Behav.* **2010**, *5*, 1575–1582. [CrossRef]

69. Pogson, B.J.; Albrecht, V. Genetic dissection of chloroplast biogenesis and development: An overview. *Plant Physiol.* **2011**, *155*, 1545–1551. [CrossRef]
70. Koussevitzky, S.; Nott, A.; Mockler, T.C.; Hong, F.; Sachetto-Martins, G.; Surpin, M.; Lim, I.J.; Mittler, R.; Chory, J. Signals from chloroplasts converge to regulate nuclear gene expression. *Science* **2007**, *316*, 715–719. [CrossRef]
71. Lopez-Juez, E.; Jarvis, R.P.; Takeuchi, A.; Page, A.M.; Chory, J. New Arabidopsis cue mutants suggest a close connection between plastid- and phytochrome regulation of nuclear gene expression. *Plant Physiol.* **1998**, *118*, 803–815. [CrossRef]
72. Miura, E.; Kato, Y.; Matsushima, R.; Albrecht, V.; Laalami, S.; Sakamoto, W. The balance between protein synthesis and degradation in chloroplasts determines leaf variegation in Arabidopsis yellow variegated mutants. *Plant Cell* **2007**, *19*, 1313–1328. [CrossRef]
73. Mei, J.; Li, F.; Liu, X.; Hu, G.; Fu, Y.; Liu, W. Newly identified *CSP41b* gene localized in chloroplasts affects leaf color in rice. *Plant Sci.* **2017**, *256*, 39–45. [CrossRef] [PubMed]
74. Albrecht, V.; Ingenfeld, A.; Apel, K. Characterization of the Snowy cotyledon 1 mutant of Arabidopsis thaliana: The impact of chloroplast elongation factor G on chloroplast development and plant vitality. *Plant Mol. Biol.* **2006**, *60*, 507–518. [CrossRef] [PubMed]
75. Qin, D.; Li, M.; Xu, F.; Xu, Q.; Ge, S.; Dong, J. Analysis of agronomic characters and preliminary mapping of regulatory genes of a barley yellow-green leaf mutant ygl. *J. Triticeae Crop.* **2019**, *39*, 653–658.
76. Ren, Y.; Zhu, Y.; Sun, D.; Deng, Y.; An, G.; Li, W.; Liu, J. Physiological characteristic analysis of a leaf-yellowing mutant in watermelon. *J. Fruit Sci.* **2020**, *37*, 565–573. [CrossRef]
77. Polivka, T.; Frank, H.A. Molecular factors controlling photosynthetic light harvesting by carotenoids. *Acc. Chem. Res.* **2010**, *43*, 1125–1134. [CrossRef]
78. Giorio, G.; Stigliani, A.L.; D'Ambrosio, C. Phytoene synthase genes in tomato (Solanum lycopersicum L.)—New data on the structures, the deduced amino acid sequences and the expression patterns. *FEBS J.* **2008**, *275*, 527–535. [CrossRef]
79. Cazzonelli, C.I.; Pogson, B.J. Source to sink: Regulation of carotenoid biosynthesis in plants. *Trends Plant Sci.* **2010**, *15*, 266–274. [CrossRef]
80. Toledo-Ortiz, G.; Huq, E.; Rodriguez-Concepcion, M. Direct regulation of phytoene synthase gene expression and carotenoid biosynthesis by phytochrome-interacting factors. *Proc. Natl. Acad. Sci. USA* **2010**, *107*, 11626–11631. [CrossRef]
81. Busch, M.; Seuter, A.; Hain, R. Functional analysis of the early steps of carotenoid biosynthesis in tobacco. *Plant Physiol.* **2002**, *128*, 439–453. [CrossRef] [PubMed]
82. Aluru, M.R.; Bae, H.; Wu, D.; Rodermel, S.R. The Arabidopsis immutans mutation affects plastid differentiation and the morphogenesis of white and green sectors in variegated plants. *Plant Physiol.* **2001**, *127*, 67–77. [CrossRef] [PubMed]
83. Aluru, M.R.; Zola, J.; Foudree, A.; Rodermel, S.R. Chloroplast photooxidation-induced transcriptome reprogramming in Arabidopsis immutans white leaf sectors. *Plant Physiol.* **2009**, *150*, 904–923. [CrossRef] [PubMed]
84. Fang, J.; Chai, C.; Qian, Q.; Li, C.; Tang, J.; Sun, L.; Huang, Z.; Guo, X.; Sun, C.; Liu, M.; et al. Mutations of genes in synthesis of the carotenoid precursors of ABA lead to pre-harvest sprouting and photo-oxidation in rice. *Plant J.* **2008**, *54*, 177–189. [CrossRef]
85. Park, H.; Kreunen, S.S.; Cuttriss, A.J.; Della, P.D.; Pogson, B.J. Identification of the carotenoid isomerase provides insight into carotenoid biosynthesis, prolamellar body formation, and photomorphogenesis. *Plant Cell* **2002**, *14*, 321–332. [CrossRef]
86. Seeram, N.P.; Nair, M.G. Inhibition of lipid peroxidation and structure-activity-related studies of the dietary constituents anthocyanins, anthocyanidins, and catechins. *J. Agric. Food Chem.* **2002**, *50*, 5308–5312. [CrossRef]
87. Zhao, D.; Tao, J. Recent advances on the development and regulation of flower color in ornamental plants. *Front. Plant Sci.* **2015**, 6. [CrossRef]
88. Zhu, Z.; Wang, H.; Wang, Y.; Guan, S.; Wang, F.; Tang, J.; Zhang, R.; Xie, L.; Lu, Y. Characterization of the cis elements in the proximal promoter regions of the anthocyanin pathway genes reveals a common regulatory logic that governs pathway regulation. *J. Exp. Bot.* **2015**, *66*, 3775–3789. [CrossRef]

89. Davies, K.M.; Albert, N.W.; Schwinn, K.E. From landing lights to mimicry: The molecular regulation of flower colouration and mechanisms for pigmentation patterning. *Funct. Plant Biol.* **2012**, *39*, 619–638. [CrossRef]
90. Petroni, K.; Tonelli, C. Recent advances on the regulation of anthocyanin synthesis in reproductive organs. *Plant Sci.* **2011**, *181*, 219–229. [CrossRef]
91. Holton, T.A.; Cornish, E.C. Genetics and biochemistry of anthocyanin biosynthesis. *Plant Cell* **1995**, *7*, 1071–1083. [CrossRef] [PubMed]
92. Gu, C.; Liao, L.; Zhou, H.; Wang, L.; Deng, X.; Han, Y. Constitutive activation of an anthocyanin regulatory gene PcMYB10.6 is related to red coloration in Purple-Foliage Plum. *PLoS ONE* **2015**, *10*, e0135159. [CrossRef] [PubMed]
93. Fukusaki, E.; Kawasaki, K.; Kajiyama, S.; An, C.I.; Suzuki, K.; Tanaka, Y.; Kobayashi, A. Flower color modulations of Torenia hybrida by downregulation of chalcone synthase genes with RNA interference. *J. Biotechnol.* **2004**, *111*, 229–240. [CrossRef] [PubMed]
94. Sun, W.; Meng, X.; Liang, L.; Jiang, W.; Huang, Y.; He, J.; Hu, H.; Almqvist, J.; Gao, X.; Wang, L. Molecular and biochemical analysis of chalcone synthase from freesia hybrid in flavonoid biosynthetic pathway. *PLoS ONE* **2015**, *10*, e0119154. [CrossRef] [PubMed]
95. Nishihara, M.; Nakatsuka, T.; Yamamura, S. Flavonoid components and flower color change in transgenic tobacco plants by suppression of chalcone isomerase gene. *FEBS Lett.* **2005**, *579*, 6074–6078. [CrossRef] [PubMed]
96. Itoh, Y.; Higeta, D.; Suzuki, A.; Yoshida, H.; Ozeki, Y. Excision of transposable elements from the chalcone isomerase and dihydroflavonol 4-reductase genes may contribute to the variegation of the yellow-flowered carnation (Dianthus caryophyllus). *Plant Cell Physiol.* **2002**, *43*, 578–585. [CrossRef]
97. Kim, E.-Y.; Kim, C.-W.; Kim, S. Identification of two novel mutant ANS alleles responsible for inactivation of anthocyanidin synthase and failure of anthocyanin production in onion (Allium cepa L.). *Euphytica* **2016**, *212*, 427–437. [CrossRef]
98. Brugliera, F.; Tao, G.-Q.; Tems, U.; Kalc, G.; Mouradova, E.; Price, K.; Stevenson, K.; Nakamura, N.; Stacey, I.; Katsumoto, Y.; et al. Violet/blue chrysanthemums-metabolic engineering of the anthocyanin biosynthetic pathway results in novel petal colors. *Plant Cell Physiol.* **2013**, *54*, 1696–1710. [CrossRef]
99. Katsumoto, Y.; Fukuchi-Mizutani, M.; Fukui, Y.; Brugliera, F.; Holton, T.A.; Karan, M.; Nakamura, N.; Yonekura-Sakakibara, K.; Togami, J.; Pigeaire, A.; et al. Engineering of the rose flavonoid biosynthetic pathway successfully generated blue-hued flowers accumulating delphinidin. *Plant Cell Physiol.* **2007**, *48*, 1589–1600. [CrossRef] [PubMed]
100. Cao, Y.; Li, K.; Li, Y.; Zhao, X.; Wang, L. MYB transcription factors as regulators of secondary metabolism in plants. *Biology* **2020**, *9*, 61. [CrossRef] [PubMed]
101. Ramsay, N.A.; Glover, B.J. MYB-bHLH-WD40 protein complex and the evolution of cellular diversity. *Trends Plant Sci.* **2005**, *10*, 63–70. [CrossRef] [PubMed]
102. Yamagishi, M.; Toda, S.; Tasaki, K. The novel allele of the LhMYB12 gene is involved in splatter-type spot formation on the flower tepals of Asiatic hybrid lilies (Lilium spp.). *New Phytol.* **2014**, *201*, 1009–1020. [CrossRef] [PubMed]
103. Dubos, C.; Stracke, R.; Grotewold, E.; Weisshaar, B.; Martin, C.; Lepiniec, L. MYB transcription factors in Arabidopsis. *Trends Plant Sci.* **2010**, *15*, 573–581. [CrossRef] [PubMed]
104. Borevitz, J.O.; Xia, Y.; Blount, J.; DIXon, R.A.; Lamb, C. Activation tagging identifies a conserved MYB regulator of phenylpropanoid biosynthesis. *Plant Cell* **2000**, *12*, 2383–2394. [CrossRef]
105. Zhou, H.; Kui, L.-W.; Wang, H.; Gu, C.; Dare, A.P.; Espley, R.V.; He, H.; Allan, A.C.; Han, Y. Molecular genetics of blood-fleshed peach reveals activation of anthocyanin biosynthesis by NAC transcription factors. *Plant J.* **2015**, *82*, 105–121. [CrossRef]
106. Yang, P.; Zhou, B.; Li, Y. The bHLH transcription factors involved in anthocyanin biosynthesis in plants. *Plant Physiol. J.* **2012**, *48*, 747–758. [CrossRef]
107. Bai, H.; Song, Z.; Zhang, Y.; Li, Z.; Wang, Y.; Liu, X.; Ma, J.; Quan, J.; Wu, X.; Liu, M.; et al. The bHLH transcription factor PPLS1 regulates the color of pulvinus and leaf sheath in foxtail millet (Setaria italica). *Theor. Appl. Genet.* **2020**, *133*, 1911–1926. [CrossRef]

108. Li, P.; Chen, B.; Zhang, G.; Chen, L.; Dong, Q.; Wen, J.; Mysore, K.S.; Zhao, J. Regulation of anthocyanin and proanthocyanidin biosynthesis by Medicago truncatula bHLH transcription factor MtTT8. *New Phytol.* **2016**, *210*, 905–921. [CrossRef]
109. Sharma, S.; Holme, I.B.; Dionisio, G.; Kodama, M.; Dzhanfezova, T.; Joernsgaard, B.; Brinch-Pedersen, H. Cyanidin based anthocyanin biosynthesis in orange carrot is restored by expression of AmRosea1 and AmDelila, MYB and bHLH transcription factors. *Plant Mol. Biol.* **2020**, *103*, 443–456. [CrossRef]
110. Xiang, L.-L.; Liu, X.-F.; Li, X.; Yin, X.-R.; Grierson, D.; Li, F.; Chen, K.-S. A novel bHLH transcription factor involved in regulating anthocyanin biosynthesis in Chrysanthemums (Chrysanthemum morifolium Ramat.). *PLoS ONE* **2015**, *10*, e0143892. [CrossRef]
111. Li, Q.; Zhao, P.; Li, J.; Zhang, C.; Wang, L.; Ren, Z. Genome-wide analysis of the WD-repeat protein family in cucumber and Arabidopsis. *Mol. Genet. Genom.* **2014**, *289*, 103–124. [CrossRef] [PubMed]
112. Xu, W.; Grain, D.; Le Gourrierec, J.; Harscoet, E.; Berger, A.; Jauvion, V.; Scagnelli, A.; Berger, N.; Bidzinski, P.; Kelemen, Z.; et al. Regulation of flavonoid biosynthesis involves an unexpected complex transcriptional regulation of TT8 expression, in Arabidopsis. *New Phytol.* **2013**, *198*, 59–70. [CrossRef] [PubMed]
113. Broun, P. Transcriptional control of flavonoid biosynthesis: A complex network of conserved regulators involved in multiple aspects of differentiation in Arabidopsis. *Curr. Opin. Plant Biol.* **2005**, *8*, 272–279. [CrossRef] [PubMed]
114. Liu, C.; Jun, J.H.; Dixon, R.A. MYB5 and MYB14 play pivotal roles in seed coat polymer biosynthesis in Medicago truncatula. *Plant Physiol.* **2014**, *165*, 1424–1439. [CrossRef]
115. Neale, D.B.; Kremer, A. Forest tree genomics: Growing resources and applications. *Nat. Rev. Genet.* **2011**, *12*, 111–122. [CrossRef]
116. Chen, S.; Lin, X.; Zhang, D.; Li, Q.; Zhao, X.; Chen, S. Genome-wide analysis of NAC gene family in Betula pendula. *Forests* **2019**, *10*, 741. [CrossRef]
117. Wang, S.; Huang, H.; Han, R.; Liu, C.; Qiu, Z.; Liu, G.; Chen, S.; Jiang, J. Negative feedback loop between BpAP1 and BpPI/BpDEF heterodimer in Betula platyphylla x *B. pendula*. *Plant Sci.* **2019**, 289. [CrossRef]
118. Voichek, Y.; Weigel, D. Identifying genetic variants underlying phenotypic variation in plants without complete genomes. *Nat. Genet.* **2020**, *52*, 534–540. [CrossRef] [PubMed]
119. Liu, Y.; Du, H.; Li, P.; Shen, Y.; Peng, H.; Liu, S.; Zhou, G.-A.; Zhang, H.; Liu, Z.; Shi, M.; et al. Pan-genome of wild and cultivated soybeans. *Cell* **2020**, *182*, 162. [CrossRef]
120. Tanksley, S.D.; Ganal, M.W.; Martin, G.B. Chromosome landing: A paradigm for map-based gene cloning in plants with large genomes. *Trends Genet.* **1995**, *11*, 63–68. [CrossRef]
121. Lin, T.; Sun, L.; Jing, D.; Qian, H.; Yu, B.; Zeng, S.; Li, C.; Gong, H. Characterization and gene mapping of a Yellow Green Leaf Mutant ygl14(t) in Rice. *J. Nucl. Agric. Sci.* **2018**, *32*, 216–226.
122. Gao, S.; Gao, W.; Liao, X.; Xiong, C.; Yu, G.; Yang, Q.; Yang, C.; Ye, Z. The tomato *WV* gene encoding a thioredoxin protein is essential for chloroplast development at low temperature and high light intensity. *BMC Plant Biol.* **2019**, 19. [CrossRef] [PubMed]
123. Han, F.; Cui, H.; Zhang, B.; Liu, X.; Yang, L.; Zhuang, M.; Lv, H.; Li, Z.; Wang, Y.; Fang, Z.; et al. Map-based cloning and characterization of BoCCD4, a gene responsible for white/yellow petal color in B-oleracea. *BMC Genom.* **2019**, 20. [CrossRef] [PubMed]
124. Yao, G.; Ming, M.; Allan, A.C.; Gu, C.; Li, L.; Wu, X.; Wang, R.; Chang, Y.; Qi, K.; Zhang, S.; et al. Map-based cloning of the pear gene MYB114 identifies an interaction with other transcription factors to coordinately regulate fruit anthocyanin biosynthesis. *Plant J.* **2017**, *92*, 437–451. [CrossRef]
125. Gao, M.; Hu, L.; Li, Y.; Weng, Y. The chlorophyll-deficient golden leaf mutation in cucumber is due to a single nucleotide substitution in CsChlI for magnesium chelatase I subunit. *Theor. Appl. Genet.* **2016**, *129*, 1961–1973. [CrossRef]
126. Mueller, A.H.; Dockter, C.; Gough, S.P.; Lundqvist, U.; von Wettstein, D.; Hansson, M. Characterization of mutations in barley fch2 encoding chlorophyllide a oxygenase. *Plant Cell Physiol.* **2012**, *53*, 1232–1246. [CrossRef]
127. Wang, Z.; Gerstein, M.; Snyder, M. RNA-Seq: A revolutionary tool for transcriptomics. *Nat. Rev. Genet.* **2009**, *10*, 57–63. [CrossRef]
128. Stark, R.; Grzelak, M.; Hadfield, J. RNA sequencing: The teenage years. *Nat. Rev. Genet.* **2019**, *20*, 631–656. [CrossRef]

129. Yang, Y.; Chen, X.; Xu, B.; Li, Y.; Ma, Y.; Wang, G. Phenotype and transcriptome analysis reveals chloroplast development and pigment biosynthesis together influenced the leaf color formation in mutants of Anthurium andraeanum 'Sonate'. *Front. Plant Sci.* **2015**, 6. [CrossRef]
130. Cao, H.; Li, H.; Miao, Z.; Fu, G.; Yang, C.; Wu, L.; Zhao, P.; Shan, Q.; Ruan, J.; Wang, G.; et al. The preliminary study of leaf-color mutant in dendrobium officinale. *J. Nucl. Agric. Sci.* **2017**, *31*, 461–471.
131. Sun, C.; Deng, L.; Du, M.; Zhao, J.; Chen, Q.; Huang, T.; Jiang, H.; Li, C.-B.; Li, C. A Transcriptional network promotes anthocyanin biosynthesis in tomato flesh. *Mol. Plant* **2020**, *13*, 42–58. [CrossRef] [PubMed]
132. Li, M.; Hensel, G.; Mascher, M.; Melzer, M.; Budhagatapalli, N.; Rutten, T.; Himmelbach, A.; Beier, S.; Korzun, V.; Kumlehn, J.; et al. Leaf variegation and impaired chloroplast development caused by a truncated CCT domain gene in albostrians barley. *Plant Cell* **2019**, *31*, 1430–1445. [CrossRef] [PubMed]
133. Wu, H.; Shi, N.; An, X.; Liu, C.; Fu, H.; Cao, L.; Feng, Y.; Sun, D.; Zhang, L. Candidate genes for yellow leaf color in common wheat (Triticum aestivum L.) and major related metabolic pathways according to transcriptome profiling. *Int. J. Mol. Sci.* **2018**, *19*, 1594. [CrossRef] [PubMed]
134. Gang, H.; Li, R.; Zhao, Y.; Liu, G.; Chen, S.; Jiang, J. Loss of *GLK1* transcription factor function reveals new insights in chlorophyll biosynthesis and chloroplast development. *J. Exp. Bot.* **2019**, *70*, 3125–3138. [CrossRef]

MDPI
St. Alban-Anlage 66
4052 Basel
Switzerland
Tel. +41 61 683 77 34
Fax +41 61 302 89 18
www.mdpi.com

Forests Editorial Office
E-mail: forests@mdpi.com
www.mdpi.com/journal/forests

www.ingramcontent.com/pod-product-compliance
Lightning Source LLC
LaVergne TN
LVHW070119170726
843515LV00007B/1715

* 9 7 8 3 0 3 6 5 0 6 3 8 8 *